Intermediate Tropical Agriculture Series

General Editor

Dr. W.J.A. Payne,
Consultant in tropical livestock production

Tree and Field Crops of the Wetter Regions of the Tropics

C.N. Williams M.Sc., Ph.D.

W.Y. Chew M.Agr.Sc.

J. H. Rajaratnam Ph.D.

Longman

Longman Group Limited
London

Associated companies, branches and representatives throughout the world

First published 1980

ISBN 0 582 60319 6

Text set in 10/11 pt Photon Times.
Printed in Hong Kong by
Wing Tai Cheung Printing Co Ltd

British Library Cataloguing in Publication Data

Williams, C N
Tree and field crops of the wetter regions of the Tropics
(Intermediate Tropical Agriculture Series)
1. Tropical crops
I. Title II. Chew, W y III. Series
632.5′0913 SB111

Other titles in the Intermediate Tropical Agriculture Series

Already published:

H. T. B. Hall, *Diseases and Parasites of Livestock in the Tropics*
Describes the causes, symptoms, treatment and control of the main diseases of livestock in the Tropics.

L. R. Humphreys, *Tropical Pasture and Fodder Crops*
Describes natural tropical grasslands, discusses the possibilities for pasture improvement and suggests grass and legume species to accomplish this. Also includes sections on the establishment of pastures, soil fertility and pasture management.

J. C. Abbott and J. P. Makeham, *Agricultural Economics and Marketing in the Tropics.*
Describes the inter-relations of agriculture, farm management, marketing and their economics, as they occur in the Tropics.

Titles in preparation:

D. S. Hill and J. M. Waller, *Pests and Diseases of Tropical Crops*
Volume 1: Principles and Methods of Control
Volume 2: Field Handbook

J. A. Eusebio, *Pig Production in the Tropics*

H. F. Heady and E. B. Heady, *Range and Wildlife Management in the Tropics*

Contents

Preface

This book is primarily intended for students of tropical agricultural science in the higher secondary school and diploma levels, but should also be useful to farmers, businessmen, administrators and the interested man-in-the-street as well. In writing it the authors have assumed an elementary knowledge of agriculture, biology and allied sciences on the part of the reader, since the wide scope of the book makes it difficult to cover the whole of the background information. A glossary is provided to help with technical terms. However, by way of introduction, a detailed treatment of the principles of crop production in the humid Tropics is given in Part I which covers 7 chapters. The reader should bear in mind the relevant agronomic principles discussed in these chapters when reading Part II of the book which deals in detail with the individual crops, grouped together in chapters according to the nature of their products; in each chapter, major crops are treated first, and in greater detail, followed by minor crops, both groups being given in alphabetical order. This arrangement of the crops seems more logical for a book dealing with the agronomy of tropical crops, rather than purely with their botany, although the important botanical features of the crops are generally described.

The book does not attempt to deal with all the numerous agricultural systems encountered in the Tropics. Rather it advocates intensive systems which, in the long run, are an inevitable development. Intensive agriculture is less destructive to the environment than extensive agriculture because the area of land under cultivation is less and the area that can be left to natural vegetation is greater. For example, cut-and-burn shifting cultivation has a poor effect on fertility, species richness and desertification. Intensive systems of agriculture aim to maintain stable permanent farms. They yield several times as much crop per unit area as extensive systems; are available to irrigation and thus multiple cropping in the Tropics; and allow about a 10-fold increase in productivity over extensive farming.

The authors are obviously indebted to many works of reference as general sources of information and facts and figures. But most of the material in this book is based on many years of experience in different parts of the humid Tropics. It is hoped that many agricultural students and agriculturists in the Tropics will benefit from it.

Acknowledgements

The publishers are grateful to the following for permission to reproduce photographs in the text:

A.A.A. for Figs 11.6, 13.5a, 13.5b, 19.5a, and 19.5b; A–Z Collection for Figs 5.6c, 5.7, 14.25, 16.19, 18.9, and 18.20; Barnaby's Picture Library for Figs 16.16b, 17.7, and 17.13; Biofotos/Heather Angel for Figs 12.21b, 12.30, 12.46, 12.55, 18.23b, and 19.6; Bruce Coleman Ltd. for Figs 9.3 (Jane Coleman), 16.7c, 18.4 (Jane Burton), and 18.15 (Eric Crichton); W. Y. Chew for Figs 12.23, 12.24, 12.26, 12.29, 12.33, 12.34, 12.35, 12.36, 12.38, 12.39, 12.40, 12.41, 12.42, 12.44, 12.45, 12.47a, 12.48, 12.49, 12.50, 12.52, 12.56, 12.57, 12.58, 12.59, 12.60, 12.62, 12.63, 14.20, 14.21, 14.23, 14.24, 15.5, 16.17, 16.18, 17.3, 17.15, 17.16, 17.17, 18.5, 18.6, 18.7, 18.8, 18.10, 18.11, 18.12, 18.13, 18.14, 18.16, 18.17, 18.18, 18.21, 18.22, 20.4a, 20.4b, 20.4c, 20.5, 20.6, 20.7, 20.8, 20.9, 20.10, and 20.11; R. A. Francis (Longman Group) for Figs 12.27 and 15.2; J. B. Free/Natural History Photographic Agency for Fig. 12.27; W. J. A. Payne for Figs 11.2, 11.9, 11.10, and 12.37; Picturepoint Ltd. for Fig. 12.51, Don Tindall for Figs 2.4, 7.2a, 7.2b, 9.2, 10.4a, 10.4b, 11.1, 11.4, 11.5, 11.8, 12.21a, 12.22, 14.26, 15.3, 15.4a, 15.4b, and 16.20; J. Waller for Figs 12.47b, 14.2, and 16.12b; C. N. Williams for Figs 2.1, 2.2a, 2.2b, 2.2c, 2.2d, 2.2e, 2.2f, 2.3, 3.2, 3.4a, 3.4b, 3.4c, 5.2, 5.5, 5.8, 5.10, 5.11, 5.12, 5.13, 5.14, 5.15, 6.4, 6.5, 6.6, 6.7, 6.8, 6.9, 6.10, 6.11, 6.12, 6.13, 6.14, 6.15, 6.16, 7.3, 7.4, 9.1, 9.4a, 9.4b, 9.5, 9.8, 9.9, 9.12a, 9.12b, 9.13, 9.14, 9.20, 9.21a, 9.21b, 9.22, 9.24, 9.25, 9.26, 10.1, 10.2, 10.3, 10.6, 11.3, 11.11, 12.1a, 12.1b, 12.3a, 12.3b, 12.4a, 12.4b, 12.5a, 12.5b, 12.5c, 12.6a, 12.6b, 12.7a, 12.7b, 12.8, 12.11a, 12.11b, 12.11c, 12.12, 12.13, 12.15, 12.16, 12.18, 12.19, 12.20, 13.1, 13.7a, 13.7b, 13.9, 13.10, 13.11, 13.12a, 13.12b, 13.13, 13.14, 13.16, 13.17, 13.18, 13.19, 13.20, 13.21, 14.3, 14.5a, 14.5b, 14.8, 14.10, 14.12, 14.13, 14.14, 14.15, 14.16, 14.17, 14.18, 15.6, 16.1, 16.2, 16.3, 16.4, 16.5, 16.7b, 16.9, 16.10, 16.12a, 16.13a, 16.13b, 16.14a, 16.14b, 16.15, 16.16a, 16.16c, 17.1, 17.2, 17.4, 17.5, 17.6, 17.8a, 17.8b, 17.9, 17.10, 17.11, 17.12, 17.14, 18.1, 18.3, 19.1, 19.3, 20.1, 20.2a, 20.2b; Zambia Information Services for Fig. 11.7.

The publishers regret that they are unable to trace the copyright holders of Figs 12.31 and 12.32, and apologise for any infringement of copyright caused.

The cover photograph was kindly supplied by Alan Hutchinson Library.

Glossary

adventitious root Side roots growing from the lower nodes of the stem.

aerobic With reference to soil, the presence of oxygen in the air-spaces of the soil.

agronomy The science of soil management and crop production.

alluvial Alluvial soils are developed in material deposited by wind or in rivers.

anaerobic With reference to soil, waterlogged soil in which the oxygen has been excluded or used up.

antagonism Refers to the inhibition of uptake of one element (e.g. magnesium) by another (e.g. potassium) in the roots of a plant.

anthesis The stage of development in a plant at which the pollen becomes ripe.

apical meristem The main growing point of a plant or the growing point of a major branch.

available water That part of the soil water retained between field capacity and permanent wilting point.

axil The junction between a leaf and the stem.

Bordeaux mixture An insecticide spray prepared by stirring 1·25 kg of hydrated lime, $Ca^{2+}(OH^-)_2$, into 40 litres of water and adding 1 kg of copper(II) sulphate(VI), $Cu^{2+}SO_4^{2-}$, dissolved in 10 litres of water to make 50 litres.

bud clone Clone from which bud wood is obtained.

budding The grafting of a bud on to a stock seedling.

bud wood Shoots bearing buds of the scion material.

bulbils Vegetative bulbs on a plant.

bunds (rice) Low earth walls to retain water in rice fields.

cambium The zone of dividing cells in a stem which gives rise to the tissues of the stem.

canopy The array of leaves produced by the crop.

catch-crop A short-term crop grown between widely-spaced tree crops; it is grown before the inter-row areas become shaded by the tree crops.

cation exchange capacity (CEC) The ability of the soil to hold cations (positive ions); expressed in milli-equivalents of cation per 100 grams of soil.

cation exchange sites Negatively-charged sites in soil which attract cations.

cherelle wilt The death of young immature fruits of cocoa from physiological causes.

chlorotic Pale coloration of foliage and other plant parts resulting from lack of chlorophyll.

chromosomes Thread-like bodies in the nuclei of cells that carry the genes or hereditary characteristics of the organism.

cleft graft A graft made by inserting a wedge-shaped scion base into a slit made in the stock seedling shoot.

clone The entire stock of plants produced from one stock seedling by vegetative propagation.

coir Fibrous material from the husks of coconuts.

colluvial Colluvial soils are developed in material which has moved down a slope under gravity.

compost Decomposed organic matter, usually made from straw or similar plant material.

copra The dried flesh of the coconut.

coralline Coralline soils derived from old coral reefs and composed chiefly of calcium(II) carbonate.

corm An underground stem usually swollen with starch; used for propagation in some plants.

cover crop A crop grown to provide ground cover between widely-spaced tree crops; it should be a fast-growing species, usually a legume.

cultivar A crop variety.

cyclone A violent tropical storm.

endocarp The inner layer of a fruit wall.

endosperm Reserve food material in a seed.

epiphytic Growing on but not parasitising a plant.

fan (cocoa) A specific branch-type of cocoa which grows in the shape of a fan.

field capacity Water held by soil after excess water has drained downwards.

flush (cocoa) A burst of leaf growth.

friable A soil which has a loose structure and permits rapid growth and ramification of plant roots.

frond The compound leaves of palms.

fumigation The use of a toxic gas to kill pests.

grafting The joining of two plants vegetatively.

hermaphrodite Having both male and female parts.

hybrid The genetic progeny of two plant varieties.

hydroponics The growing of plants (usually vegetables) in water or sand treated with a nutrient solution.

hygroscopic Having the property of taking up water from the atmosphere.

infestation An accumulated attack by pests.

inflorescence A flower shoot or group of flowers.

intercrop A crop grown in the spaces between the plants of another crop; e.g. cocoa under coconuts.

inter-planting The planting of a second crop in the spaces between the plants of the main crop.

jat (tea) A cultivar of tea.

jorquette (cocoa) A specific branch-type of cocoa.

larvae The juvenile stages of insects (e.g. caterpillars, grubs, etc).

laterite Soil or soil material composed of either soft or indurated iron rich soil material or soil rich in hard ironstone concretions.

latex The milky sap of a plant.

latosol Free-draining, deeply weathered soils, usually red or yellow, stable, and high in iron oxides and kaolinitic clays.

leaching Washing out.

leaf area index (LAI) The ratio of the leaf area of a crop to the area of ground covered by the crop, e.g. an LAI of 4 means that leaf area is 4 times ground area.

loam Soil textural class in which the sand, silt and clay fractions occur in balanced amounts without any one of them being very dominant.

lodging The falling over of plants, usually as a result of wind or flood.

macerate To chop up.

marcotting Vegetative propagation of a shoot achieved by tying a ball of soil or other material (such as coconut husk) around the base of the shoot until roots form at that place. The shoot is then cut below the soil ball for planting.

maturation Ripening.

mesocarp The central tissue of a fruit wall.

metamorphic Rock which has been fundamentally changed by heat and pressure.

mucilage A slimy plant sap characteristic of certain plants (e.g. okra).

mulch A layer of non-living material on the soil surface to conserve moisture and prevent the growth of weeds, for example, chopped grass and other plant material or plastic sheet.

mutant An organism that is genetically changed from its original form.

necrosis Death of a cell or tissue while part of a living body.

nematode Soil-inhabiting worm some kinds of which are parasitic on plants.

nitrophoska blue special A compound fertiliser containing N, P, K and Mg in the ratio 12:12:17:2 with added micronutrients; useful where little is known of the nutrient status of soils.

node The regular areas on a plant stem from which side branches and leaves arise.

nucellar Refers to seeds derived solely from the female parent; they carry the full chromosome number; they are common in citrus.

nutriculture Growing crops in sand or other inert material, supplied with a solution of plant nutrients.

optimal yield The highest yield obtainable under given environmental conditions.

out-cross or out-pollination Results when one plant is pollinated by another with a different genotype.

parthenocarpy Development of a fruit without genetic union.

perennial Long lasting; plants which go on growing through many seasons.

pericarp The wall of a fruit.

permanent wilting point The condition of the soil at which plants wilt permanently; generally equated with the amount of water held at a tension of 15 atmospheres.

primordia Young, undeveloped organs such as shoots and roots.

photosynthesis Formation of carbohydrates from water and carbon dioxide using the energy of the sun.

puddling The condition of the soil in rice culture when waterlogged to give mud.

ratoon A second growth occurring after harvest of the first crop.

retting Soaking fibres to break down the tissues.

reducing soils Waterlogged or submerged soils from which oxygen has been excluded.

rhizobium Bacteria living in association with legume roots; they are usually capable of fixing nitrogen.

rhizome An underground stem.

rhizosphere Zone of soil immediately surrounding a plant root.

rotavate Cultivation of the soil with a rotavator.

rotenone A plant alkaloid useful in insect control.

saline solution A solution of common salt (sodium chloride).

salt index An expression of the solubility of inorganic fertilisers; sodium nitrate(V) is very soluble and is assigned a salt index of 100.

scarifying Mechanical scratching of the surface of seeds to increase water absorption and germination; this may be done with sandpaper and is particularly useful for leguminous seeds.

scion A vegetative bud or shoot grafted on to a root stock for vegetative propagation.

senescent Refers to plant parts in the process of aging

and dying.

sink strength The flow of nutrients to a growing organ.

soil colloids Very fine clay and organic particles of the soil.

spot weeding Selective spraying, rubbing or removal of particular weeds or clumps of weeds.

sticker A chemical, usually a water-miscible oil, which can be mixed with chemicals for spraying to make them adhere to the plant surfaces.

stigma Female organ of a flower which the pollen sticks to before fertilising the ovary.

strip weeding Killing weeds in a defined area, usually the inter-row area between field crops, or the area along the planting line in large tree crops.

surface active Reducing surface tension so that wetting of the surface occurs (e.g. as with soapy water).

sward An area of grass or pasture.

systemic Systemic compounds are those which enter the tissues of the plant.

tajak A sharp weeding instrument used for cutting weeds at the soil surface.

tetraploid Having double the normal (diploid) number of chromosomes.

tillage Cultivation of the soil, e.g. ploughing, rotavating, etc.

tiller The branched shoot of a grass or cereal plant.

toxic reduction products Toxic substances formed in reducing soils.

translocation The movement of water through the plant, carrying dissolved food; some herbicides are also transported in this way.

transpiration The movement of water through the plant from roots to leaves, caused by the evaporation of water via the stomata of the leaves.

triploid Having three times the haploid chromosome number.

turgid Filled with fluid, e.g. plant cells filled with sap giving rigidity to the plant.

ultra-low volume (ULV) sprayer A machine which delivers very small volumes of spray.

understory A canopy of plants beneath the main dominant plants e.g. bushes growing under taller trees in rainforest.

urease The enzyme that denitrifies urea.

volatilisation When a substance becomes a gas.

waterlogged Saturated with water.

Part 1 General considerations

1 Introduction

This book is about the tree and field crops of the humid Tropics where rainfall is generally in excess of evaporation and where the natural vegetation is rain-forest of one sort or another. The regions of the world classified as humid and tropical are indicated in Fig. 1.1.

The most economically important crops in the humid Tropics are perennial and essentially non-deciduous, many are trees. This results from the climatic conditions prevailing: the high ambient temperatures and excess of rainfall. Irrigation practices are minimal since rainfall is rarely a limiting factor. Most of the humid Tropics are located near the equator (between 10°N and 10°S latitude), a region characterised by an absence of cyclones. In the Tropics in higher latitudes and in sub-tropical regions, crops are often destroyed by cyclones.

Some general sections on the agronomy of crops in the humid Tropics are provided before details of specific crops are discussed.

2 Land preparation: cover crops

In the wetter regions of the Tropics land preparation usually involves clearing high forest. Where land has been planted with tree crops which are old and require replanting, clearing involves the removal of the old tree crop as

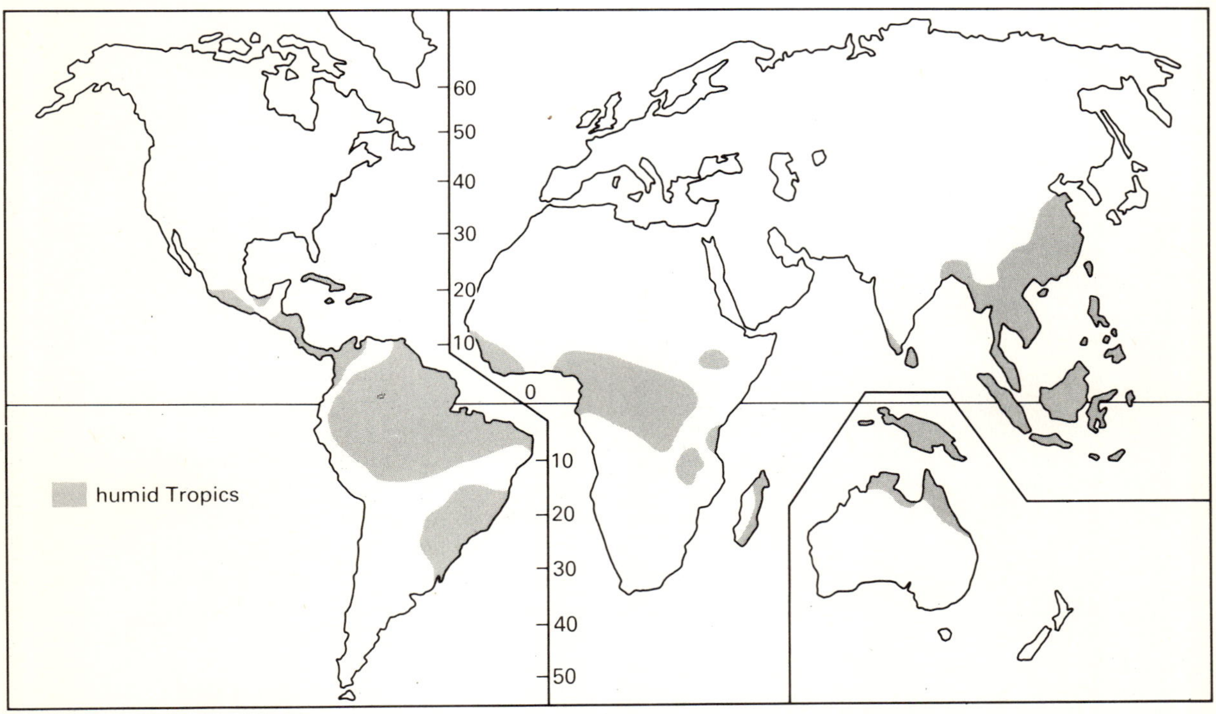

fig. 1.1 Distribution of humid tropical zones. (*Source:* **Blummenstock, D. A.** and **Thornwaite, A. V.** (1941). *Yearbook.* US Department of Agriculture.)

opposed to natural forest. In either case, clearing involves the destruction of a stable system in which the forest, or the established perennial crop, is more or less in a state of balance with the soil. The roots of the trees and cover plants draw water and nutrients upwards, preventing or greatly reducing leaching. Water and nutrients are cycled in the vegetation itself. When leaves fall and old parts of the vegetation die, the nutrients contained in them are released slowly and taken up again by the living plants. The trees also encourage the accumulation of high levels of organic matter in the top soil which stabilises nutrients. Trees provide a ground cover of leaf litter preventing raindrop impact and erosion. The roots of trees also bind the soil and keep it stable. When forest is cleared for cropping, the balance changes.

Different standards of clearing are used for different crops. In clearing high forest, which is massive, complete removal is not usually attempted as this would be expensive. Forest is usually cleared by felling and burning. In the humid Tropics burning is often incomplete because of the wet climate. Fig. 2.1 shows an area of cleared land prepared for oil palm and rubber cultivation. Further clearing will not be attempted and the field will be planted with cover legumes and the seedling trees in this state. Neither oil palms nor rubber will require harvesting for at least 3 years. In the years before harvesting starts, the field will be entered only during weeding rounds (about once every 2 months or less often) and when fertilisers are applied. Workers will have to scramble over unburnt trunks but as this is only, at the most, 6 times each year, this will not be too difficult. By the time harvesting starts, and frequent entry is required, the unburnt logs and stumps will be decayed and broken down in the moist climate of the wet Tropics. The standard of clearing shown in Fig. 2.1 is suitable for most tree crops which take a long time to mature. This standard of clearing is also used in shifting cultivation when the spaces between the trees will be planted with such crops as maize, beans, cassava and hill rice.

fig. 2.1 Land cleared by felling and burning. This standard of clearing is suitable for tree crops such as rubber and oil palms. Bulldozer tracks made in removing valuable timber trees form the first access roads for this area which is part of a large estate scheme

In clearing old plantation tree crops (rubber, coconuts, coffee, etc.) for replanting, a higher standard of clearing is used. This is mainly to avoid root diseases which may have built up in the old crop and could affect the new crop. This occurs especially when the new crop is the same species as the old one, or when oil palms are planted on old coconut land. It is much easier to clear old tree crops than to clear high forest.

Methods of clearing tropical forest

High forest is cleared by felling and burning. The trees are felled with a chain saw; the trunks are cut above the spreading buttress roots. A skilled operator will cut large trees so that they fall and bring down smaller trees. After felling, the trees are allowed to dry for as long as possible. In humid climates, it may be possible only to get a short drying season, often not more than a few weeks.

Burning in a wet climate is generally aided by pre-killing large trees up to 6 months before felling using a strong (10 per cent) solution of 2:4:5-T. This is painted around the bole or into slashes made in the bark. For killing bamboos, a strong (7·5 per cent) solution of dalapon can be sprayed on the lower parts of the stems. Pre-killing causes trees partially to dry out prior to felling. Any valuable timber may be removed, at the time of felling, by making tracks with bulldozers to the trees; these tracks may form the basis of the first road network into the cleared area. The result of this type of clearing is generally similar to that shown in Fig. 2.1.

Sometimes explosives instead of chain saws are used for felling trees. Ammonium nitrate(v) explosive (which is also a fertiliser) is used for this purpose. The main steps

fig. 2.2 Land clearing by blasting: (a)—(c) drilling and placing the charge in a tree; (d) the explosion; (e) and (f) by placing suitable charges the trunk can be severed and the roots shattered. (*Source:* **Williams, C. N.** and **Hsu, Y. C.** (1970). *Oil Palm Cultivation in Malaya.* University of Malaya Press.)

in felling trees by explosive are shown in Fig. 2.2. Explosives will also shatter the main roots. To clear forest with explosives the assistance of someone experienced and licensed to use them should be obtained.

With secondary and less massive forest, complete clearing with root and stump removal may be carried out as when clearing old crop trees, as discussed below.

Clearing old tree crops

Because old tree crops are less massive it is possible completely to remove them from the soil. This is done with a heavy tractor or a bulldozer using a dozer-blade and knock-down boom (Fig. 2.3). Trees such as rubber are bulldozed into rows and burnt. Old palms (coconuts and oil palms) are treated in the same way but are much more difficult to burn and usually require cutting into lengths, re-stacking and burning on more than one occasion. In clearing old tree crops using a bulldozer the stumps and large roots are mostly uprooted during the pushing down operation. After this, the area may be ploughed with a heavy disc plough to cut up smaller roots, and then harrowed. When clearing of this type is undertaken, the area will be suitable for the use of mechanical equipment.

Cultivation of field crops as well as tree crops becomes possible.

Planting cover crops

When an area is cleared for agriculture, whether it be virgin forest or an old plantation area, the soil will be exposed to leaching, erosion and the breakdown of organic matter. It must therefore be used without delay. When perennial crops (tree crops) are to be grown, the soil is usually planted with cover legumes when the trees are young or before the trees are planted. This prevents accelerated erosion, combats weed growth and adds nitrogen from the legumes to the soil. In the wet Tropics, the legumes most commonly grown are *Centrosema pubescens* (centro), *Pueraria phaseoloides* (puero) and *Calopogonium mucunoides* (calapo). These are usually sown together since puero and calapo establish first and centro takes over later. They are usually sown in rows (1–2 m apart) at a seed rate of 5 to 10 kg per hectare, in ratios varying from about 1 : 2 : 2 in West Africa to 4 : 1 : 5 in Southeast Asia. Legume seeds are treated by scarifying (rolling them in a drum lined with sand

fig. 2.3 Clearing old oil palms with a bulldozer fitted with a knock-down boom

fig. 2.4 An area of young oil palms with cover legumes planted between the trees

paper) or with acid to make them germinate better. Acid treatment is carried out using concentrated sulphuric(VI) acid. A little acid is mixed with the dry seeds using just enough to wet them (keeping it off hands and clothing) and after a period of 5 to 10 minutes the seeds are dropped into a large drum of water in small quantities and stirred vigorously (to prevent the build-up of heat). The water is then poured off and the seeds are surface-dried and planted. Legume seeds are planted by shallow drilling after inoculation with a *Rhizobium*. Fig. 2.4 shows an area of young oil palms planted with cover legumes. These legumes have been shown to fix nitrogen thus benefiting the soil. The correct *Rhizobium* may be obtained from a number of sources, including government and commercial organisations.

Sometimes intercropping or catch-cropping is undertaken instead of growing cover legumes. When this is done, the same considerations of soil cover apply and the ideal intercrop is one which provides good cover and also fixes nitrogen. Groundnuts have been used in rubber as a successful intercrop (p. 38). Cowpeas also make an excellent intercrop as they are long lasting and have a semi-creeping or scrambling habit.

3 Weed control

Methods of weed control

Early and regular weeding, before weeds are growing strongly, both increases crop growth and reduces the work required in weeding.

Weed control by hand

In many tropical countries weed control in both tree and field crops is still carried out by hand. In the case of tree crops, this usually means keeping a circle round the tree free from weeds, and in field crops planted in rows, by weeding between the rows. When circle weeding by hand, care must be taken not to hoe the circle too deeply. This produces a saucer-like depression around each tree and damages surface roots. Hoeing should be shallow and the hoed soil should be spread back into the circle. The best type of implements have sharp blades which cut just beneath the soil, such as the *tajak* used in Southeast Asia.

The size of the circle to be kept free of weeds should be about 1 m in diameter in the case of newly-planted seedlings. It should then be kept about the same size as the developing crown of the tree (the spread of the branches or fronds). Sometimes, when trees are planted closely spaced in rows, as in avenue planting of rubber, strip weeding is carried out. Both circle weeding and strip weeding are shown in Fig. 3.1 on p. 6.

In plantations of mature trees, in which the canopy has closed over, shading will reduce weed growth. Weed control consists of periodical spot weeding in the cleared area, particularly when harvesting and access paths are to be kept clear. Control of weeds in inter-row areas is usually carried out by the use of herbicides (see p. 7). Slashing of weeds by hand in inter-row areas should be avoided as this will encourage a grass-dominant weed cover which is not desirable. Shallow rotavating or disking is satisfactory as a substitute for control by herbicides.

Mechanical weed control

Mechanical weed control, using tractors and other types of cultivation equipment including buffalo-drawn cultivators (Fig. 13.11 on p. 156), is much used in field crops which are grown in rows. Pre-cultivation of the field with plough and rotavator kills off most existing weeds and for this reason, planting should be carried out immediately after cultivating. This makes use of the first weed-free period for establishment of the crop. When crops are planted in rows, the rows should be straight and parallel to allow for mechanical weeding in the early stages of growth. When seeds are planted mechanically there is no problem because the planter or seeder will ensure straight planting; but when crops are planted by

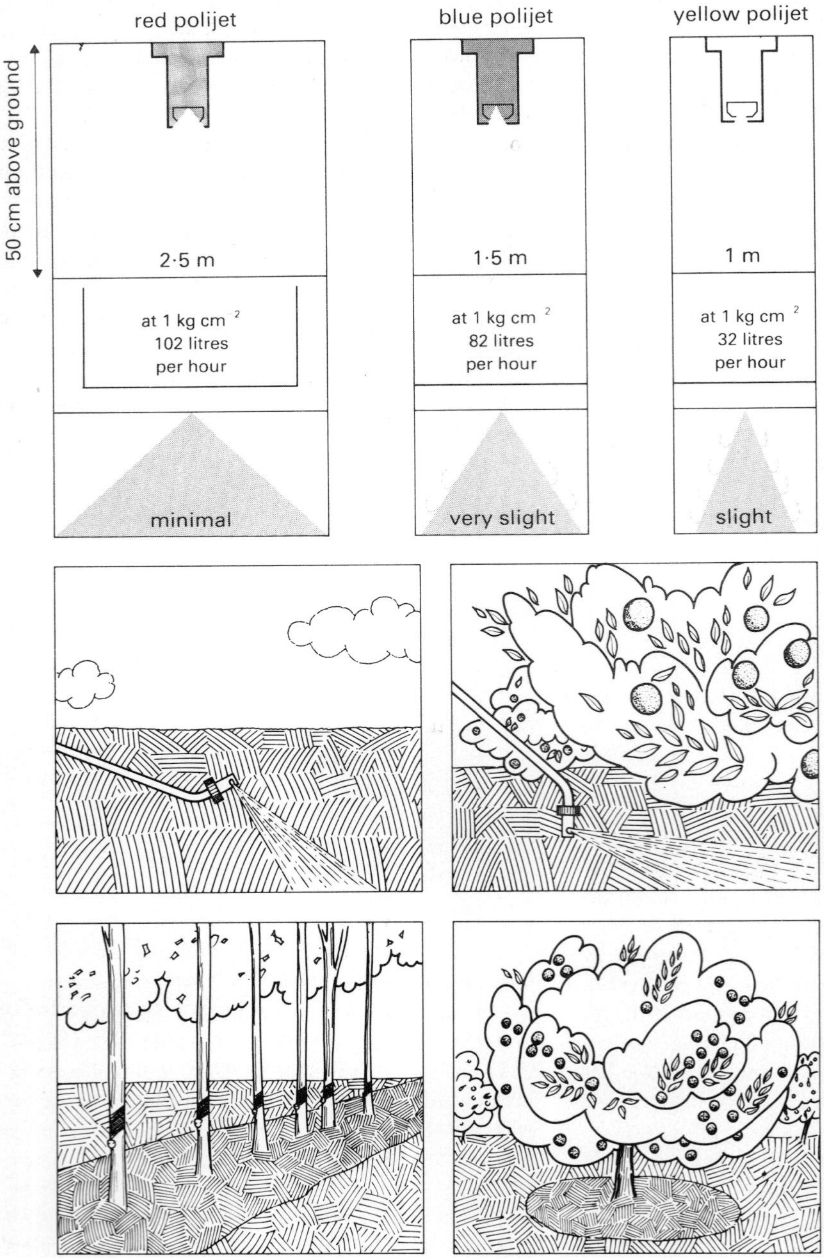

(a) 'Polijet' spraying which produces a continued swath with only slight drift

(b) Normal position of spray

(c) For spraying under low-hanging branches invert spray lance and project spray fan almost horizontally

(d) Weed killing between the rows in rubber

(e) Weed killing in a circle around the tree

fig. 3.1 (*Source:* a pamphlet published by ICI Ltd: UK)

hand (cf. rice throughout much of the Tropics) care should be taken to plant in straight rows.

Other methods of weed control

Cover crops control weeds by covering the inter-row areas and shading out the growth of weeds. The cover crop should be a legume which fixes nitrogen thus benefiting the soil. In humid regions periodical spot weeding is necessary to eliminate or reduce invading weeds. Under sub-humid conditions, where the dry season exceeds about 2 months, it is sometimes desirable to disc the cover at the onset of the dry season to reduce competition for water. Under drier conditions, regeneration of the cover generally takes place from seed in the following wet season, but under very humid conditions, seed production of cover legumes is much reduced, and frequent rounds of spot weeding are necessary to maintain a legume-dominated cover.

Intercropping and **catch-cropping** (see p. 39) also effectively control weeds in the main crops because of weed control measures used on the intercrop.

Mulching consists in spreading dead plant material in the weeding circle and between the rows. Prunings or residues of the crop may be used unless these harbour diseases or pests. Prunings may be used in tea, coffee, oil palms, bananas and other crops. In some cases (e.g. coffee) mulch crops, usually grasses, may be especially grown. Mulch also reduces soil erosion and conserves soil moisture. Mulching is a desirable practice which should be encouraged.

fig. 3.2 Weed control by cattle grazing; left-hand plot ungrazed, right-hand plot grazed at 2·4 head per hectare for a period of 10 days

Grazing livestock between the trees is a method of weed control used in coconut cultivation. This eliminates the need for weeding by hand and provides animal products. When cattle are used, however, they should be properly managed and not allowed to damage the crop. They must be moved to different parts of the estate at regular intervals to graze the weeds back. Cattle are only used to any great extent in coconuts, at the present time, but may also be used in other crops providing they do not cause damage. Fig. 3.2 shows how cattle can be used to control weeds in oil palms. Sheep could be useful in rubber plantations.

Flooding is important in controlling weed species of dry land in rice cultivation (see p. 157).

Chemical weed control

The use of chemicals (herbicides) is becoming more and more important in weed control, particularly in tree crops for circle and inter-row weeding. The use of herbicides in circle weeding is considered to cause less root disturbance to the crop than mechanical or hand weeding. The uses of specific herbicides are mentioned in the sections concerned with the various crops; here we consider some general aspects of herbicide use and describe the properties of some important herbicides.

Herbicides may be taken up from the soil by the roots of the weed which is thus killed. Others kill germinating seeds or small weeds directly. These are insoluble and remain in the top layers of the soil for long periods. This is important for perennial crops where weeding is a continual requirement over a long period. Simazine and other triazines, diuron and monuron are important soil-surface herbicides active for periods up to 6 months. They are commonly used as pre-emergence sprays, that is, immediately after cultivation or after the use of contact herbicides and before new weeds germinate.

Other herbicides are absorbed by the foliage of the weed. The herbicide may kill by direct action on the foliage or by a general action after translocation throughout the plant. When they kill the foliage directly, they are called contact herbicides. For effective weed control, contact herbicides are generally used in combination with soil-surface herbicides; the former to kill off standing

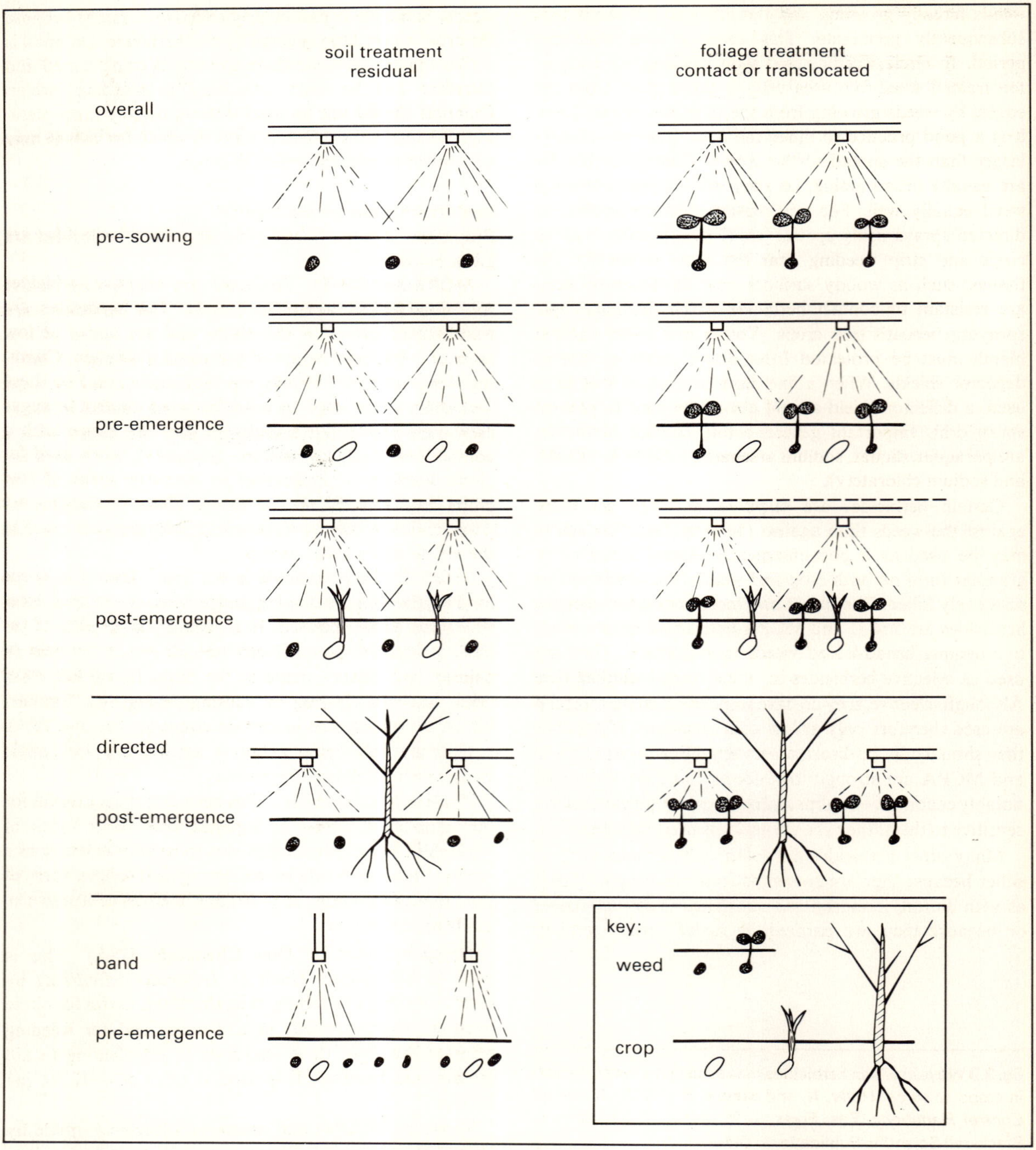
soil treatment
residual
foliage treatment
contact or translocated
overall
pre-sowing
pre-emergence
post-emergence
directed
post-emergence
band
pre-emergence
key:
weed
crop

weeds already growing, and the latter to kill seeds that subsequently germinate. This gives a long weed-free period. In circle weeding and strip weeding, invasion of the treated weed-free area usually takes place from the edges, by weeds growing from the inter-row areas. Thus, it is a good practice to make the weed-free area slightly larger than the spread of the crown. Contact herbicides are usually broad acting, i.e. they will kill both crop and weed equally well. For this reason they are applied as directed sprays using special jets (Fig. 3.1) when used in circle and strip weeding near the crop. Generally, old tissues, such as woody surfaces near the bases of trees, are resistant to contact herbicide which facilitates safe spraying beneath tree crops. Young and short-statured plants must be protected from direct spray by use of deflector shields. When a fine atomising jet or mist jet is used, a deflector shield should always be used to prevent spray drift. Important general-acting contact herbicides are paraquat, diquat, sodium arsenate(V), DSMA, MSMA and sodium chlorate(V).

Certain herbicides are selective, i.e. they act more against the weeds than against the crop. Such herbicides may be used as a post-emergence general spray or in granular form on both crop and weeds; the weeds will be selectively killed. The oldest and most important selective herbicides are 2:4-D and MCPA both of which are selective against broad-leafed weeds (non-grasses). They are used as selective herbicides in cereal crops including rice. Although selective, they do have some effect on the crop and are used therefore only at low concentrations. If possible, they should only be directed between the crop rows. 2:4-D and MCPA are strongly translocated and certain crops, notably coconuts, oil palms and other palms are particularly sensitive to them; they should never be used in these crops.

Many other herbicides are selective to a certain extent either because they are prevented from reaching the crop, as with diuron, monuron and simazine mentioned above, or because they are particularly active against certain weeds. Simazine is particularly effective in maize because the crop is very little affected by this herbicide. Dalapon is active against grass weeds in which it is translocated and therefore can be used selectively in weeding circles. Dalapon should not be used directly on any crop plant. Fig. 3.3 shows the different ways in which herbicides may be applied to control weeds in crops.

fig. 3.3 Ways in which herbicides may be used for weed control in crops. (*Source:* **Holly, K.** and **Steele, B.** (1968). In: *Weed Control Handbook.* Eds. Fryer, J. D. and R. J. Makepeace, Blackwell Scientific Publications: Oxford.)

Useful herbicides for the Tropics

Properties of some of the more important herbicides are given below.

MCPA and 2:4-D These are very selective herbicides for broad-leaved weeds in cereals. The herbicides are translocated (move in the plant) and are active at low rates of 1 litre per hectare of commercial solution. Climbing weeds such as *Mikania* are very easily killed by these herbicides. They may be used for weed control in sugar cane, rice and maize. Sometimes they are mixed with a contact herbicide, e.g. sodium chlorate(V), when used for circle weeding or application to inter-row areas of tree crops (not directly on the crop). These herbicides are made commercially as sodium and potassium salts and as the amine and various esters.

2:4:5-T This herbicide is not used directly in crops but may be used in clearing jungle trees or old crop trees which are to be removed. It is painted on the side of the trees or injected (poured) into holes drilled in the trees or painted into slashes made in the bark. It will kill most trees, and aids clearing by causing drying out. The herbicide is now banned in certain countries because of its persistence and toxicity but is satisfactory for jungle clearing when followed by burning.

Amiben (chloramben) This herbicide is very useful for the control of weeds in legumes and many types of vegetables. It has been found also to be very effective as a pre-emergent herbicide in wet rice against sedges, sprayed just after transplanting at 5–10 litres of 20 per cent active ingredient per hectare.

Dalapon (made by Dow Chemicals, USA) This is used to kill grasses (such as *Imperata cylindrica*) by direct spraying or rubbing. It is, therefore, useful for circle weeding of grass weeds in tree crops and for weeding between the rows. It is also used in pre-clearing to kill grasses and bamboo. It is used at rates of 4–12 kg per hectare.

Simazine, atrazine and ametryne (triazines) (made by Geigy, Switzerland) These are used as selective herbicides in maize for treatment of inter-row areas, and in

(a)

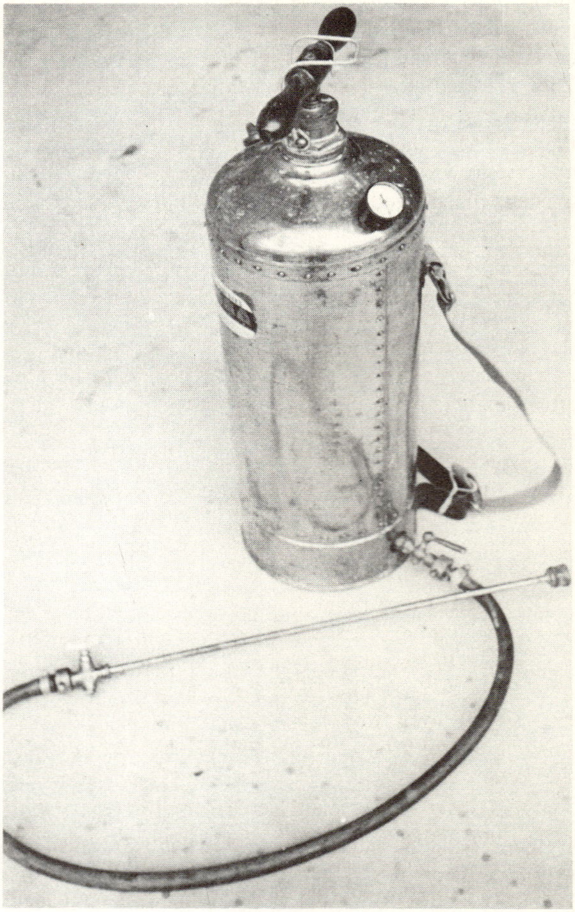

(b)

fig. 3.4 (a) Knapsack sprayer operated by a low-pressure pump, suitable for herbicides (b) This sprayer is pumped up to 3–5 kg cm^2 pressure before use and gives a fine spray suitable for fungicides (c) Low-volume sprayer which uses chemicals mixed with oil rather than water, the powerful fan gives good coverage over the leaves of the crop, it is suitable for fungicides and insecticides

(c)

many other crops for circle weeding; they are used for pineapples. They are long lasting and should be sprayed only after cultivation or spraying with a contact herbicide. They are used at 1–4 kg per hectare active ingredient.

Monuron (CMU) and diuron (DMU) (made by Du Pont, USA) These herbicides are used in circle weeding and inter-row weeding in many tree crops. They kill mainly small young weed seedlings and germinating seeds and should only be used after cultivation or after spraying with a contact herbicide. They persist for up to 6 months; used at about 0·5–1 kg per hectare active ingredient.

Paraquat (Gramoxone) (made by ICI, UK) This is a very effective contact herbicide for directed application in circle and inter-row weeding. It should be used after planting but before the crop seedlings emerge in clayey soils (not sands). The herbicide is inactivated on contact with soils containing clay; it is much used also in 'minimal cultivation'. This herbicide is very effective against grass weeds. It is used at 0·25–2 litres of commercial solution per hectare. It is fairly toxic to animals and should be used with care.

Diquat (Reglone) (made by ICI, UK) It is similar to paraquat but more effective against broad-leaved weeds.

Glyphosate (Roundup) (made by Monsanto) It is a slow-acting translocated herbicide used for clearing grass and broad-leaved weeds before planting, and for strip weeding in tree crops such as rubber. It is inactivated in the soil.

TOK (CMU) This is a selective herbicide widely used in rice. It is active against a wide range of weeds including grasses but is relatively harmless to fish which may live in rice fields. It is used in granular form which contains about 7 per cent of active ingredient. It is applied at rates of about 20 kg per hectare immediately after transplanting and before weed germination has occurred.

Arsenical herbicides Both sodium arsenate(V) and sodium arsenate(III) are very effective general contact herbicides. They are used for circle weeding and for weeding between the rows in rubber and oil palms. They are very toxic to animals and must be used with extreme care. They must not be sprayed near streams and water. These herbicides are banned in many countries.

DSMA and MSMA These are similar to sodium arsenate(III) but much less toxic, nevertheless they should be handled with extreme caution.

Sodium chlorate(V) This is a general contact herbicide for inter-row and circle weeding. It can be mixed with 2:4-D and MCPA shortly before use. It is effective at 10–15 kg per hectare under dry conditions. Avoid contact with clothing as it is inflammable when dry.

Spraying equipment and spraying methods

In tropical countries the most commonly-used spraying equipment is the knapsack sprayer (Fig. 3.4(a)). These are usually of 8 to 16 litre capacity and are used with various types of jet depending on conditions. For general weed spraying before the crop is planted, a mist jet will give better results (when there is no danger of spray drift to nearby crops). When circle weeding or inter-row weeding is carried out, a fan jet such as the 'polijet' (Fig. 3.1), which produces a confined swath of spray of larger droplet size which is less likely to drift to the crop, should be used. The fan jet may also be used for spraying under low-hanging bushes as shown in Fig. 3.1.

Commonly the chemicals are applied with a surface-active compound (if this is not already in the preparation) and sometimes also with a sticker (e.g. Tenac, Mobilcer, etc). About 400 litres of water are used per sprayed hectare. For fungicides and insecticides low-volume spraying is very effective (see Fig. 3.4(c)); these sprayers are not yet used very much for herbicides mainly because of drift problems.

Spraying equipment should always be well cared for and cleaned after use. Never use equipment for fungicides or insecticides that has been used for herbicides. Periodically, about 4 times a year, all equipment should be taken apart and lightly oiled, replacing any washers that are worn. Always keep spare washers handy and when buying equipment make sure that spare parts can be obtained locally.

Herbicides may be applied in granular form, which eliminates spraying and the possibility of spray drift. This method has been particularly useful in rice with soil-active herbicides such as TOK granular. Other soil-active herbicides may also lend themselves to granular application; 2:4-D and MCPA, both selective, translocated herbicides, have recently been used in granular form for control of broad-leafed weeds in rice. Granular herbicides have been used in rice because the flooded cultivation lends itself to spread of the herbicide; for efficient use the fields must not dry out. However, dalapon has been used in granular form for weed control in forest trees and it seems likely that many herbicides, particularly soil-active ones, might be used in granular form under conditions of high rainfall.

4 Soil, nutrients and fertilisers

The value of a soil for agriculture depends on the following factors:

1 Nutrient content and nutrient holding capacity.
2 Availability of those nutrients.
3 Acidity and alkalinity as it affects availability of nutrients.
4 Physical properties (whether it is friable and well drained or heavy and poorly drained, degree of aeration).
5 Water-holding capacity (the quantity of water the soil can hold between field capacity after rain and permanent wilting point, or the water content at which plants do not recover from wilting at night or in the early morning).
6 Depth.
7 Stability to erosion.

The value of a soil also depends on the properties imparted to it by the organic matter of the soil. This improves drainage in heavy soils and water and nutrient-holding capacity in sandy soils. The content of organic matter in natural soils varies greatly. Under virgin forest the level of organic matter may be about 4 per cent of the dry weight of the top soil; under savanna and secondary forest it may be between 1 and 3 per cent.

The amount of organic matter in cultivated soil depends much on how it is farmed. When farmed correctly the level of organic matter can be maintained and increased. One approach is to leave a minimum of open exposed surface. This can be achieved by minimising the time that bare soil is left exposed after clearing or cultivating; by planting cover legumes, in the case of tree crops; or appropriate intercrops or nurse crops, with correct spacing, in the case of field crops. The organic matter of the soil can also be maintained by the correct use of fertilisers to replace nutrients. Correct use of fertilisers encourages good crop growth. The crop root and other residues which remain after harvest add organic matter to the soil. If cleared land is farmed without the use of fertilisers, to replace nutrient losses, and other techniques to maintain fertility, then organic matter decreases. This results in a decrease of fertility; the soil becomes poor and yields poor crops.

Usually, the first crop after clearing will be good because of all the nutrients released from the natural vegetation after burning. Great losses can occur over a short period of time and continued high production will depend on good farming practices and use of fertiliser. The ways of maintaining fertility in cropping, whether with perennial tree crops or short term field crops, will be considered in more detail below, but first we must briefly discuss some of the more important soils of the humid Tropics and how they may be used safely for agricultural purposes.

Soils of the humid Tropics

Latosols

Latosols are highly weathered soils and consequently stable; they are often red or yellow. They are formed from a variety of parent rocks and usually contain iron oxides. Generally, those which are formed from granitic and basaltic rocks are the most stable and will not easily erode or slip (cause land slides). They can be used for agriculture even if they are very steep. Such soils occur widely in the humid Tropics; for example, in Malaysia, Sumatra, parts of East Africa (the Kikuyu soils) and so on. They are friable soils, often deeply profiled and they drain well.

Other latosols formed from various sedimentary rocks such as shales, sandstone, mudstones, etc., are generally not stable. When cleared for cultivation they will be easily eroded and in steep areas may cause land slides. These soils can be used for agriculture if special precautions are taken. If the slope is low (up to 15°) then they may be used for field crops with proper contour planting, etc. (see p. 36). However, if they slope at more than 15°, they should only be used for perennial tree crops, with a minimum of cultivation or for pasture. On these soils slopes of more than 20° should be left with natural vegetation. Other types of latosols derived from metamorphic rocks are variable but generally intermediate in stability.

Most latosols are capable of holding about 100–150 mm available water per metre of soil (between field capacity and permanent wilting point). Shale-derived soils will generally hold about 150 mm available water per metre. Some latosols have layers of stony laterite beneath the surface and these are generally not suitable for deep-rooted tree crops because the depth of soil available to the roots is small. Generally latosols tend to be low in nutrient content in the humid Tropics.

Volcanic soils

Other soils encountered in the wetter regions of the Tropics are recent volcanic-ash soils (for example in parts of the Philippines, New Britain and Central America). They may be good for agriculture because they are nutrient-rich and are also porous and freely drained so that they are stable. They do not hold much water, however, and at a later stage volcanic-ash soils may become powdery (andosols) and then may be easily eroded by wind and rain.

Alluvial and colluvial soils

Various alluvial and colluvial soils also occur widely in the humid Tropics, chiefly in coastal areas. The soils may be beach deposits of sand, generally considered poor for agriculture but often supporting coconuts. These soils are stable and may be used profitably for intensive vegetable production, with irrigation and correct use of fertiliser. Generally, beach deposits (and also ash soils) hold much less water than latosols, in the range 50–75 mm per metre of soil. Consequently they require even and frequent rainfall or irrigation for good crop yields. Other alluvial soils are much more clayey and are usually considered good for agriculture. They are much used for crops such as rice, sugar cane and oil palm. Their chief problems are flooding and drainage. In Malaysia, alluvial soils formed under a marine environment are rich in nutrients while those formed under freshwater conditions are generally poorer in nutrients and require more fertilisers. Clayey alluvial soils hold about 150–200 mm of available water per metre of soil depth.

Sulphide soils

Certain coastal and low-lying soils called sulphide or acid-sulphate soils are found throughout the humid Tropics. These soils have high levels of sulphide when they are waterlogged and poorly drained; they may give off a smell of hydrogen sulphide (bad eggs). When drained, they become very acidic (pH down to 2·0) and are poor for cropping. They require so much lime to reduce acidity that they are often considered unsuitable for agriculture. Drainage of such soils is not recommended and they should be used for rice growing under flooded culture. Sulphide soils occur in the region of mangroves and nipa palms.

Other soils

Two other humid tropical soils which deserve mention are peat and coralline soils. Peat soils are formed in lowland, poorly-drained areas and are often very extensive. Malaysia for example has about 3 million ha of peat soil. Peats contain up to about 90 per cent or more of organic matter and are very friable and loose. When limed and fertilised they can be used for growing crops such as pineapples, cassava, vegetables and many others. They are especially suitable for cassava and other root crops because of easy harvesting from the loose and light-textured peat; they should not be used for tree crops because they give poor support and trees fall over. Deficiencies of micronutrients, particularly copper, zinc and boron, occur on peats.

Coralline soils occur widely throughout the Pacific, the Caribbean and Central America and are used mainly for coconut production. The main problem here is correct use of fertiliser and attention to deficiencies of micronutrients, particularly iron and manganese.

Using soil profitably and maintaining fertility

If a soil is to be kept in good condition it can be considered a good rule for most agricultural purposes that the level of organic matter should be maintained. An exception to this is nutriculture on pure sands. Organic matter here refers to humus, undecomposed organisms and organisms. The effects of organic matter may be summarised as follows:

1 The physical addition of the residues improves the drainage of heavy clay soils and the water-holding capacity of coarse-grained sandy soils by the formation of humus.

2 Organic matter acts as a stabilising agent by encouraging micro-organisms (fungi and gum-forming bacteria) which bind the soil particles together.

3 Surface organic matter such as leaf litter reduces raindrop impact which is one cause of soil erosion.

4 Organic matter is highly active in storing the plant nutrients nitrogen, phosphorus and sulphur; potassium, magnesium and calcium are also stored. Increased humus increases the cation exchange capacity (CEC) of the soil. When the organic matter level is very high, as in peat soils, micronutrients such as copper and zinc are immobilised. The living cells of organisms in organic matter (e.g. bacteria) also store nutrients.

In crop production, there should be a balance between

the formation and breakdown of organic matter, since the breakdown or mineralisation of organic matter is necessary to release bound nutrients, particularly nitrogen, phosphorus, sulphur and micronutrients.

Organic matter is best maintained by reducing soil exposure and correctly replacing lost nutrients. Not, as often supposed, by adding tonnes of cow or other dung to the soil, though this, when available, should also be used.

Soil cover is necessary to prevent erosion since it is the impact of raindrops on the soil that is the main cause of erosion. Table 4.1 summarises an experiment carried out by Hudson in Rhodesia to show the importance of soil cover in preventing erosion.

Table 4.1

Type of soil cover	Soil erosion (tonnes per hectare)
Permanent grass cover (protection from raindrop impact)	3·3
Bare soil	350·0
Soil covered with mulch	3·0

Soil cover may be achieved by living plants (e.g. cover crops) or mulch. Fig. 4.1 shows the factors involved in soil erosion and stability.

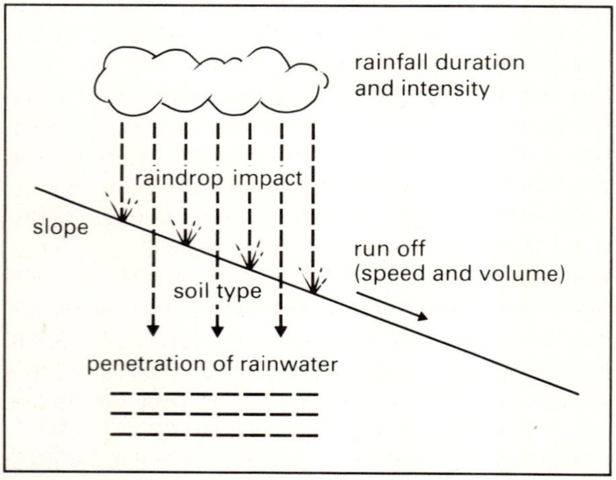

fig. 4.1 The factors affecting soil erosion

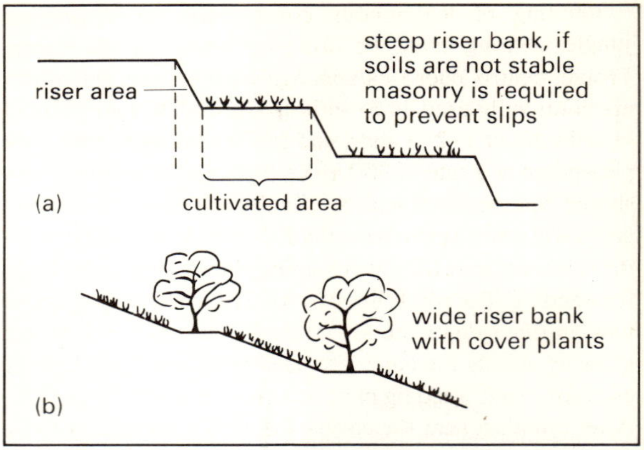

fig. 4.2 Two types of terrace used in the cultivation of steep land (a) Bench terraces for vegetables (b) Bench terraces for tree crops

Thus, in good farming the aims are to reduce soil exposure and maintain organic matter and these aims can be achieved in different ways for different types of crops. Briefly, we may divide crops into perennial tree crops and short-term field crops. These are considered separately.

Perennial crops

Perennial crops, most of which are trees, are grown for long periods, usually in excess of 20 years and sometimes for more than 100 years. In correct land use for perennial crops the old natural vegetation (usually forest) which is stable should be replaced with another artificial forest of the perennial crop, as quickly as possible. With unstable soils such as shales, slopes above about 20° should not be used for agriculture but left with natural vegetation. With lesser slopes small terraces and long riser banks should be used as indicated in Fig. 4.2.

With stable soils (such as granite soils) even slopes of more than 35° may be planted without terracing, though terracing should still be encouraged to reduce rainfall run-off and loss of fertiliser. With steep soils the establishment of legume covers becomes difficult as the seeds tend to be washed out. It may become necessary to encourage a natural grass cover. Grasses are encouraged by slashing. On steep slopes disturbance of inter-row areas by weed control should be minimised to allow growth of a strong soil cover.

Planting of leguminous cover crops in incomplete jungle clearing must be done by hand. In the humid Tropics centro, puero, calapo, *Stylosanthes spp.* and others are commonly used as cover crops (p. 7). They are planted in drills about 1–2 m apart and fertilised with ground rock phosphate at a rate of 200 kg per hectare. At planting they should be inoculated with a suitable *Rhizobium*. The cover crops will start to grow from the drill line and spread between the rows. Because they are creeping plants, they can be lifted for weeding. The whole line of cover is lifted from the ground first on one side of the line, then on the other, and weeds growing among the cover are hoed out or sprayed. During establishment, weeding of covers should be done regularly. After establishment the covers will reduce weed growth by competition. Slowly, over the course of time, weeds will invade the cover which will become a mixture of natural species and the planted legumes. At this stage only certain weeds (weeds known to have a bad effect on the crop) should be killed by selective weeding, using suitable herbicidal or hand methods. Later still, the crop trees will shade out ground cover so that maintenance weeding in mature tree crops will be greatly reduced.

In the humid Tropics disc-cultivation to kill off covers prior to a dry season, is not practised. However, in subhumid regions (for example, in parts of West Africa, Thailand, Burma, Sri Lanka, South India and elsewhere) this is done to reduce competition for ground water between the cover and tree crop. Under these conditions a light disc-cultivation is used and the covers generally reseed themselves and reappear during the following wet season.

Perennial crop seedlings should be planted at the same time as the cover crop, using the correct spacing and methods. The tree crop is protected from weed competition by circle weeding. This is the main weeding operation carried out with perennial crops. Under certain circumstances it may become necessary also to control cover and weed growth between the rows. This occurs when the growth becomes too profuse, and particularly when the proportion of weeds to cover crops is high.

Unfortunately, there are no selective herbicides which favour cover legumes and kill other weeds. Thus, controlling the weeds between the rows involves either expensive selective weeding by hand (in the case of grasses such as *Imperata cylindrica* the leaves may be rubbed with mineral oil using a cloth), or general control that cuts back both cover legumes and invading weeds. Two types of general control are used. The first is slashing. This encourages grasses to grow (because their growing points are protected) and will generally only be used in parks, agricultural gardens, and steep areas where grasses are needed to bind the soil. The second method is the use of a general herbicidal spray between the rows. The type of herbicide used must depend on the types of weed present and the sensitivity of the species of tree crop grown. These matters are considered in the section on herbicides (p. 7), and in the sections dealing with individual crops in which particular herbicides are recommended. Spraying between the rows should not be so frequent that a bare soil surface is produced, but should be timed so that the weeds are controlled at a low level and there is always a mulch of dead weeds over the soil. It will be necessary to change the herbicides from time to time as different weeds become dominant.

When the crop is young, circle weeding is usually done by hand. At later stages, weeds may be controlled by spraying with herbicidal chemicals. These must be known not to harm the crop and care should be taken not to spray them directly on to the foliage of the crop.

In some perennial crops and under some circumstances, mulch, instead of cover legumes, may be used to suppress weeds very effectively. For example, in certain parts of East Africa where the rainfall is not quite enough for coffee, mulching greatly increases yields of this crop. Mulching is valuable because it provides ground cover and prevents erosion, increases water acceptance and reduces evaporation (see p. 7) but does not, like cover legumes, compete for moisture with the crop. The only problem with the use of mulch is availability. It must be borne in mind that a considerable amount of material is required to mulch each hectare and, further, it is a continuing requirement. A variety of materials can be used, for example, prunings of the crop itself, but these must only be used when they do not harbour pests and diseases. Artificial material such as polythene sheet may be used. Because of high costs, this is generally only practicable in nurseries or in high-value crops (for example, pineapples in Hawaii). Thirdly, the cut stems of grasses may be used (such as Guatemala, Napier, etc.). These will have to be grown separately and cut at intervals of 2 to 3 months. About 20 tonnes of green material per hectare should be used, to cover the area between trees or bushes.

Sometimes, instead of growing cover crops, or using mulch, the spaces between tree crops may be cultivated for short-term annual crops. General methods of growing these are considered further in Chapter 6, but here it should be noted that intercropping is generally only profitable when the main tree crop is young. Also, the intercrop should not interfere with the growth of the main crop (this usually means growing shallow-rooted intercrops, restricting their spread and supplying extra fertilisers so that the intercrop does not compete with the main crop for nutrients). Legumes are favoured as intercrops because they help provide the main perennial crop with nitrogen.

Field crops

Field crops are annual or seasonal crops that occupy the soil for a short period. They include short-term annuals and crops which are a year or more in the ground, such as cassava, sugar cane and others. The same principles as for perennials apply to the maintenance of fertility in field crops; namely, minimising soil exposure and adequate use of fertilisers to replace nutrient losses. With field crops the degree of soil exposure is much greater than with tree crops because of the more frequent need to replant. Generally, it can only be recommended to use less steep soils for growing field crops. Slopes up to 5° are suitable for cultivation with little risk of erosion but slopes of 5–10° generally require contour planting. Steep slopes can be used safely with extensive terracing and construction of retainer walls to avoid land slips. Terraces should be suitably graded to allow sufficient run-off and reduce infiltration. Excessive infiltration in unstable soils, such as shales, causes them to become fluid and subject to land slips. With steep, unstable soils, mulch can also be effectively used to prevent erosion without any attempt at terracing or contour planting. This allows a high proportion of run-off and minimises land slips. It is usually better to use steep slopes on unstable soils for perennial crops or forest. Stable soils when steep also require terracing for convenience of cultivation, but retainer walls are not necessary. Control of flooding is important on flat coastal land which is commonly used to cultivate rice.

When a crop is grown occasionally, as in shifting cultivation or, for example, when cassava is grown as a fallow crop and occasionally harvested, there is an extensive fallow period between cropping during which secondary ground cover invades the soil and some regeneration of fertility occurs. This recovery of fertility from nutrients in the sub-soil is not very great, and the soil will never regain its virgin state of fertility without addition of fertilisers. However, the occasional use of soil is a way of extending its natural fertility so that it lasts longer.

A better way of growing field crops is to farm one area of land intensively in such a way that it maintains its fertility and produces high yields. This requires the use of fertilisers at appropriate rates. Taiwanese farmers manage to grow 5 crops a year including 2 of rice, various legumes and other vegetables. This is only possible by interplanting the old maturing crop with the seeds or seedlings of the following crop. For example, beans may be planted between the rows of ripening paddy; during harvesting the paddy, care is taken not to trample the young bean plants. Asian farmers have been intensively cultivating fields for centuries without much loss of fertility. They use animal and human dung and other natural fertilisers and grow legume crops in the rotation. Nowadays artificial fertilisers are also used especially with new, more demanding and higher-yielding varieties of crops. In general a rotation of crops should include a legume as frequently as possible. In intensive cultivation, cereals, legumes, other vegetables and root crops such as sweet potatoes should be rotated to prevent build up of pests and diseases. Two crops belonging to the same family should not be grown in succession. When drainage is difficult, crops adapted to flooded culture should be grown in rainy seasons.

Plant nutrients and fertilisers

The key to crop improvement throughout many parts of the Tropics is the correct use of fertilisers and the use of newer, more fertiliser-responsive varieties. In the case of some crops fertiliser requirements are well established, with other crops certain general rules must be followed to obtain benefits from fertilisers. Correct use of fertiliser involves supplying the correct amounts of nutrients for optimal yield of the crop and replacing those lost during cultivation of the soil by leaching and in other ways. As a general rule, it might be considered that the application of about 1·5 times the quantities of nutrients removed by the harvesting of the crop will maintain the fertility of soil and crop yields. Certain other general facts about fertilisers and plant nutrition are important in the efficient use of fertilisers and these are considered below.

Essential plant nutrients

Altogether thirteen mineral elements are known to be necessary for plant growth. These are nitrogen (N), phosphorus (P, sometimes calculated as phosphorus(V) oxide, P_2O_5), potassium (calculated as K, or potassium oxide, K_2O), calcium (Ca), magnesium (calculated as Mg or magnesium(II) oxide MgO), sulphur (calculated as sulphate(VI) ion, SO_4^{2-} or S), manganese (Mn), boron (B), copper (Cu), iron (Fe), molybdenum (Mo), zinc (Zn) and chlorine (Cl). Because of the fact that these nutrients are required in different quantities, they are divided into three groups called primary, secondary and micro plant nutrients, or trace elements as indicated in Table 4.2.

Table 4.2 Essential plant nutrients

Primary	Secondary	Micro
nitrogen	calcium	boron
phosphorus	magnesium	iron
potassium	sulphur	copper
		chlorine
		manganese
		zinc
		molybdenum

In addition to these, silicon (Si), though not an essential nutrient, is useful to crops because an adequate supply gives thicker cell walls and increased resistance to diseases (e.g. blast disease of rice).

Nutrients play many roles in plant growth. Primary and secondary nutrients are required in relatively large amounts. Most of them form parts of the structural components of plant cells. Micro elements are generally concerned with cell metabolism and enzymes and are thus required in very small amounts.

Throughout the development of a crop, a considerable movement of nutrients occurs. Some move to the regions of greatest growth activity such as shoot or root tips and others in the direction of water movement from roots to tops (the transpiration pathway). Thus, young developing organs, such as leaves, remove some nutrients from senescing leaves as development occurs. During flowering and fruiting, the strong sink strength (drawing off of nutrients) developed by this growth activity brings about a limitation in the growth of other organs as nutrients are diverted to the inflorescence. Some nutrients are more mobile than others. Usually nitrogen and phosphorus are relatively mobile while suphur is relatively immobile. Thus, sulphur deficiency symptoms first appear in the youngest leaves since older leaves are unable to translocate sulphur to younger growing regions. Of the cationic nutrients, potassium and magnesium are usually highly mobile in plants while calcium appears not to be redistributed to any great extent following uptake. Of the minor elements, copper and zinc are relatively mobile. There is evidence that manganese, boron, iron and molybdenum become fixed in the leaves, but move freely in other parts of the plant. Like sulphur deficiency, iron, boron, manganese and molybdenum deficiencies also develop first in the youngest leaves.

Nutrient deficiency symptoms

Nutrient deficiency symptoms for individual elements vary considerably from plant to plant. Nevertheless there are certain classical symptoms which are commonly associated with shortages of the individual elements, and these are described briefly below:

Nitrogen and **sulphur** deficiencies may result in smaller, pale green (chlorotic) plants. Sulphur deficiency appears in young leaves and causes tip and marginal leaf death, and blackening (necrosis); nitrogen deficiency produces a more general yellowing and in the early stages can often be recognised by the yellowing of the leaf veins as opposed to the region of the leaf blade.

Phosphorus deficiency also produces stunted plants but in this case the leaves often become very dark green in colour with some leaves becoming purple.

Potassium deficiency characteristically results in a yellow mottling of the leaf surface and in later stages this develops into a patchwork of orange and necrotic areas both at the leaf margin and on the blade; the older leaves are affected first.

Calcium deficiency is usually confined to the leaf margin which may develop a white, bleached appearance. Cup-shaped and deformed young leaves, due to interrupted development in the early stages of cell division are also common in calcium-deficient plants. Usually the younger leaves are affected first in calcium deficiency.

Magnesium deficiency causes a bronze yellowing of leaves which may be distinguished from other deficiencies causing leaf yellowing by the fact that the older leaves or those leaves which are fully exposed to the sun are always affected first. The veins usually remain green with the interveinal tissues becoming chlorotic, the opposite of the symptom of nitrogen deficiency. Magnesium is highly mobile in plants and is transported to the younger developing leaves.

Manganese deficiency produces similar leaf symptoms. These can be distinguished from those caused by magnesium deficiency by the fact that younger leaves are affected first and the bronze coloration is usually absent.

Iron deficiency, because of the role of iron in chlorophyll formation, causes a leaf yellowing; the youngest leaves are affected first, and the yellowing is uniform over the entire leaf. Iron is relatively immobile in plants, so that as the supply of iron is depleted progressive symptoms of deficiency develop in the newer tissues. The symptoms are easily confused with sulphur deficiency.

Zinc and **boron** deficiencies are both known to produce unexpanded malformed leaves, and the condition is often given the name of 'little-leaf'. Zinc deficiency also results in interveinal chlorosis, while boron deficiency in the more advanced stages results in necrosis of tissues, root rot, heart rot and death. Boron deficient plants have an increased susceptibility to disease.

Copper and **molybdenum** deficiencies both affect the tips and margins of leaves which show yellowing and scorching.

Types of fertiliser and their use

It is necessary to understand the properties of fertilisers in order that they may be safely and profitably used in crop production, and for this reason a short account of each of the major groups: potassium, phosphate and nitrogen fertilisers, is given below. Features which are of particular importance are:

1 The degree of solubility in the soil solution (or salt index).

2 The acidification of the soil which results from use of the fertiliser.

Salt index is a measure of fertiliser solubility which is expressed with reference to sodium nitrate(V), which is highly soluble and is taken to have a salt index of 100. Risk of damage from excessive application rates is thus high with sodium nitrate(V). Other fertilisers with high salt index are ammonium nitrate(V) (105), potassium chloride and potassium nitrate(V) (114 and 74), urea (75) and ammonium sulphate(VI) (69). Fertilisers with a low salt index are the phosphatic fertilisers.

Ammonium fertilisers acidify the soil. Potassium nitrate(V) and sodium nitrate(V) tend to be neutral while mineral phosphate and of course lime, make the soil less acidic. Calcium(II) nitrate(V) and calcium(II) ammonium nitrate(V) also tend to make the soil more alkaline and may be used as a nitrogen source in soils of high acidity. Choice of nitrogenous fertilisers depends on conditions of soil acidity, on the stage of crop growth and on the crop.

An advantage of urea is that it contains a high proportion of nitrogen (45 per cent) and is therefore a useful nitrogen source where transport is costly. Urea has been found to be a useful nitrogen source under tropical conditions and is often mentioned in connection with specific crops. It is not so strongly acidifying as ammonium fertilisers.

Ammonium fertilisers and urea should not be mixed with alkaline fertilisers such as lime, rock phosphate or basic slag because a chemical reaction takes place resulting in the release of ammonia. Losses of ammonia also occur in soils from urease activity especially with urea. Using small quantities more often, reduces this ammonia loss.

Moisture uptake by fertilisers containing hygroscopic chemicals is a practical problem in their use and various methods have been used by fertiliser manufacturers to overcome this. These include protective packaging in polythene bags and the coating of fertiliser granules with a substance which does not take up water, such as lime, clay or a film of oil. Fertiliser bags should only be opened immediately before they are used. Tables 4.3, 4.4 and 4.5 indicate the main properties of the common fertilisers used in agriculture.

Nutrient supply problems

In the flooded culture of rice only urea or ammonium fertilisers should be used because nitrate fertilisers are lost as nitrogen gas.

Phosphate fixation is a special problem in acid tropical soils, especially latosols. Experience has shown that the application of comparatively inexpensive ground rock phosphate is preferable because fixation of the more expensive soluble phosphates is so rapid. In acid soils, rock phosphate liberates phosphate(V) ions due to reaction of the slowly soluble calcium(II) phosphate(V), and the residual effect of high applications has been found suitable for the phosphate nutrition of tree crops such as rubber, oil palm, etc. over long periods. With fast-growing field crops, however, and on less acid soils, superphosphate should be used. In general, phosphate supply is not a problem in the wet cultivation of rice, and rock phosphate can be used as a fertiliser if the soil is moderately acid. In calcareous soils, such as the coralline soils (p. 13), it has been found that the soluble calcium phosphates tend to revert to less soluble forms and should therefore be used at somewhat higher rates.

There are no particular problems associated with the use of potassium fertilisers except antagonism with magnesium and other cationic nutrients (see below). Potassium is relatively strongly retained on to soil colloids so that many soils may be treated before planting without risk of leaching.

Table 4.3 Properties of nitrogen fertilisers

Fertiliser	Origin	Approximate analysis	Particular condition of use
Ammonium nitrate(V)	From reaction of nitric(V) acid and ammonia	32–33 per cent nitrogen	A commonly used source of nitrogen well retained by soil. Appropriate care should be taken to avoid leaf burn. Continued use can lead to development of undesirable acidity.
Ammonium phosphate(V)	Reaction of phosphoric(V) acid and ammonia	16–21 per cent nitrogen 48–53 per cent phosphorus (as P_2O_5)	Has advantage of supplying both nitrogen and phosphorus. The phosphate fraction is readily available to plants but in red tropical soils may be converted to unavailable forms. It should not be mixed with alkaline fertilisers
Ammonium sulphate(VI)	From reaction of sulphuric(VI) acid and ammonia and by-product of gas industry	20–25 per cent nitrogen 59 per cent sulphate(VI)	Has advantage of supplying both nitrogen and sulphur. Should not be mixed with alkaline fertilisers. Continued use can lead to development of undesirable acidity.
Calcium(II) nitrate(V)	From reaction of lime and nitric(V) acid	15 per cent nitrogen	Used to prevent build up of acidity
Urea	Made synthetically from carbon dioxide and ammonia	45 per cent nitrogen	A concentrated source of nitrogen with good keeping properties but slow-acting. Convenient for remote areas.

Table 4.4 Properties of phosphate fertilisers

Fertiliser	Origin	Approximate analysis	Particular conditions of use
Basic slag	By-product of steel industry	16–20 per cent phosphorus (as P_2O_5)	Relatively low solubility. Must be applied in advance of the crop. Calcium content makes it suitable for leached acid soils.
Calcium(II) phosphate(III)	Prepared from rock phosphate or calcium(II) phosphate(V)	53 per cent phosphorus (as P_2O_5)	Relatively low solubility. Must be applied in advance of the crop. Also suitable for leached acid soils or for iron-containing soils.
Rock phosphate (Calcium(II) phosphate(V)	Mined and pulverised	28 per cent phosphorus (as P_2O_5)	Very low solubility. Must be applied in advance of crop but has often been found to be more economical than the soluble phosphates in phosphate-inactivating soils. Can generally be recommended for tree crops in acidic tropical soils.
Superphosphate	Reaction of rock phosphate and sulphuric(VI) acid	18 per cent phosphorus (as P_2O_5)	Soluble, rapidly acting phosphate fertiliser but on latosols tends to be uneconomical because of fixation. Also contains calcium and sulphur.
Triple super-phosphate	Reaction of rock phosphate and phosphoric(V) acid	45 per cent phosphorus (as P_2O_5)	A concentrated soluble phosphate source convenient for remote areas or for preparing mixtures. Also rapidly fixed in latosols.

Tables 4.5 Properties of potassium fertilisers

Fertiliser	Origin	Approximate analysis	Particular conditions of use
Potassium chloride	Mined and purified	50–60 per cent potassium (as K_2O)	Most widely used potassium fertiliser.
Potassium nitrate(V)	Manufactured from Chile saltpetre and potassium chloride	48–52 per cent potassium (as K_2O) 13 per cent nitrogen	Used for crops sensitive to chloride.
Potassium sulphate(VI)	Manufactured from sodium sulphate(VI) and potassium chloride	48–52 per cent potassium (as K_2O)	Used for crops sensitive to chloride.

Sulphur and calcium are seldom found to be limiting in crops. Sulphur is usually supplied as sulphate in fertiliser mixtures, and calcium is supplied in phosphatic fertilisers and as lime. The increasing use of urea to replace ammonium sulphate(VI), however, may lead to problems of sulphur deficiency.

Magnesium is often included in bag mixtures as kieserite (magnesium(II) sulphate(VI)) or is supplied as dolomitic limestone. The former is more useful for fast-growing annual crops and in less acid soils, and the latter for crops grown on acidic soils and with slower-growing tree crops.

Nutrient ions which are similar in charge antagonise each other at the site of uptake into plant cells so that a high level of one will lead to reduced uptake of the other, and this may lead to development of deficiency symptoms even in the presence of normally high levels of a particular element. The cationic nutrients, calcium, magnesium, manganese, iron and potassium all antagonise each other. Heavy applications of either potassium or calcium are known to induce magnesium deficiency. For example, it was established soon after World War II that certain soils used for oil-palm plantations in Malaysia were deficient in potassium. Subsequent heavy applications of potassium led to an induced magnesium deficiency which reached serious proportions before it was detected. This is one of the most commonly encountered antagonisms between fertilisers. Antagonism between ions, of course, works in any direction since it depends on their relative concentrations. It is seldom, however, that heavy applications of magnesium are made to soils. Calcium and other cations also antagonise manganese and iron uptake both through site antagonism and by raising soil pH (reducing acidity) which decreases their solubility. Sodium antagonises the uptake of all cationic nutrients in saline soils and the practice of applying gypsum (calcium(II) sulphate(VI)) to reclaimed saline soils, serves the purpose of creating a more favourable balance of the plant nutrient ions, calcium and magnesium, to sodium.

The micronutrients are not normally included in bag mixtures of fertilisers or single fertilisers except as impurities. Molybdenum is sometimes included in superphosphate. This is because it is needed by legumes and phosphatic fertilisers are often applied to legume crops. A few bag fertilisers, of which the best known is 'nitrophoska blue special', also contain added micronutrients. When deficiencies have been determined, micronutrients are applied as necessary, usually as single elements. Quite often foliar sprays are used to supply micronutrients (see p. 23). Copper and zinc are commonly supplied as foliar sprays, often as fungicides. Iron and manganese deficiencies seldom occur in acid tropical soils and manganese, along with aluminium, may be present at toxic levels on very acid soils (see below). Deficiencies occur mostly on alkaline soils (e.g. coralline soils) when iron and manganese may also be supplied as foliar sprays. Copper, zinc and boron are less available to plants on alkaline soils, and deficiencies of these nutrients also occur on organic peat soils because the elements become bound in the organic matter. Molybdenum is more readily available on alkaline soils.

Boron is best supplied to the soil as sodium borate(III). Chlorine is seldom supplied as a nutrient because it is generally present in sufficient quantities as an impurity in fertilisers and as potassium chloride (KCl).

Toxicity of some mineral elements

Some of the nutrient elements mentioned above are only required in relatively small amounts and concentrations in excess of low levels may be distinctly toxic to plants. Manganese, iron and aluminium toxicities are most commonly encountered in acidic tropical soils. In very acid soils, aluminium becomes soluble and is taken up by plants thus giving rise to toxicity symptoms. The beneficial effect of lime in some cases is due to reduction in the solubility and uptake of aluminium. Aluminium toxicity often appears similar to phosphorus deficiency. Manganese and iron toxicity symptoms usually involve some form of leaf-yellowing and necrosis, spotting or chlorotic mottling.

Fertiliser application and timing

There are two factors which affect the timing of fertilisers.

1 The minimising of leaching losses by fractional application.

2 The changing of crop requirements for fertiliser with developmental and seasonal changes.

In humid tropical soils mineralisation of nutrients occurs rapidly. In combination with high rainfall, this leads to rapid losses of nutrients by leaching. Generally, with perennial crops it may be said that fractional application of fertilisers will be more economical than applying large quantities once a year, which is common practice with many tree crops.

Changing crop requirements may generally be related to developmental changes. The composition of fertilisers required may change as well as the total quantities required, as for example, an increased demand for potassium and a reduced demand for nitrogen with approaching maturity or with fruit production. This occurs in coconuts, pineapples, cassava and many other crops. Crop requirements for fertiliser may also change with the season. For example, during a season of moisture sufficiency a high nitrogen status may be more desirable while during a dry season a high potassium status may be required. During an overcast cloudy season nutrient requirements may be generally lowered because of lowered photosynthesis and growth. These considerations have been very little studied in practical agriculture.

With annual crops such as rice or maize, however, it quite often appears that there is one main period in development when fertiliser is most required. That is when leaf area is increasing most rapidly after planting. Fertilisers for rice should be mixed into the soil after flooding to prevent losses by denitrification.

Methods of application of fertilisers are of practical significance in the economics of fertiliser use. Broadly, there are three ways of applying fertilisers.

1 Fertilisers may be broadcast more or less evenly over the soil surface. Sometimes they are worked in, to reduce loss by erosion and to reach the root zone.

2 The fertiliser may be applied in a confined band along or between rows of plants in field crops, or in a circle around tree crops.

3 Certain nutrients may be applied as a foliar spray or by application to other aerial parts of the plant.

Each of these methods has advantages in certain respects. Much evidence suggests that feeding roots in a crop are more or less evenly distributed throughout the upper soil layers; the roots of neighbouring plants intermingle, consequently in many situations, an even distribution of fertiliser over the whole surface would bring the nutrients in contact with the greatest number of plant roots. With large plants such as trees, there is conflicting evidence regarding the occurrence of concentric zones of feeding roots around each tree. In oil palms for example, it has been claimed that the zone of feeding roots occurs 2–3 m from the tree. However, it has been observed that death of roots occurs in the region of concentrated particles of fertiliser and that this is followed by the growth of numerous new roots, so that zones of roots may result from previous band application of fertilisers.

With tree crops it is usual to allow weeds or cover plants to grow in between the trees to some extent. Circles around each tree, or lines or paths following rows of trees, are kept free of weeds. Consequently, application of fertiliser is probably best confined to these weed-free areas. The fertiliser may be spread evenly or applied in bands within the circle. In certain circumstances, band application may be better than broadcasting; for instance, in the uptake of phosphorus from fertilisers in soils where phosphorus fixation occurs. In the case of perennial crops roots will generally become established in the fertiliser zone. When finely ground phosphatic fertiliser is broadcast, the more complete contact will favour reactions between fertiliser and soil which may inactivate phosphorus.

Foliar nutrition

Foliar nutrition has traditionally been used for supplying micronutrients. Foliar sprays may be used to supply a proportion of nitrogenous fertiliser, particularly for strategic applications when it is required rapidly. Urea is usually the source of nitrogen in foliar sprays. At rates of up to 5 per cent solution, urea is generally found to cause no damage to plants. Foliar application of urea is used extensively for pineapples, bananas and sugar cane. The main advantage of foliar sprays of nitrogen is that they correct deficiencies rapidly, an important point with short maturing crops. Urea is also commonly supplied to the leaf axils of pineapples.

Iron has been supplied as iron(II) sulphate(VI) at a 1 per cent concentration. Young plants are damaged by stronger solutions. The relative immobility of iron sometimes makes repeated applications necessary. Foliar application of iron has special importance in alkaline soils (e.g. coralline soils) where this element is not easily available from the soil. Iron deficiency has been corrected in pineapples under circumstances where the element was not available on account of the high manganese levels in the soil. Iron can also become immobilised in the presence of phosphate.

Zinc is often supplied as a foliar spray in various forms (sulphate(VI), oxide, carbonate, etc.) both in solution (e.g. 0·5–1·5 kg per 1 000 litres per hectare) and as a dust. Zinc application, as a foliar spray or dust, appears to be quite successful in correcting deficiencies and has been used in controlling mottle-leaf caused by zinc deficiency in citrus. Zinc is commonly supplied along with fungicides to control blister blight in tea.

Copper and manganese sprays at about 1 kg per 1 000 litres per hectare, with an equal quantity of lime, are also successful in controlling deficiencies in many species. Sufficient copper may be supplied in the form of copper fungicides.

Molybdenum deficiency is probably less common in the Tropics but responses have been reported to foliar sprays; for example, in citrus at a rate of 50 g per 400 litres per hectare of sodium molybdate. Foliar sprays of boron, either as boron or boric acid, have proved successful in many species. In palms, the application of boron to the leaf axils has been used to correct deficiencies.

With all micronutrient sprays, care must be taken to avoid overdoses causing toxic effects.

The best time to spray is when temperature and humidity conditions favour the formation of a liquid film on the leaf; drying conditions promote leaf damage.

Organic fertilisers and manures

Organic waste products have long been used as fertilisers to enrich the soil, particularly those which contain a high

Table 4.6 Approximate analyses of various organic fertilisers and manures

Manure	Nitrogen (per cent)	Phosphorus (as P_2O_5) (per cent)	Potassium (as K_2O) (per cent)
Bone meal		22	
Blood meal	12		
Fish meal	5	5	
Groundnut cake	7		
Coconut cake	1–3		
Farmyard manure	0·4–0·6	0·1–0·3	0·4–0·6
Poultry and pig manure	0·5–1·5	0·5–1·5	0·6–0·8

proportion of nitrogen. Sometimes it is thought that organic fertilisers are superior to artificial fertilisers, although most evidence suggests that the fertilising value of any particular compound is proportional to the content of its constituent nutrient elements. It is a fact, however, that the risk of crop damage is low with organic fertilisers. The use of organic fertilisers depends largely on cost and availability in any particular area. The approximate analyses of some common organic fertilisers are given in Table 4.6.

Animal manures should be used fairly promptly because nitrogen is lost as ammonia on storage, but a few weeks of rotting is required to prevent crop damage. The nutrients in animal manures are not all immediately available since the manure must be broken down in the soil to release the nutrients. It may be assumed that 10 tonnes of animal manure adds about 2 tonnes of organic matter to the soil.

Straw should not be used as manure or to increase the organic-matter level of the soil, except to improve a very heavy or very sandy soil. If straw is ploughed in, it will immobilise nitrogen and affect subsequent crops; it will be necessary to add about 50 kg of ammonium sulphate(VI) or 30 kg of urea for each tonne of straw ploughed in, to prevent nitrogen immobilisation. The old stubble of a cereal crop for example may require as much as 500 kg of ammonium sulphate(VI) per hectare to break it down over a period of 1 year. On the whole it is better to burn such material as rice straw and maize stalks or to use them as mulch over the soil, and to increase organic matter in the soil by growing a legume crop for green manure.

Nitrogen and legumes

Under natural conditions, rock weathering accounts for the presence of all plant nutrients except nitrogen. Nitrogen comes largely from the fixation of atmospheric nitrogen by legumes and other organisms. These may be either free living, in the soil or water, or in symbiotic association with higher plants, notably those higher plants belonging to the legume family.

Nitrogen added to the soil is removed continuously by various processes. These include leaching and erosion; volatilisation as ammonia (some of this is returned again in rainfall); denitrification of nitrate(V) in reducing soils (e.g. in wet soils for rice) and removal in the harvested crop.

All these losses must be made good in a stable agricultural system, either by the use of fertilisers, or by legume fixation.

Not all tropical legumes have been found to fix nitrogen and, unfortunately, too little is known about the relative value of the different legume crops and cover species. Nitrogen is fixed in nodules, formed by bacteria from the soil, on the roots of legumes. There is often a good correlation between the total weight of nodules formed and apparent nitrogen fixation. More particularly, nodules which are pink or red in colour from the pigment haemoglobin are most effective in nitrogen fixation. As far as tropical crop legumes are concerned, nodule weight is sometimes found to be relatively high. Varieties of *Vicia, Phaseolus, Arachis, Glycine,* and *Vigna* may produce from between 100 and 1 000 nodules per plant, weighing up to 200 mg each. Reports of nitrogen fixation by tropical legumes include *Stylosanthes gracilis* in Australia—85 kg per hectare per year; *Pueraria thunbergiana* in association with grass in Puerto Rico—169 kg per hectare; the leguminous shrub *Leucaena leucocephala* in Mauritius—191 kg per hectare per year. It has been found that nitrogen levels in the soil increased by about 500 kg in 2·5 years under the cover crop *Centrosema pubescens* in association with star grass in Nigeria. In green manuring, nitrogen bound in the crop itself is also incorporated into the soil. Species of *Calopagonium, Pueraria, Centrosema, Indigofera* and *Phaseolus* are reported to yield nitrogen in quantities ranging from 50 to 250 kg per hectare after 4–6 months growth. A study of nitrogen fixation by *Glycine javanica* over 9 years in East Africa on tropical red loam soils showed an increase in soil nitrogen content from 0·206 per cent to 0·270 per cent and, including the nitrogen in the crop itself, this was calculated to represent an annual fixation of 175 kg per hectare per year.

Nitrogen fixation by food legumes is of particular interest in tropical countries both because of the protein value of the harvested crop in human and animal nutrition, and for the benefit of soil nitrogen fixation. It has been shown that soy-bean can produce as much as 180 kg nitrogen per hectare of which about 80 per cent is removed as the harvested beans. Estimates of nitrogen fixation by pigeon pea (*Cajanus cajan*) average about 80 kg per hectare without fertiliser, and 100 kg per hectare with the addition of phosphate. Recently cowpeas (*Vigna sinensis*) have been reported to be very efficient in fixing nitrogen. In all legumes the appropriate strain of

Rhizobium should be used and generally this can be obtained through agricultural extension services.

In general, it would seem that tropical legumes are as capable of fixing nitrogen as temperate legumes, but they appear to be much more variable in their activity. It has recently been found that some tropical grasses and other non-leguminous plants are also capable of fixing nitrogen, presumably from bacteria living in the rhizosphere or root zone. Further studies are obviously necessary on nitrogen fixation by both legumes and non-leguminous plants, particularly on the occurrence and inoculation of suitable rhizobial bacteria in the soil and on the effects of nutritional, soil and other factors on nitrogen-fixing organisms.

5 Propagation and nursery techniques

A nursery phase is an important part of the planting operation in the cultivation of many tropical crops. This is true of most tree crops and some field crops, for example, rice. With tree crops, the planting distance is relatively wide compared to field crops and for this reason there is a large area of ground between planting points. Small tree-seedlings are more subject to weed competition and it is difficult to protect them from pest and diseases. If kept in a nursery until they are larger, seedlings can be better cared for with less effort and can also be conveniently watered to increase growth and development.

Many tropical crops are grown from cuttings or buddings and these require special attention involving frequent watering and shading. They are consequently better and more cheaply raised in a nursery.

In the case of perennial tree crops it is advisable to plant only strong healthy seedlings to the field, because the individual trees will be kept for many years. Weak and poor-yielding trees will be very unprofitable to the grower over an extended period of time. For this reason also the nursery is useful because it allows weak and abnormal seedlings to be rejected before planting out.

fig. 5.1 Use of watering-can in a small nursery

Nursery practice

The site

A nursery site should be chosen carefully, bearing in mind the following conditions:

1 It should be near a source of water, either wells or a stream. In the case of well water it should be noted that about 50 000 litres of water will be required daily to irrigate each hectare of nursery.

2 It should be as near as possible to the site of field planting.

3 The area should not be subject to flooding.

4 The ground should be level or terraced.

Watering

A small nursery up to about 0·1 ha in size can be watered by one man and his family using watering cans. The best way is to have cans mounted on a shoulder pole as shown in Fig. 5.1. A larger nursery will require more workers or the use of a pump and hose system. A small pump of about 3 kW will be suitable for a nursery of 1 ha and for larger nurseries appropriately heavier equipment will be

fig. 5.2 A large oil-palm nursery under sprinkler irrigation

needed. When a pump is used, a sprinkler system may be more convenient than a hose system but initially more costly. Fig. 5.2 shows a large nursery of oil palm seedlings under sprinkler irrigation. Normally once-daily watering at a rate of 50 000 litres per hectare on rainless days will be sufficient for most nursery seedlings in the humid Tropics.

Containers for seedlings and soil

Polythene bags have become the almost universal container for nursery seedlings, although when available, woven frond or bamboo baskets may also be used. Polythene however, is cheap and durable and makes handling of seedlings at planting out very easy. Usually black perforated polythene bags are used, perforated with 5 mm holes which should be about 5 cm apart and should cover the whole lower half of the bag. A good loamy soil should be used for filling the bags for most crops. Jungle top-soil is a favourite, but well-drained (structured) red soil may be used when jungle top-soil is not available; in this case rock phosphate should be mixed in at a rate of 1 kg per 100 kg of soil. If only heavy clays are available they should be pre-mixed with sand at a ratio of about 1 part sand to 2 parts of clay soil. The method of bag filling is shown in Fig. 5.3. Bags should be filled completely to the top to prevent the edges from folding in during watering.

The size of bags used depends on the seedlings. For instance, large seedlings like oil palms require bags of size about 60 × 40 cm laid flat; cocoa seedlings raised from seed require small bags of size about 15 × 10 cm, laid flat.

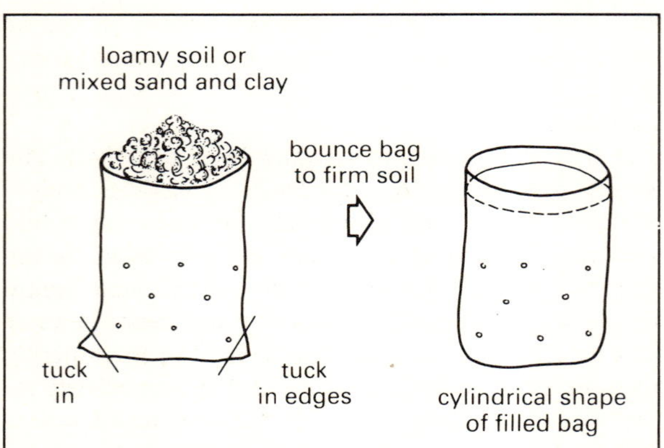

fig. 5.3 Method of filling a polythene bag with soil

Shading and spacing

Most young seedlings do better under partial shade at the early stages. When they get larger they will become self-shading and shade should be removed. The spacing of bags depends on the time they are to be kept in the nursery. Generally, however, spacing will either be close packed (as in Fig. 5.1) in which case seedlings will be ready to plant out as soon as they have completely filled the bags, or spaced up to about 3 times the bag diameter between seedlings (Fig. 5.2). When they are close packed, paths must be left between the seedlings for watering, weeding, etc. and the beds should not, for convenience, be more than about 1·5 m wide. Shade is usually provided by means of a framework covered with fronds (Fig. 5.4). A single layer of fronds will produce about 50 per cent shade which should be adequate for most purposes. When

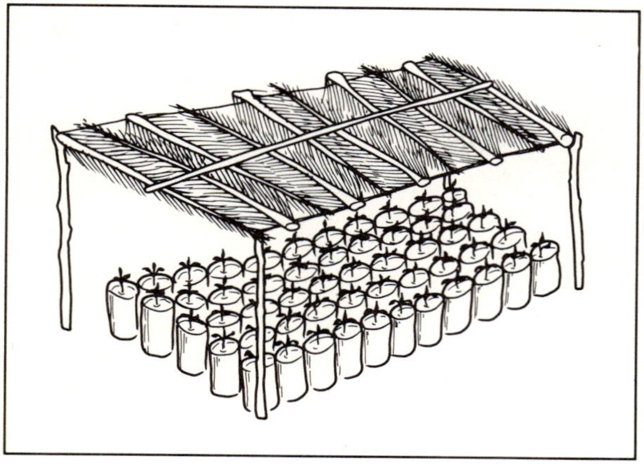

fig. 5.4 Shade frame using fronds

fig. 5.5 Stripping the polythene bag from a young papaya seedling ready for transplanting

the seedlings are well grown, but before they completely fill the bags, the shade should be removed. At first, half the fronds should be removed for a few weeks, and then, a few weeks before planting out, the seedlings should be completely exposed for hardening.

Transplanting

Transplanting of seedlings to the field should be done during a sufficiently wet season and the production of nursery seedlings should be timed for this. Use of rainfall records is necessary to determine the best time for field-planting. Nursery seedlings should just fill the bags at the time of planting out, (that is, the shoots of adjacent seedlings should just touch). If they have to be kept for one reason or another (e.g. because of unusual weather or bad planning) they should be widely spaced with at least 3 times the bag diameter between them. Seedlings established in soil contained in perforated polythene bags are very useful in this respect and can be kept for extended periods until required.

Before planting out, any weak, slow-growing, diseased, deformed or otherwise abnormal plants should be discarded. This is very important as it affects the long-term yield of the field-planting.

Seedlings grown in polythene bags are easy to transport to the field and, depending on the situation, may be carried by lorry, tractor, trailer or hand. A convenient method of carrying seedlings over rough ground is to use a box and double poles carried on the shoulders of 2 workers.

Before planting, the polythene bags must be stripped off by slitting lengthwise with a razor blade or very sharp knife. The roots of the seedlings should have formed in a firm core with the soil which can be separated easily from the bag (Fig. 5.5); but nevertheless the bag should be stripped off with care so as not to fracture the soil core.

Fertilisers

Usually nursery seedlings will grow faster if fertilised. The easiest way to fertilise seedlings is to dissolve a high-nitrogen fertiliser containing all the major nutrients: nitrogen, phosphorus, potassium and magnesium, in the ratio of about 15:10:10:5 in water, and to water lightly once every week. Fertiliser should be well mixed in water at a rate of about 1 kg per 400 litres (25 g per watering can). Fertiliser scorch can be avoided by immediately re-wetting with water alone. Sometimes micronutrient deficiencies will occur and these are best dealt with by watering with a complete fertiliser such as 'nitrophoska

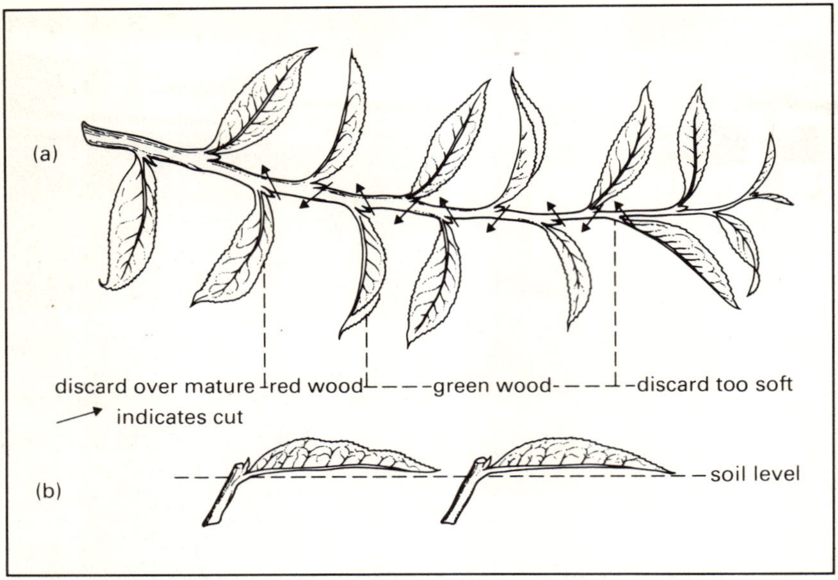

(c)

fig. 5.6 (a) Method of preparing single-leaf cuttings of tea for vegetative propagation (b) Method of inserting cuttings in the soil (c) Single-leaf cuttings of tea rooting in soil in a nursery (*Source:* **Eden, T.** (1965). *Tea.* Longman: London.)

blue special' dissolved in water as above, and given periodically throughout the nursery life of the seedlings.

Vegetative propagation

Many species of tropical crops are raised from vegetative parts. The 3 main types of vegetative material are:
1 side shoots (used for example, with crops such as bananas and pineapples);
2 cuttings;
3 graftings.

Side shoots do not usually require nursery care and can be planted straight to the field. Similarly, **cuttings** which are planted beneath the soil (e.g. sugar cane and cassava) are also planted directly to the field. Cuttings which are planted exposed to the air need special nursery care and protection from drying out. This can be achieved by shading and by enclosing the cuttings in a polythene 'tent' over a frame to reduce water loss. Cuttings may consist of short lengths of stem bearing several buds and several leaves, or a single leaf. Alternatively, a length of stem from which the leaves have fallen or have been removed may be used. The method of preparing single-leaf cuttings of tea is shown in Figs. 5.6 and 5.7. When considering stem cuttings of several nodes, it is generally found that green or semi-green wood is better material than old brown wood.

With all cuttings, part of the stem and cut surface

fig. 5.7 Single-leaf cuttings of tea under a plastic shade to retain humidity

should be planted beneath the soil when roots will form at the cut surface and elsewhere after a period of time (usually several weeks). Sometimes a rooting hormone is needed to promote rooting of cuttings. This is usually naphthalene-acetic acid or a similar compound mixed with a neutral carrier of talcum powder. Many brands are available commercially. To use the hormone, the cut ends are simply dipped in the powder before planting.

In preparing cuttings, a sharp knife or blade must be used to make a clean cut and the cuttings should be planted without delay. A well-drained rather sandy soil with little or no organic matter is often best. The use of fertilisers should be delayed until rooting has taken place. Watering, however, should be frequent to prevent drying out and hence the requirement for a well-drained soil which prevents excessive accumulation of water. Cuttings must always be planted the right way up.

Cuttings can be used to produce small trees and bushes but root-depth is less than that obtained from the propagation of seedlings since there is no tap root. Cuttings are suitable, therefore, for small fruit trees and

fig. 5.8 A lime tree grown from a cutting, note the much-branched main stem

hedges, such as those used in tea production. Trees tend to branch frequently from the base for support as shown in the lime tree grown from a cutting in Fig. 5.8. Large trees cannot be grown from cuttings because of the lack of support from the root system and the tendency to damage by wind.

Closely related to the cutting method of propagation is the method known as **marcotting**. This is a slow and tedious method of propagation but nevertheless much used in Asia. To produce a marcott, a healthy side twig or branch of about finge· thickness is selected, the bark is removed for about 0·5 cm with a sharp blade leaving the wood exposed, the twig is then bound in a ball of good top soil held in place with a plastic sheet and string. The method is illustrated in Fig. 5.9. After a time roots grow into the soil and the twig is cut and planted after removing the plastic cover.

The other main method of vegetative propagation is known as **budding**. With this method the need to protect the material against water loss is less than with cuttings; normal nursery care (i.e. daily watering and shade, p. 26) is satisfactory. Propagation by budding involves grafting the bud from a good-yielding, healthy tree (with desirable properties of disease and pest resistance, quality etc.) on

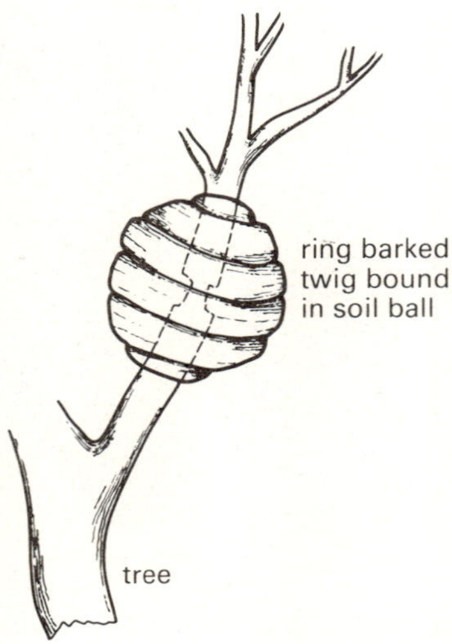

fig. 5.9 Method of preparing a marcott

fig. 5.10 A nursery-grown rubber plant pruned to produce bud wood

to a normal seed-grown stock seedling. Usually budding is best done at the green-wood stage when the bud wood and the stock seedling may be about 1 cm thick. The technique of making an inverted 'T' patch budding is shown in Fig. 12.10 on p. 104. Essentially all that is done is to cut off a bud with a sliver of wood still attached, using a sharp knife. Sometimes the wood is carefully peeled off leaving only the bark. The bud is then inserted under the bark of the stock seedling and taped with transparent plastic tape. Care must be taken not to bend the bud patch since this will bruise tissues and cause exudation of

fig. 5.11 Bud grafted rubber seedlings; the grafts have grown into tall seedlings

sap (this is especially important with latex-bearing species such as rubber). The bud must be the right way up and the inner face of the bud patch must be firmly in contact with the outer wood surface of the stock seedling. Sometimes a piece of bark of the stock seedling, the same size as the bud patch, is completely removed and the patch inserted in the space and bound with tape.

Budding should only be done on healthy stock seedlings which are known to make healthy combinations with the bud clone being used. The budding is usually made at about the centre of the stock seedling and after a few weeks, when the bud has taken, the top of the stock seedling is cut back and the tape removed. If successful the bud will then begin to grow and the only special care necessary is to pinch off any side shoots forming on the stock until the bud patch has grown strongly and become the dominant shoot. Sometimes, for large numbers of buddings, the tree or bush from which the buds are taken is especially grown in a nursery and pruned back to stimulate the production of numerous side shoots for bud wood. Fig. 5.10 shows a rubber nursery tree pruned to produce many bud-wood shoots and Fig. 5.11 shows a

fig. 5.12 A cleft graft of citrus. The shoot above is inserted into a V-shaped notch in the stock below. The graft is then taped for a few weeks to prevent separation

fig. 5.13 Equipment for extraction of soil-grown seedlings from the ground, *from left*: piston, core-extractor, key and hammer

fig. 5.14 The core is hammered into the ground around the seedling and then prised from the soil with the key

nursery of rubber buddings in which the patch buds have already grown into tall shoots.

Another method used to produce grafted seedlings is known as **cleft grafting**. Here the stock is cut back and a young shoot of the desired tree is cut wedge-shaped and inserted in a cleft in the stock as shown in Fig. 5.12.

Ground nurseries

Although polythene bags are widely used for nurseries and are very convenient, some planters and farmers may still prefer to plant seedlings directly in the soil. This often means, however, that considerable damage is done to the seedling on extraction thus delaying its development; in addition, more seedlings are lost. A convenient method of lifting seedlings which reduces losses, is to use a core extractor such as the one developed by the Rubber Research Institute of Malaysia. Figs 5.13–5.15 show the extraction of a rubber seedling. The core extractor is made of steel pipe of convenient size (in this case 200 mm diameter) with heavy steel wings. It is driven into the soil using a double-headed hammer and prised loose with the steel key as shown, the core is then extruded with the piston, bound in newspaper and sent for planting.

fig. 5.15 The piston is used to extract the core. After extraction the core is wrapped in paper and tied firmly for transport to the planting site

6 Light use and spacing

The importance of soil and soil fertility in crop production has been considered in some detail because the soil is one phase of the crop environment which can be changed readily by agronomic practices such as cultivation, drainage, the use of fertilisers, liming, the building up of organic matter and so on, to improve crop growth and yield. This emphasis on the soil often leads us to forget that the major part of plant growth and the agricultural yield, comes, not from the soil, but from the air through the process of photosynthesis. Much of the actual material of plants and crops belongs to the group of chemical compounds known as carbohydrates. These are formed from the gas carbon dioxide (CO_2) which is found in the air and in water, and include sugars, starch and cellulose. There are hundreds of other compounds formed directly or indirectly from carbohydrates or from the carbon dioxide in the air, for example, the phenolics which go to form wood, are formed indirectly from carbohydrates. The primary process of fixation of carbon dioxide into the materials of the plant is called **photosynthesis** and this takes place when the carbon dioxide from the air enters the green parts of the plant (chiefly the leaves) and is changed (reduced) into carbohydrate by the action of sunlight. This is illustrated diagrammatically in Fig. 6.1.

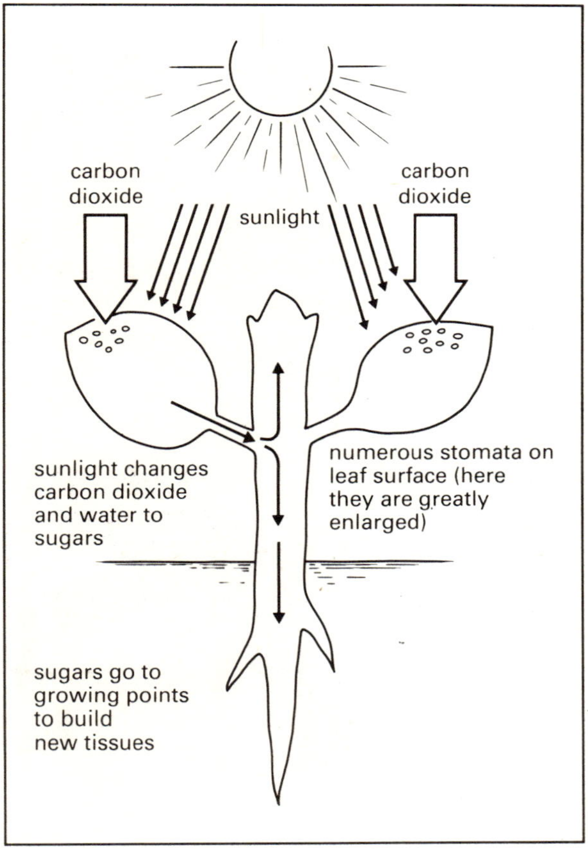

fig. 6.1 Diagrammatic illustration of the processes of photosynthesis

Carbon dioxide enters the leaf through minute openings called stomata which open during daylight hours and close at night (Fig. 6.2). Since photosynthesis requires

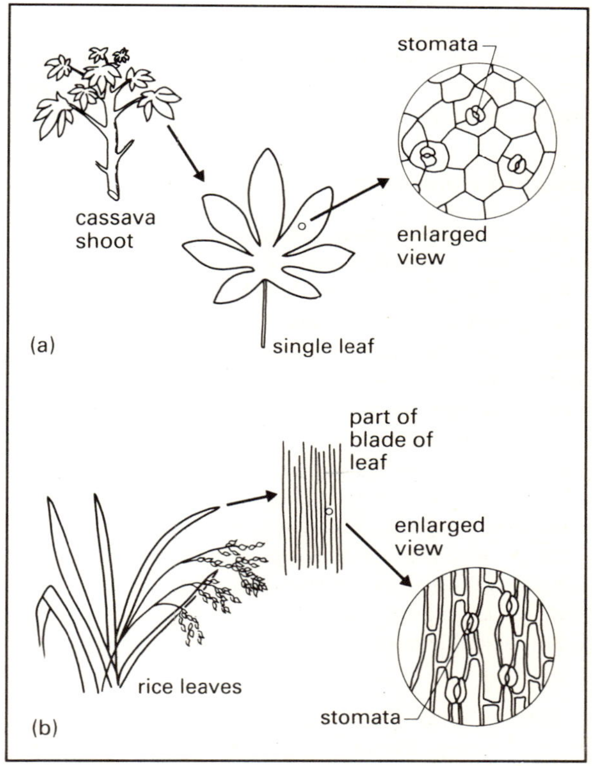

fig. 6.2 Position and microscopic appearance of stomata of (a) cassava, (b) rice

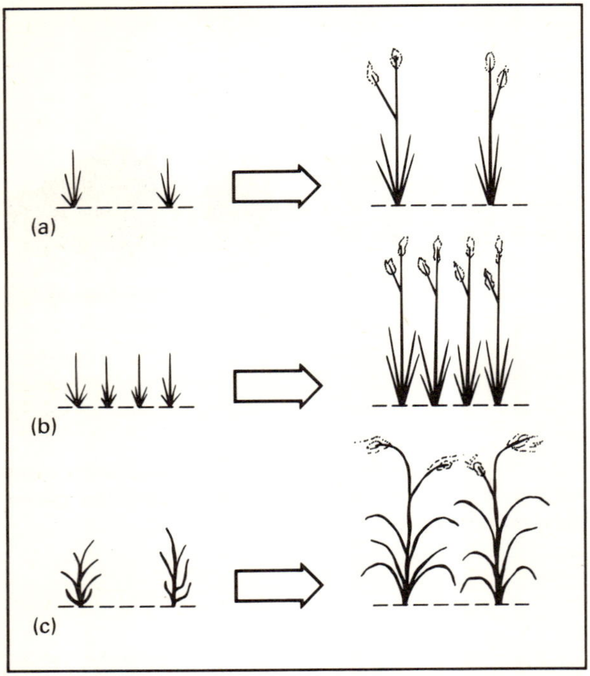

fig. 6.3 Plant spacing and yield (a) small erect variety at wide spacing gives low yields per hectare (b) close spacing gives very high yields (c) large leafy varieties require wider spacing to give the best yield (generally lower than in (b))

light, this process only takes place during daylight hours. Stomata also close and limit carbon dioxide uptake and photosynthesis during periods of moisture stress. This is one of the primary reasons why drought reduces the growth and yield of crops.

Living plants use up carbohydrates in the process of respiration. The growth of the crop depends on the balance of photosynthesis, which builds up carbohydrates and plant materials, and respiration, which uses them up. When photosynthesis exceeds respiration, which is normally the case in the developing crop, then growth occurs. Under deep shade, however, respiration may equal photosynthesis and no growth will occur. Since photosynthesis takes place primarily in green leaves, it follows that, up to a limit, the greater the area of green leaf the greater will be the amount of photosynthesis and growth. When the area of leaf becomes too great however, self-shading takes place and growth declines. In crop production it is best to grow the crop so that it attains the correct amount of leaf for maximum growth and production; this is usually achieved by using the correct spacing and the correct fertilisation. Agricultural scientists have referred to the amount of leaf formed on a crop in terms of **leaf area index** (LAI). Leaf area index is defined as the area of leaf of the crop in relation to the area of ground on which the crop is established. The optimal LAI is that at which the greatest growth and productivity occurs.

Optimal LAI varies with different crops and different varieties of crops. For broad-leaved short-statured species such as sweet potatoes, groundnuts and similar crops, optimal LAI may be between 4 and 5 depending on the variety (that is, the most productive leaf area will vary between 4 and 5 times the ground area on which the crop is growing). For cereal crops such as maize or rice, or other grass crops such as sugar cane, optimal LAI will be between about 7 and 12 depending on the variety. The main reason why grasses (i.e. cereals, sugar cane etc.) have higher optimal LAI values is because their leaves are more erect and permit better light penetration into the canopy and thus better illumination of the lower leaf surfaces. The optimal LAI of tree crops is usually between about 6 and 10 depending on the type of leaf and the height of the tree.

fig. 6.4 A correctly-planted small field of cassava. The spacing is close enough to give a good ground cover and eliminate unnecessary competition from weeds. Yield per hectare will be high

fig. 6.5 A good stand of sugar cane correctly planted; weeds are eliminated between the rows by shading out

fig. 6.6 Widely-planted fruit trees require unnecessary weed control and wastes ground. This spacing is suitable for planting in parks and gardens

It is difficult to determine LAI under field conditions so the farmer must use common sense in deciding on spacing. As general rule, it can be said that when the crop is fully developed, there should be a minimum of open spaces between individual plants of the crop (Fig. 6.3). With most annual crops the foliage should overlap at full development because a reduction in canopy size occurs on flowering. In crops like sugar cane, a reduction in canopy size occurs as the cane matures. Figs 6.4 and 6.5 illustrate correctly-planted cassava and sugar cane, respectively. With most perennial crops, the inter-row spaces should be minimal but excessive spread of the canopy is often prevented by pruning. Others, like palms, are self-restricting and do not spread. With certain crops (e.g. citrus and tea) in which the framework of the tree or bush is low, spaces are needed between the individual trees or rows for harvesting; pruning is carried out to prevent the spread of branches. When this is done, the size of the inter-plant or inter-row space should be kept minimal (that is, just wide enough for entry). Wide open spaces as illustrated in Figs 6.6 and 6.7 should be avoided. Fig. 6.8 shows a field of closely-planted tea in which the spaces between the bushes are just wide enough for the pluckers to enter.

Another general rule is that small erect-leaved varieties can be closer spaced (or more densely planted) than large varieties with drooping leaves. Such small erect-leaved varieties often respond to higher levels of fertiliser, particularly nitrogen. Examples of different varieties are shown in Figs 6.9 and 6.10. It often happens that small, erect-leaved varieties are higher yielding because of the high plant populations possible per unit of ground surface. This is illustrated diagrammatically in Fig. 6.3. Examples of incorrect plant spacing are shown in Figs 6.6, 6.7, 6.11 and 6.12. At the other extreme, if plants are too closely spaced, yield will also decline because individuals become excessively thin, susceptible to diseases and may also lodge (fall over).

Often, in tropical countries, the spacing adopted for crops is too wide. This is probably because in village and

fig. 6.7 The trees of this small variety of papaya are too widely planted. The exposed ground will suffer erosion and loss of organic matter; weed control will be difficult; they should be planted at about twice the density shown

fig. 6.8 Correctly-planted hillside tea with only narrow spaces between the bushes

backyard agriculture the plants are considered as individuals, rather than as members of a large community such as a field of corn.

It is often best to grow a number of different crops at the same time, as this gives the farmer a range of produce which will provide variety for home consumption and which may be easier to market than a large quantity of a single crop. The growing of a number of species also allows crop rotation, by moving crops from one plot to another in successive cultivations; this reduces the build up of soil pests and diseases of any particular species.

fig. 6.9 Two varieties of cassava growing side by side. The one on the left has broad horizontal leaves, it should be planted at a fairly wide spacing in rows 1 m apart. The one on the right has narrow erect leaves and should be planted closer

fig. 6.10 Two varieties of rice. On the left, the straight-leaved variety should be close spaced; alternatively, it may respond to high levels of nitrogen fertiliser. The fertiliser encourages production of side shoots which increases plant density in the same way as closer spacing

fig. 6.11 Maize planted in rows 1 m apart. This is too wide for this variety and light entering between the rows has encouraged the growth of weeds

fig. 6.12 Young sugar cane shoots too widely planted allowing growth of weeds between the plants

fig. 6.13 Young rubber seedlings interplanted with groundnuts

fig. 6.14 Young coconuts interplanted with sugar cane

fig. 6.15 Coconuts interplanted with coffee. The tall variety of coconut is particularly well suited to intercropping because it is usually widely spaced and permits a relatively high penetration of light to the intercrop

Crops suitable for rotation in small farms could include maize, various legumes, sweet potatoes, pumpkins and other curcurbits and many others. With perennial tree crops, however, short-term rotation is not possible and full use of the land involves interplanting as discussed below.

Correct plant spacing demands that the foliage of individuals overlaps and excludes spaces between the plants. This is true as a general rule but sometimes individual plants require special attention: when growing prize vegetables, for instance. In the special case of plants from which bud wood, cuttings, side shoots, or seeds are required the plant should be superior genetically to obtain the highest yield. Such plants should be isolated to receive maximum sunlight, kept clean, weeded and otherwise well tended; plants for bud wood must be pruned to encourage the growth of side shoots (e.g. see Fig. 5.10 on p. 30).

Interplanting in tree crops

With large species such as trees, there is a long period of time, while the trees are young, when there is extensive ground area between planting points which requires attention. In plantation cropping, this space is often planted with cover legumes. Cover legumes enrich top soil with nitrogen and reduce weed invasion but do not provide any economic product which can be sold or consumed unless grazed by livestock. For this reason intercropping is sometimes practised. The best kind of intercrop to grow is, of course, a legume; Fig. 6.13 shows young rubber interplanted with groundnuts. With all intercrops, extra fertilisers are needed so supply the needs of both the main crops and the intercrop. Figs 6.14 and 6.15 show young coconuts interplanted with sugar cane and coffee respectively.

With most tree crops, the early years are most profitable for intercropping. In the case of rubber, for

fig. 6.16 A taller species, banana, planted between young rubber to act as a nurse crop. This provides some shade for the developing seedlings

example, the first 2 or 3 years can be used; after the main crop has grown over, there will be too much shade for intercropping to be profitable. For this reason the intercrops chosen are usually short-term species such as groundnuts and sugar cane. The tall variety of coconut, however, is an exception and lends itself well to the intercropping of permanent crops such as coffee (Fig. 6.15), cocoa and some fruit trees. Coconuts are sun-loving and need to be fairly widely planted for optimal yield (usually about 10 m apart for the tall varieties). Consequently, there is sufficient light under tall coconuts to permit intercropping with shade-tolerant species. Pastures for grazing cattle and/or sheep are often grown under coconut.

Another type of intercropping is the use of nurse crops, tall-growing, temporary crops such as bananas, to shade perennial tree species when they are young, as illustrated in Fig. 6.16. This is often used to aid the establishment of cocoa in West Africa.

7 Farming practices and major crops of the humid tropics

The agriculturist is concerned with designing practical agricultural systems. They must be ecologically suited to the soil, climate, topography, etc., and economically sound in that they provide food and other products for the subsistence of the farmer and his family, or products which can be readily sold.

The prevalence of pests and diseases is a major problem only briefly considered in this book. In developing countries crop production may not be sufficiently well organised for widespread control measures involving quarantine, for instance, to be carried out. The small farmer growing a wide range of crops chiefly for subsistence, has a better chance of controlling pests and diseases both because he can give more attention to his crops and trees and because pests are generally not as serious on small plots of mixed crops. When he tries to expand his cropping and increase the size of his farm the farmer may run into trouble. The small farmer is best advised to grow a larger number of crop species and to rotate all his fields with different crops in successive seasons where the crops are annual short-lived species. Advice on crops of national importance, for example, groundnuts in Malawi or rubber in Malaysia, is generally provided by government or other agencies; the farmer can get advice and help with pest and disease control, as well as with problems of manuring, husbandry, etc.

The economic side of farming practice in the Tropics is complicated. Firstly population density and hence the market for the farmer's products is very variable. Farmers located near towns and large population centres can often enjoy markets for a wide range of products but farmers located away from such centres are very restricted in the choice of crops which can be sold for cash income; consequently they must concentrate on food and other crops useful for subsistence. Crops such as rubber and coconuts, the products of which can be stored for transport and sale when convenient, are very important to the isolated farmer as they provide cash income. It is a mistake for the farmer to concentrate all his energies on such crops and to neglect subsistence crops, because the prices obtained from crops like rubber and coconuts may fluctuate and the farmer may find himself in receipt of a very small cash income and faced with paying high prices for food. The first consideration of any small farmer should be the production of his basic subsistence crops.

Even where a farmer is located near a large population centre, there is no guarantee that he can find markets for any other than well-established products. This can be attributed to the conservatism and low buying power of the consumer and to lack of marketing drive and control by the farmer, due principally to insufficient capital. Often the farmer is in the hands of unscrupulous and usurious middlemen who offer him low prices for his crops and restrict the expansion of trade by demanding high prices, even in nearby towns. The organisation of efficient farmers' cooperatives capable of handling their own distribution and marketing problems provides the answer. In developing countries the trained agriculturist may be required to advise on this subject as well as on the growing of crops. The chief ways in which cooperatives can help the farmer are set out below.

1 Providing lorry transport at operating costs for distribution of farm products.

2 Organising marketing outlets such as market stalls which may be cooperatively owned.

3 The timing of crop production to suit the market.

4 Collective ownership of farm machinery such as cultivators, irrigation equipment, etc.

5 Through the employment of a good agriculturist or through access to extension services, obtaining the correct advice on fertiliser use, pest and disease control, seed and planting material, etc.

6 Providing credit for fertilisers, etc. and cash advances on the sale of crops.

7 The generation of political pressure with regard to land tenure and reform, protective tariffs, subsidies, extension services, etc.

8 Involvement in adult education concerning technical farming matters, health and hygiene, family planning and so on.

All these functions depend on good management of the cooperative and it could be said that farmers' cooperatives in tropical countries often fail because of the lack of this factor.

The qualities to look for in the management of a farmers' cooperative are:

1 Common sense. (In fact one of the rarest senses bestowed on man.)

2 Honesty and the ability and willingness to do hardwork.

3 A knowledge of farming and farmers' problems.

A farmers' cooperative, if it is to be successful, should also truly represent the interests of the farmer and not the interests of processing, manufacturing, distributing and selling organisations making use of farmers' products.

Crop systems

Farming systems may be intensive or extensive. Intensive systems, which may include both perennial crops and annual crops, make full and continuous use of the land using fertilisers and other inputs to maintain production, prevent soil deterioration and so on. Extensive systems, e.g. shifting cultivation, some forms of grazing, etc. make use of fallow periods during which the soil regenerates, to some extent, its fertility. Extensive systems of agriculture aim to ration out the nutrient and water reserves of the soil to make them last longer or to ensure that productivity declines at a slower rate. Intensive agriculture will become more and more necessary and is emphasised in this book. The methods of agriculture encountered in the Tropics are extremely varied and range from the most primitive to highly-evolved systems. The following account very briefly considers the most important farming

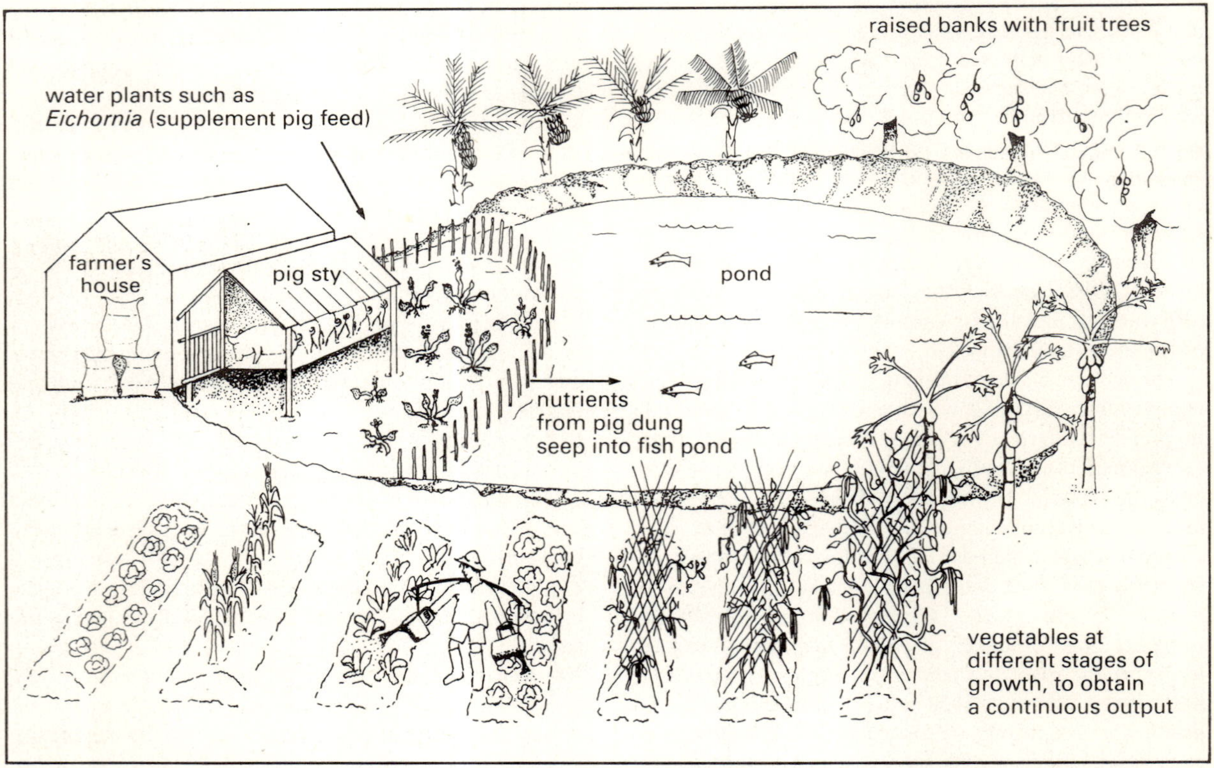

fig. 7.1 Diagrammatic representation of a Chinese market farm. No space is wasted and no material thrown away. Pig houses are usually located over the water and the dung first drops into an enclosed shallow area of the pond where water plants such as water hyacinth are grown and are used as pig food. Remaining nutrients seep to the deeper parts of the pond and cause algae and other organisms to grow; the fish feed on this growth

fig. 7.2 Vegetable beds are extremely neat. Crops are staggered so that vegetables at all stages of growth are available for marketing

systems of the humid tropical world which for convenience is divided into the regions of Southeast Asia, Africa, the American Tropics including the Caribbean, and the Pacific.

Southeast Asia

Probably the most highly-evolved and intensive farming systems of the world, other than hydroponics or nutriculture, are encountered in Southeast Asia. For thousands of years Chinese and Indo-Chinese peoples have employed stable agricultural systems for growing rice and vegetables and sometimes fruit trees; they use waste products and excrement of farm animals to maintain satisfactory levels of nutrients in the soil. Pig dung is often cycled through fish ponds, which may be an important feature of the farm. Seed-bed preparation is of a very high standard and in some of the tropical regions crop husbandry is so highly evolved that 5 or more crops of rice and various vegetables may be obtained in any one year. Nothing is wasted. In rice-growing the paddy straw is used to grow mushrooms. The nitrogen content of the soil is maintained at satisfactory levels by the inclusion of legumes in the cycle of crops in drained soils. In rice fields the level is maintained by the fixation of nitrogen by various algal organisms. Traditionally, prawndust which consists of parts of prawns and other small sea animals, has been used as a fertiliser to augment soil fertility and crop production. In recent years more and more chemical fertilisers have been used to grow

fig. 7.3 Bench terraces with retaining walls are designed so that the flow from a slow stream passes slowly across them. Manures are introduced at the top and provide nutrients to the beds of watercress

fig. 7.4 This farm on steep land is carefully terraced with strong retaining walls. Temperate vegetables are grown. It is typical of Chinese farms at elevations of 1 000 to 2 000 m in tropical Southeast Asia

the higher-yielding crops available through crop breeding programmes, since these crops remove large quantities of nutrients from the soil. With the development of intensive poultry keeping, fowl manure has become widely available as a fertiliser. Some typical Chinese vegetable farms are shown in Figs 7.1 to 7.4. The inputs of the farm in Fig. 7.1, apart from the work of the farmer and his family, are pig food, fertilisers and chemicals needed to control pests and diseases. The products are vegetables, pork, fish and fruits. The farms illustrated in Figs 7.3 and 7.4 utilise bench terraces in order to cultivate hilly land. Stone retaining walls support the terraces and water flow is controlled by the terrace slope. Both these farms are at high elevations, one specialises in growing temperate vegetables and the other in producing watercress. There is now a tendency towards the use of machinery such as irrigation equipment and small pedestrian cultivators to replace some of the manual work involved in market farming.

The use of night soil as a fertiliser is acceptable if it is subjected to anaerobic fermentation which eliminates many of the harmful parasites that it may contain. Subsequently, the aerobic condition of the soil and pond water kills off pathogenic anaerobic organisms which may be present in the fermented night soil. Methane digesters may be used to ferment night soil and the dung of farm animals, to give methane gas which may be used for cooking, refrigeration and in other ways. The effluent from a methane digester is normally fed into an algae or water-plant pond before going to the fish pond and vegetables; water coming from this series of processes normally conforms to public health standards. The design of a methane digester is given in Fig. 7.6 and the flow diagram for a farm digester is shown in Fig. 7.5.

The Chinese market farm is the most highly-evolved in Southeast Asia. The farmers of Indonesia also generally practise a standard of agriculture similar to that of the Chinese. At the other extreme are the very loosely organised farms of other indigenous peoples. Foremost is the Malay kampong farm with mixed fruit and coffee trees, a few coconuts, an odd patch of cassava, sugar cane, maize and so on, all rather poorly tended but often artistically arranged around a clean decorated house, with considerable uncultivated areas in between. Rain-fed rice is usually grown once a year on both upland and swamp areas adjoining the dwelling. Usually fowls are kept which are fed on household scraps but otherwise forage for themselves. Manure from the fowls tends to be concentrated on trees nearer the farmers dwelling and these fare considerably better than those further away. This is extensive agriculture and the farmers will often supplement their farming by gathering a range of wild vegetables, fishing in streams and paddy fields and by collecting edible snails, etc. Least organised is the sporadic and shifting cultivation of people who derive much of their living from the jungle by collecting rattan, bamboo and other forest products. The small areas cultivated are used to grow cassava, maize, hill rice and a variety of other crops.

Another system of agriculture predominating in Southeast Asia is plantation agriculture which in Malaysia, Sumatra and parts of Thailand, the southern Philippines, Indo-China and elsewhere has reached the highest levels of organisation. Particularly in Malaysia and Sumatra,

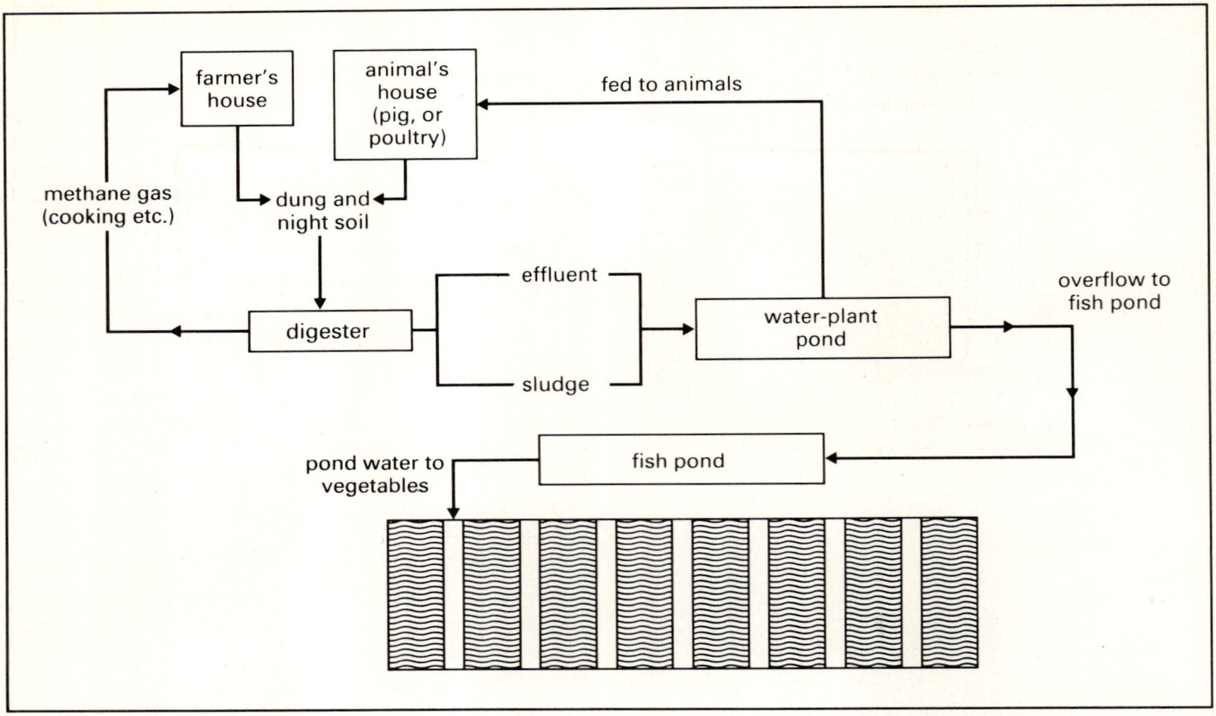

fig. 7.5 The flow diagram of a farm similar to that in Fig. 7.1 using a methane digester (Fig. 7.6). The effluent water from this digester will normally be pure enough to conform to public health standards (*adapted from* a design by Mr George Chan)

plantation industries based on rubber and oil palms have prospered. These are often organised on a large scale involving thousands of hectares by large private or government organisations. In Malaysia there are some very successful co-operatively owned settler schemes. Rubber is also important as a crop for smallholders particularly in Thailand. The agronomy of both these crops is advanced, correct fertilisers are used and research and extension services are good. However, the rubber industry rests heavily on the shoulders of the long-suffering rubber tapper whose living conditions are often very poor. Plantation agriculture of this type, involving tree crops, aims to produce a stable man-made forest to replace natural rain forest.

In the Philippines, Indonesia, south India, Malaysia and Sri Lanka coconuts grown both on plantation scale and as a smallholder crop are important. Coconuts are often interplanted with other crops or intergrazed by cattle thus supplementing earnings from copra. Coconuts, with an open canopy unlike oil palms and rubber, are particularly suited for intercropping with perennial species such as coffee, cocoa and some fruit trees. Oil palms and rubber can only be intercropped in the early establishment phase. Other plantation crops grown in Southeast Asia are sugar cane, particularly in the Philippines, Indonesia and Thailand; manila hemp in the Philippines; tea at high elevation sites in Sri Lanka and elsewhere; and bananas and pineapples which are generally grown both by smallholders and on estates. Robusta and liberica coffee are of some importance in Malaysia and Indonesia. The spices, cloves and nutmegs, are indigenous and are grown chiefly in small plantations in Indonesia and West

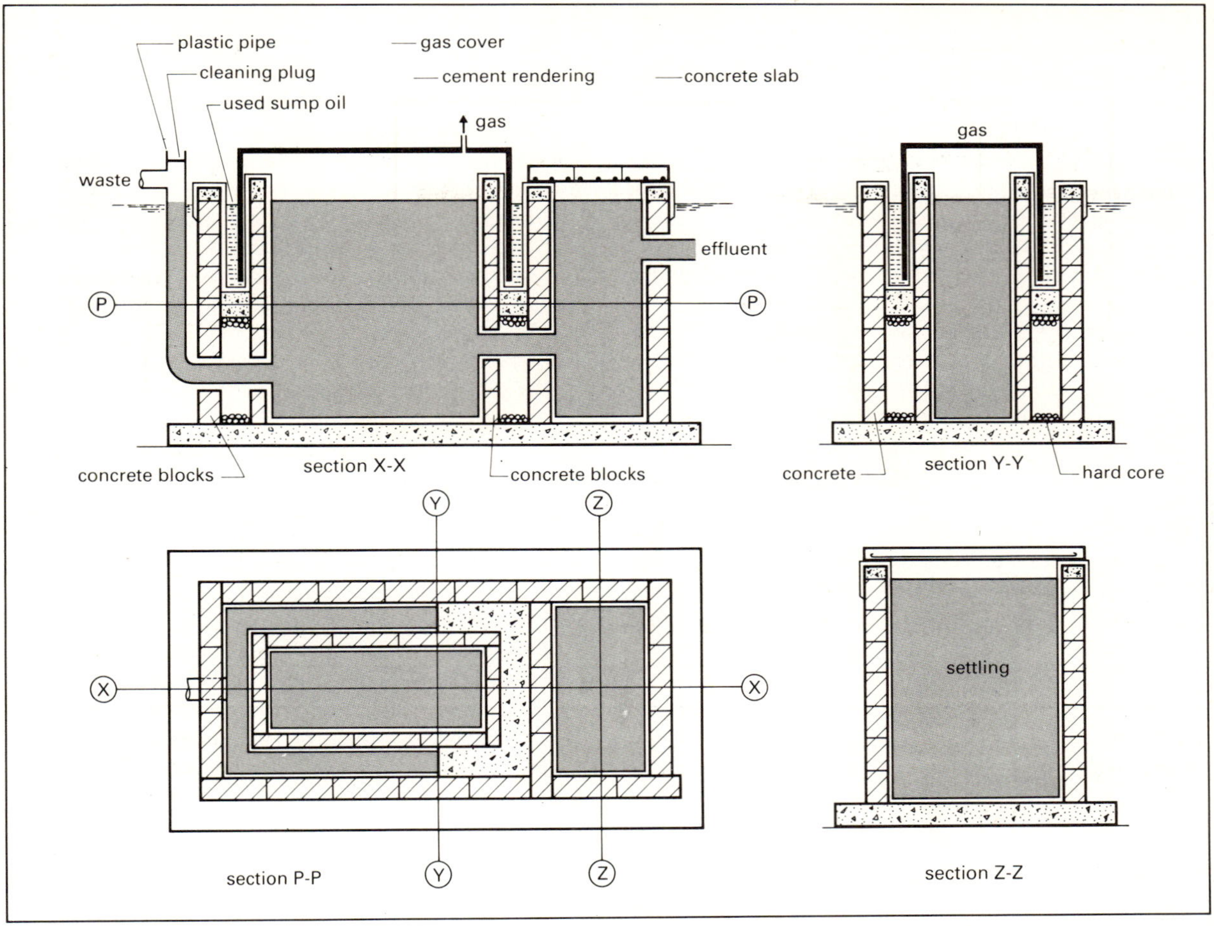

fig. 7.6 A farm digester of the type designed by Mr George Chan. A digester of this size would serve for a farm with about 30 pigs or an equivalent quantity of fowls or other livestock

Malaysia; pepper is grown in Sarawak, Sumatra and West Malaysia. In south India the areca palm is of importance as a stimulant producer and it is grown in quite large plantations.

In general soil fertility in plantations of tree crops is maintained by the use of cover legumes and fertilisers. Plantation crops are often grown on inland soils, chiefly latosols (p. 12). However, throughout Southeast Asia certain swamp soils are also used for plantation crops. Of particular importance are the very extensive areas of peat soils which are commonly regarded as 'problem soils'. Peats have been successfully used for growing pineapples. In addition, liberica coffee and a variety of other crops are grown on peat soils by smallholders. Cassava is particularly well suited to peat soils because of their loose friable structure and the consequent ease of harvesting. Other low-lying coastal and riverine soils throughout Asia are used for growing flooded rice. The standard of rice growing has changed dramatically in the past decade on account of the introduction of high-yielding, indica varieties bred mainly through programmes of the International Rice Research Institute. In addition a greater use of fertilisers has contributed to the production of higher yields and more profit; this has led to the development of extremely large irrigation schemes practising double and multiple cropping. Many of these schemes, such as the one at Tanjong Karang in Malaysia in which three crops of high-yielding rice are grown, or alternatively two crops of rice and one of maize or another crop when the fields are drained, are very successful and are technologically advanced.

Apart from rice, the only other cereal crop of importance in Southeast Asia is maize, grown mainly in the Philippines, Indonesia and Thailand.

A very recent development is the production of fodder grasses, of high crude-protein content, used for the manufacture of grass-meal and pellets. Very high yields of grass with protein levels in excess of 17 per cent may be obtained in the humid Tropics. High levels of nitrogen application (of the order 2 tonnes per hectare) have been used on napier grass (*Pennisetum purpureum*) for the production of high yields of grass with crude protein contents in excess of 15 per cent.

Africa

The main humid tropical zones of Africa are in the west and in the Congo basin (Fig. 1.1, on p. 1). Eastern Africa is very much drier with the exception of certain parts of Uganda and other areas near the great lakes. Of the crops considered in this book, bananas are important in Uganda and some coconuts fringe the eastern seaboard. Tea is grown in the highlands of East Africa and Malawi and arabica and robusta coffee in East Africa. The latter is grown both on plantations and smallholdings. For both these crops research and extension services have been good. However, marketing has been ineffectual and profitability low in the past and this has restricted the inputs of fertilisers, etc., necessary to maximise yield. The problem has been accentuated by the occurrence of coffee berry disease (CBD). In some marginal rainfall areas the application of grass mulch is important in raising coffee yields.

In West Africa, large plantations occur as in Southeast Asia: rubber in Liberia; oil palm in Zaïre, the Ivory Coast, Guinea, the Cameroons and Nigeria; and bananas in the Ivory Coast and the Cameroons. These plantations are organised in a similar manner and attain the same standards as those in Southeast Asia. Oil palms are, however, widely exploited for local use both for the manufacture of cooking oil and palm wine (the fermented sap from the inflorescence stalk). Wild palm groves and scattered semi-cultivated palms are chiefly exploited at the village level and the kernels form an important product for export.

Generally village farming systems are loosely organised and extensive, involving little or no use of fertilisers. Fallowing and some cultivation of legumes reduce fertility loss. There is very little integration of livestock with village agriculture. Annual food crops are yams (*Dioscorea* spp.), cassava, sweet potatoes and cocoyams (*Colocasia* spp. and *Xanthosoma* spp.). The major legumes are cowpeas and groundnuts. Plantains are also grown to some extent and important tree crops include grapefruit, lime, orange, pawpaw and kola.

Tree crops grown chiefly by smallholders as cash crops are cocoa (mainly in Ghana), robusta coffee (Ivory Coast, Zaïre and the Cameroons), and kola nuts. Of these cocoa is the most important. Ghana is one of the world's largest exporters of 'Amelonado' cocoa. Although some high-yielding unshaded cocoa is grown with the use of fertiliser, most smallholder cultivation depends on natural soil fertility. In establishing cocoa, understory trees which are not valued for their timber are cut and burned leaving considerable numbers of shade trees. 'Nurse plants' of plantain and cocoyam are grown between the cocoa seedlings.

Tropical America and the Caribbean

One of the unfortunate features of many parts of tropical Central America concerns land tenure. Many land owners are absentees and the use of the land is entrusted to unskilled 'managers'. This results in a low standard of productivity and living for agricultural workers. In other areas, increasing population and sub-division of holdings has led to the proliferation of very small farms. This is particularly so in the Caribbean islands. There has been a general lack of research and extension services and it is perhaps significant that the two best-known agricultural research and extension centres, the Inter-American Institute of Agricultural Sciences in Costa Rica, and the College of Tropical Agriculture, Trinidad, were in the past both organised from outside the region. However, there are now attempts at land reform, and more research and extension services are being provided both by local government organisations and by foreign-aid schemes. The agronomy of some crops (e.g. citrus) is most advanced in tropical America. In Cuba, a large number of moderately successful cooperative farms have been established.

Of the food crops, maize and beans (*Phaseolus* spp.) together with cassava (manioc or yuca) are the most important. Standards of husbandry, fertiliser use and availability of improved varieties vary considerably. The rotation of grain and legume crops helps to maintain soil fertility. In the Caribbean islands, yams and pigeon peas (*Cajanus cajan*) are more important. Rice production has also become of significant importance; Brazil is the largest producer. This rice is grown predominately under flooded conditions.

Sugar cane is one of the oldest cash crops of the region, it is grown both on smallholdings and on large estates. Bananas are also grown in large plantations, they are an important export crop of the region. Coffee is the most important smallholder crop supplying 60 per cent of the world's trade coffee. This is mostly arabica coffee, with some robusta in Trinidad and Jamaica. Cocoa is a crop of importance, native to tropical America, and although at present secondary to West Africa in production, Brazil, Ecuador, Colombia, Venezuela and Mexico are important producers. Standards vary greatly for both cocoa and coffee. Very often there is a poor standard of cultivation with low yields and soil erosion.

Citrus fruits are among the other important crops in tropical America and citriculture is more advanced, particularly in regard to varieties of orange, than in other tropical regions of the world.

The Caribbean islands are noted for certain specific crops; arrowroot from St Vincent, allspice from Jamaica and nutmegs and mace from Grenada.

Recently, considerable jungle clearing has been undertaken in the Amazon basin, and in other areas of tropical rain-forest, for the establishment of pasture and ranching.

The Pacific

The Pacific is an extremely large area comprised of numerous small islands, the large island of Papua New Guinea and the tropical region of Australia.

In most small islands, coconuts are the most important cash crop. The coconut is also an important food crop and its products are used in many other ways. Coconuts often fringe the sea shores and receive very little aid from man. Use of fertiliser is almost unknown, though trees growing near dwellings benefit from the dung of animals and other waste products. A gradation in the condition of the trees and their yield of nuts can be related to their distance from dwellings or villages. In remote islands the coconut is, together with fishing, the mainstay of the economy. This is true to a greater or lesser extent in some of the coastal regions of Southeast Asia; for example, the east coast of Malaya. Although yield of coconuts in this system is not very high compared to that which can be obtained by improved fertiliser use, etc., the trees are a stable part of the ecological system. In steep mountainous islands surrounded by a flat coastal shelf, coconuts derive subterranean water and nutrients from rainwater percolating from the inland hills. In flat islands without relief, coconuts obtain nutrients from occasional inundation by sea water causing fluctuating salinity or the roots may grow through the sandy porous soil to reach saline water. Much of the agricultural extension activity in the Pacific regions has been concerned with the up-grading of coconut culture by judicious fertiliser use and by overcoming limiting deficiencies of minerals such as potassium, iron and manganese on coralline soils.

Food crops grown throughout the Pacific are taro (*Colocasia* spp.), yams, sweet potatoes and bananas, including the Australimusa bananas of Tahiti, Papua New Guinea and elsewhere. The sago palm is exploited, and throughout Melanesia and New Guinea the areca palm produces a nut that provides an important stimulant.

In Papua New Guinea, the main food crop of the populous central highlands is the sweet potato which is grown on steep hillsides to a very high standard of husbandry. Commonly rows run up and down the hillside. A practice regarded with horror by early expatriate agriculturists. Nevertheless, the wisdom of this practice has now become obvious since it has been discovered that it prevents land slides. Soil erosion may be controlled by mulching.

The wing bean *Psophocarpus tetragonolobus* is an important food legume in New Guinea.

In Hawaii, the pineapple industry has been developed to a very high standard of perfection, although this has greatly declined of late. This was one of the first industries to use an artificial mulch of tar-impregnated paper, now replaced by polythene, to conserve soil moisture and reduce weed growth.

In tropical Australia sugar cane production is of importance, together with the livestock industry.

Further reading

Acland, J. D. (1971). *East African crops.* Longman: London.

Ahn, P. M. (1974). *West African soils.* OUP: London.

Allan, W. (1965). *The African husbandmen.* Oliver and Boyd: London.

Alvim, P. de T. and Kozlowski, T. T. (1977). *Ecophysiology of tropical crops.* Academic Press: London.

Black, C. A. (1968). *Soil-plant relationships,* 2nd Ed. Wiley: New York.

Descriptions of pathogenic fungi and bacteria. Commonwealth Mycological Institute: UK.

Farmers Guide. Jamaica Agricultural Society: Kingston.

Frohlich, G. and Rodewald, W. (1970). *Pests and disease of tropical crops and their control.* Pergamon: London.

Geus, Jan G. de (1953). *Fertiliser guide.* Centre d'etude de l'azote: Zurich.

Herklots, G. A. C. (1972). *Vegetables in Southeast Asia.* Allen and Unwin: London.

Ignatieff, V. and Page, H. J. (1958). *Efficient use of fertilisers.* Revised Ed. FAO Agricultural Studies, **43.** FAO: Rome.

Irvine, F. R. (1969). *West African agriculture,* Vol. 2 West African crops, 3rd Ed. OUP: London.

Jacob, A. and Uexhull, H. von (1963). *Fertiliser use.* West Germany.

Lee, K. E. and Wood, T. G. (1971). *Termites and soils.* Academic Press: London.

McIlroy, R. J. (1963). *An introduction to tropical cash crops.* Ibadan University Press: Ibadan, Nigeria.

(1972). *An introduction to tropical grassland husbandry.* OUP: London.

Mortensen, E. and Bullard, E. T. (1970). *Handbook of tropical and sub-tropical horticulture.* Agency for International Development: Washington, D.C.

National Academy of Sciences. (1972). *Soils of the humid tropics.* National Academy of Sciences: Washington.

Ochse, J. J., Coule, M. J. (Jr.), Djikman, M. J. and Wehlburg, C. (1961). *Tropical and sub-tropical agriculture,* Vols 1 and 2. Macmillan: London.

Pesticide manual. British Crop Protection Society.

Queensland agriculture and pastoral handbook Vols I & II. Department of Agriculture and Stock: Queensland, Australia.

Russell, E. W. (1973). *Soil conditions and plant growth,* 10th Ed. Longman: London.

Sanchez, P. A. (1976). *Properties and management of soils in the tropics.* Wiley: New York.

Sanchez, P. A. and Tergas, L. E. (1979). *Pasture production in acid soils of the tropics.* CIAT: California.

Thompson, W. T. (1970). *Agricultural chemicals Book II, Herbicides.* Thompson Publications: USA.

Tisdale, S. L. and Nelson, W. L. (1966). *Soil fertility and fertilisers.* 2nd Ed. Macmillan: New York.

Williams, C. N. (1975). *Agronomy of the major tropical crops.* OUP: London.

Williams, C. N. and Joseph, K. T. (1970). *Climate, soil and crop production in the humid tropics.* OUP: London.

Wills, J. B. (1962). *Agriculture and land use in Ghana.* OUP: London.

Wood, R. C. (1957). *A note book of tropical agriculture.* Trinidad.

Part 2 Specific crops

8 Method of crop classification

There are many ways in which tree and field crops can be classified. One method is to list the crops alphabetically according to their common English or botanical names. This is a simple method of classification but there are major disadvantages; there are often several synonyms for the common English name of a crop and we cannot expect all readers to be familiar with the botanical names of all tropical crops. Furthermore, if we adopted this method, the crop 'abaca', a fibre plant, would be followed by the crop 'annatto', a crop originating from a different botanical family and used to produce a colouring material. Crops of the same botanical family or usage cannot be grouped in any rational order by alphabetical arrangement.

A second method is to list the crops taxonomically according to the botanical family to which they belong. This would perhaps be a suitable method if all our readers were botanists.

A third method is to list the crops according to their type and usage. This is, we believe, the most satisfactory classification for the average reader and it is the one which we have adopted. The major disadvantage of this method is that it is possible to classify the same crop under two or more chapter headings. For example, the soya bean (*Glycine max*) may be listed as a 'leguminous vegetable and pulse' or as an 'oil crop'.

Within the chapters concerned with crops used for a specific purpose we have listed two groups of crops alphabetically; the first group includes those crops which are more important in the humid tropical regions and to which we devote considerable attention. The second group are those crops which we consider to be of secondary importance in these regions. While it is obvious that throughout the tropical world some crops are of more general importance than others, we realise that certain crops may be of major importance in particular locations but of minor importance in the humid Tropics as a whole. We can only seek the indulgence of our readers if we have devoted too little attention to any of these.

With regard to the range of crops considered in this book, we have attempted to provide some information on the majority of tree and field crops grown in the wetter Tropics. There will inevitably be a small number of crops grown in specific areas that are not included. The very large number of horticultural crops grown in the Tropics will be described and discussed in a companion volume on tropical horticulture whilst information on crops of the drier regions of the Tropics will be included in a third volume.

9 Beverage crops

Major crops

Cocoa (*Theobroma cacao*)

Family Sterculiaceae.

Other names Sometimes cacao is used when referring to the tree and cocoa when referring to the product.
cacao (French); cacau (Portuguese); cacao (Spanish)

Cocoa is obtained from a small tree which in its natural habitat, in the Central and South American jungles, is an understory tree. It grows usually in shaded conditions. Cocoa is a major crop in Brazil and other tropical American countries, the Caribbean and West Africa; it is a minor crop in the tropical Asian and Pacific regions.

There are a number of varieties of cocoa but because much commercial cocoa is planted from seed and cross-pollination often occurs, genetic variation in cocoa is much greater than in clonally-produced crops such as rubber and bananas. The original cocoa varieties used in trade from Venezuela, Ecuador and the West Indies, as early as the sixteenth century, were of a type known as

'Criollo' which is a very high quality cocoa but is now hardly grown at all because of its low yield. The cocoas grown today are primarily of the type originally known as 'Forasteros' which are characteristically much more hardy and vigorous. Most commercial 'Forasteros' now grown are of an acceptable quality for the trade and some, particularly the 'Amelonado' cocoa grown widely in West Africa and the state of Bahia in Brazil, are much favoured for the manufacture of chocolate. The West African 'Amelonados' are considered to be relatively genetically uniform because they are derived from a very few plants introduced to West Africa via the island of Fernando Po in the latter part of the nineteenth century.

The cocoas of Trinidad in the Caribbean region have been important in cocoa production for several hundred years and the cocoa trees of this island, often referred to as 'Trinitarios', are very mixed genetically, being derived from many introductions over a long period of time.

Certain hardy cocoa varieties from the upper Amazon have been used extensively in improving cocoa yields and hardiness in recent years. These cocoas have relatively small seeds which is a disadvantage in marketing but the seed size has been improved by breeding. The upper Amazon cocoas are self-incompatible, that is, they must be out-crossed to produce viable seed, and for this reason are considered to be natural hybrids.

Botany

Cocoa is a small tree reaching a height of about 10 m under natural conditions with a main trunk up to 6 m high. Under cultivation the tree is pruned to a much lower height. The branching of cocoa is somewhat unusual. A seedling produces a straight stem for about 1–1·5 m and then ceases growing with the formation of 3 to 5 side branches known as a fan or 'jorquette' (Fig. 9.1). After a time trunk growth continues with the formation of a new terminal shoot or 'chupon' which grows for a further 1–1·5 m before forming a second layer of side branches. The lower side branches die off, leaving a straight trunk and crown of fan branches. Usually the main trunk is pruned at the first or second chupon to limit the height of the tree in order to aid harvesting. This development pattern occurs in cocoa grown from seedlings. In cocoa grown from cuttings, fan branches are usually used for propagation and the tree produces a branched habit with several lower major trunks. Sometimes by pruning, a chupon may be induced to grow with a single main trunk resembling a seed-grown tree. Fig. 9.2 shows young

fig. 9.1 The first chupon (main stem) of cocoa has formed the first 'jorquette' of side branches or fan branches; in this case three

fig. 9.2 Young cocoa (2–3 years) planted under mature coconuts

cocoa trees pruned to a height of about 1·5 m growing under old coconut palms. The girth of the main trunk and lower main branches is related to yield; generally, the thicker the wood the greater the number of pods produced.

Cocoa produces large simple leaves in flushes. The leaves, when young, are sensitive to strong sunlight and for this reason the crop is usually established under shade. Temporary shade, using plants such as bananas, is to be recommended rather than permanent shade, except in very poor soils. When planted at the correct spacing and correctly fertilised, unshaded cocoa does better when mature and gives higher yields than shaded cocoa. In young cocoa 50–70 per cent shade is required.

Under tropical conditions of high temperature and moisture the production of leaves in flushes may occur at intervals of 4–7 weeks, more frequently in young trees. Flushing is reduced at temperatures below about 26°C and in a dry season. Generally a succession of flushes occurs after a dry spell.

Flowers are produced abundantly in clusters on the sides of the trunk and larger side branches (Fig. 9.3). Flowers are produced continuously under uniform warm and humid conditions; production is reduced by cool temperatures and drought, and is even more sensitive than leaf flushing. Thus, any cool season or dry season will greatly reduce flowering and cause fluctuations in the production of mature fruit. Pollination of the flowers is carried out by various small insects, chiefly midges (Ceratopogonidae), which are small enough to enter the narrow necks of the flowers. Of the many thousands of flowers produced by a tree, only a few will develop into mature pods. This is shown in the following table which records data from four trees in Ghana.

As can be seen from Table 9.1, failure of flowers to produce pods occurs as a result of low pollination. It also occurs as a result of wilting of young pods which occurs mainly in the period about 2 months after flowering when the pods are increasing in growth rate. This wilting occurs widely in cocoa and is known as 'cherelle wilting'. The young pods cease growing, become black and wither on the tree (see Fig. 9.4). Pods take 4 to 5 months from the time of pollination to mature. The first 1–2 months is characterised by slow growth after which growth speeds up for a period until about 3 months from pollination when pods may be 9–10 cm long. After this growth slows down and filling occurs. This is the phase of seed growth.

fig. 9.3 Clusters of flowers on the trunk of a cocoa tree next to a fully developed pod

Table 9.1 Survival of flowers of cocoa (from McKelvie, 1962)

	Flowers	Sets (pollinated)	Wilted	Harvested
Tree 1	15 144	173	123	50
Tree 2	18 308	178	115	63
Tree 3	700	26	20	6
Tree 4	9 196	75	57	18

fig. 9.4 (a) Healthy cocoa pods and small wilted pods (cherelle wilt) on the same trunk

(b) Healthy cocoa pods on a young tree, 2–3 years from planting out as a polythene bag seedling

When ripe, pods are oval about 15–20 cm long (Fig. 9.3) and yellow in colour. The seeds, when ripe, are flattened and violet-coloured in most modern cocoas and embedded in a mass of pulp. The pods are plucked from the trunk when ripe.

Cocoa produces shallow superficial roots; most of the feeding roots are in the top 15 cm of soil. In seedling-planted trees but not those grown from cuttings, a short tap root may penetrate to a maximum of up to 1·5 m.

Yields of cocoa grown on smallholdings under shade and without fertilisers are only of the order of 500 kg of dry beans per hectare per year. However, with correct fertiliser use (see p. 57) yields of about 1 500 kg per hectare may be obtained with mature cocoa with some shade removal. The newer, more vigorous cocoas grown unshaded when mature, and with correct fertiliser use, may yield 4 000 kg of dry beans per hectare.

Climate and soils

Although it is grown over a wide latitudinal range, for example in the Americas from the equator to as far as 18°S in the state of Bahia, Brazil, cocoa is rather exacting in its climatic requirements for optimal growth and production. Both leaf production and flower production are reduced by low temperatures. Mean temperatures of around 29°C are best for cocoa. A slight cold season will cause uneven bearing, i.e. there will be a period of no bearing about 5 months after the cold season, but annual yields may be rather heavier than when bearing is even. The mean temperature of the coldest month should not be less than about 15°C and the minimum temperature about 10°C. Temperatures of 5°C, even for a short period, will certainly kill cocoa.

Cocoa is also sensitive to high temperatures, particularly when young and temperatures above 32°C cause damage to young leaves. This is why shade is necessary to establish young cocoa.

Cocoa is dependent on rainfall, too little or too much may cause yield loss. A dry season, if short, has more or less the same effect as a cold season. Young cocoa will be set back seriously by as little as 2 weeks drought. Cocoa is also sensitive to dry atmosphere and in a long dry season set-back will occur even if the cocoa is irrigated. For optimal growth 100 mm of rain per month is minimal and 200 mm per month is the upper limit. Rainfall above 200 mm per month may depress yield, particularly on heavy poorly drained soils, because cocoa roots are sensitive to poor aeration. Thus, the ideal rainfall for cocoa in the Tropics is round 175 mm per month, evenly distributed. Excessive humidity (above 90 per cent) also affects cocoa.

It is apparent that cocoa is most sensitive to climatic stress. Under less than optimal conditions, particularly in drought or in poor soils, cocoa should be grown under shade. The best type of shade is that provided by another crop. For temporary shade, bananas and for permanent shade, coconuts are recommended.

Traditionally, it was preferable to grow cocoa on superior soils, because the crop was grown without fertilisers. With proper husbandry, however, a wide range of soils can be used successfully. Ideally, if the rainfall is evenly distributed, well-drained inland soils should be used. If they are too acid (below pH 6·0) they should be limed with magnesium limestone to bring the pH to about this level. Heavy clay coastal soils may also be used but they must be improved with organic matter and field drains constructed at appropriate spacing to remove standing water (Fig. 9.2); in these conditions cocoa should be grown under coconuts. Very free-draining soils, such as some volcanic ash soils, can be used for cocoa if the rainfall distribution is even. Problems have also been encountered on very freely-draining basaltic soils because of the relatively superficial root system of cocoa.

Cultivation and management

The nursery Cocoa should be planted from seed preferably, since trees produced from seed have a desirable growth form. In cocoa-producing countries seed of suitable origin is generally obtainable from government agencies. If not, seed may be imported from similar climatic regions; this seed must be imported by air and protected from low temperature. Cocoa seeds will remain viable only for about 10 days when stored in charcoal powder at 30 per cent moisture, and therefore have to be planted without delay. Seed for planting is often produced by hand-pollination, from trees known to produce superior seed. Out-pollination by insects is prevented by pinning fine insect mesh (20 mesh per centimetre) around the trunk where the flowers are produced. A seed garden of 1 ha will provide seeds for about 400 ha per year by about the 5th year rising to more than 1 000 ha after the 10th year. Hand-pollination, however, is very labour-intensive and seed produced by this method is expensive. Another way of producing hybrid seed is to use naturally incompatible parent trees (such as the Upper Amazon varieties mentioned before that will only out-cross). Thus in such a seed garden alternate rows of the different parent trees are planted and natural insect cross-pollination occurs. The garden must be isolated from other cocoas to prevent the wrong kind of crosses.

fig. 9.5 Cocoa seedlings in polythene bags in a nursery. Bags of size 15 × 10 cm when flat are used. The beds are about 1 m wide with lanes for watering. About 70 per cent shade is achieved with palm fronds

The best way of raising cocoa seedlings is in polythene bags of size about 15 × 10 cm, planting 1 seed per bag. The seeds should be planted about 1 cm deep and laid flat on the side to prevent root twisting. The bags may be close packed in beds in a shaded nursery (about 50–70 per cent shade) as illustrated in Fig. 9.5.

The seedlings will be ready to plant out after 1–2 months but if more widely spaced may be held in the nursery for longer periods. A well-drained friable topsoil is most suitable for use in the bags. Alternatively a sand/clay mixture can be used. The soil should be limed with magnesium limestone to achieve a pH of 6·0. After about 3 weeks the soil should be manured using a N:P:K:Mg mixture of composition about 15 : 10 : 10 : 5 at a rate of 50 g per bed of size 1 m × 5 m (Fig. 9.5). Manuring should be repeated approximately every 2 weeks. The fertiliser should contain a soluble form of phosphate(V) (not rock phosphate) and the above quantity should be mixed in 10 litres of water and applied by watering-can.

If seeds are not available for planting, or if it is desirable to propagate cocoa vegetatively, seedlings may be raised from cuttings. This has occurred in the island of New Britain where die-back disease caused very heavy losses and certain resistant trees are propagated clonally

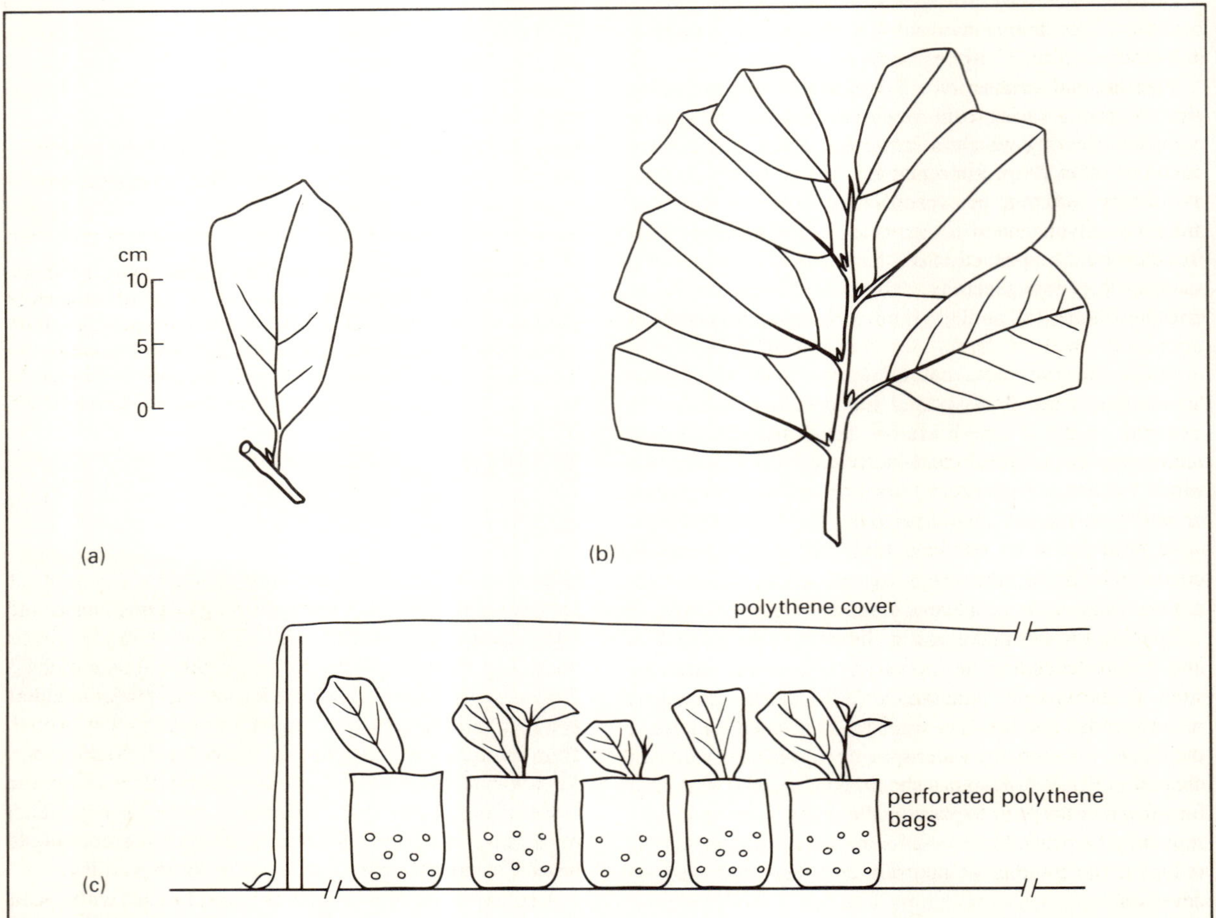

fig. 9.6 Vegetative propagation of cocoa (a) Single-leaf cuttings (b) Stem cutting (c) Arrangement of cuttings under a polythene tent

to provide growers with resistant planting material.

Cuttings are usually taken from newly-matured flushes or fan branches (the jorquette), simply because insufficient chupon branches are produced. These cuttings give rise to a strongly-branched tree usually with several lower trunk branches; this is not so convenient for harvesting. Propagation may be by the single-leaf method if planting material is in short supply or by stem cutting, if more is available. The cuttings are trimmed to about one-third the leaf size as shown in Fig. 9.6 and raised in polythene bags of size about 20 × 15 cm under a polythene cover to prevent water loss. The beds should be shaded as before and opened as little as possible, only when watering is necessary.

Buddings are very little used in cocoa growing, mainly because of the labour involved in a crop which is planted at densities up to 1 750 per hectare.

Planting and fertilisation The standard of clearing for planting cocoa varies from rough clearing and burning to complete clearing which is important when old areas of cocoa or other crops are replanted. Cocoa is best planted as nursery seedlings in appropriately sized holes, and to aid early establishment a temporary cover of 2 or 3 leaf fronds should be pushed into the ground to arch over the seedling. It is now generally recognised that mature cocoa does best under full sunlight if correctly planted, except in poor soils with a heavy texture and naturally poor drainage. In the past, natural forest trees or planted leguminous trees have been used as shade plants. In general, only 2 methods of growing cocoa are recommended: establishment under temporary shade, for which bananas or plantains may be used as nurse crops; or under permanent shade provided by coconuts which have been found to combine well with cocoa. Planting cocoa under many other crops, e.g. rubber or oil palms, has not been successful because of excessive shade.

Spacing for unshaded cocoa should be between 2·2 m and 3·0 m depending on the variety. For small varieties, such as the typical 'Amelonados', the closer spacing is recommended and for large vigorous types such as those of the Upper Amazon, the wider spacing should be used, as indicated in Fig. 9.7. Bananas should be allowed to establish for a few months prior to planting the cocoa and should subsequently be pruned to 1 or 2 followers per growing point, reducing the number at appropriate times as the cocoa develops.

The use of fertiliser is essential for growing cocoa without shade; where bananas are used as a nurse crop appropriate applications of fertiliser must be made, taking into account the reduced density of plants. Cocoa itself removes about 20 kg of nitrogen, 5 kg of phosphorus and 10 kg of potassium for each 500 kg of made cocoa (assuming the husks are returned to the soil as a mulch). Tables 9.2, 9.3 and 9.4 indicate fertiliser levels for bananas and cocoa together which should give satisfactory growth although a further response to higher levels could occur. Table 9.4 indicates rates when a compound fertiliser mixture is used instead of straight fertilisers. Cocoa is a heavy user of magnesium and in acid soils, magnesium limestone should be used to raise the pH to about 6·0 prior to planting.

A soluble form of phosphate(V) must be used with cocoa because the soil pH should be high. If ammonium sulphate(VI) is used instead of urea, twice the quantities indicated should be used. Following the 5th year, nutrient

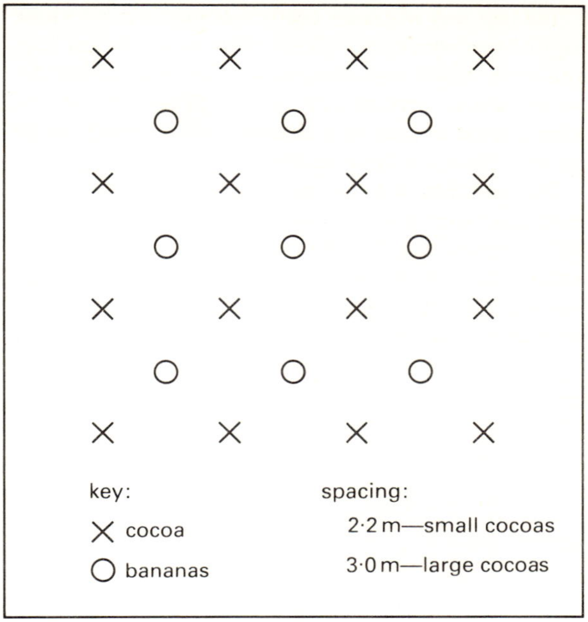

fig. 9.7 Recommended spacings for planting small cocoas, e.g. 'Amelonado', with a temporary nurse crop of bananas or plantains. Alternate banana plants are thinned at appropriate times or after 1–2 years with complete thinning when the cocoa is growing strongly (3–5 years). (*Source:* **Williams, C. N.,** (1975). *Agronomy of the Major Tropical Crops.* OUP: London.)

Table 9.2 Fertiliser recommendation for unshaded vigorous cocoa varieties e.g. Upper Amazon hybrids (rates in kg per hectare per year)

	Urea	Potassium sulphate(VI)	Magnesium(II) sulphate(VI)	Superphosphate
Year 1, newly cleared jungle or replanted soil bananas thinned	175			200
Year 2, bananas removed	250	200	100	200
Year 3	250	250	150	200
Year 4	200	200	200	150
Year 5	250	250	200	150

Table 9.3 Fertiliser recommendation for unshaded 'Amelonado' (rates in kg per hectare, per year)

	Urea	Potassium sulphate(VI)	Magnesium (II) Sulphate(VI)	Superphosphate
Year 1, newly cleared jungle or replanted soil bananas thinned	175			200
Year 2, bananas removed	225	175	75	200
Year 3	225	200	100	200
Year 4	100	100	100	100
Year 5	150	150	100	100

Table 9.4 Compound fertiliser (12:12:17:2), supplemented with urea (rates in kg per hectare, per year)

	Urea	(12:12:17:2)
Year 1–2 bananas interplanted- lower rates for new burning	100–200	200–500
Year 3–4 bananas thinned	200–250	500–650
Year 5 onwards without bananas	150–250	300–600

requirements should be correctly determined from foliage analysis if extension services are available. Fertilisers should be broadcast evenly over the soil surface rather than placed around each tree and the total amount used should be split into several applications per year, avoiding excessively wet or dry spells. A larger amount of fertilisers is often given at the beginning of the major rains.

Planting cocoa under coconuts has proved very profitable in recent years. The spacing used for coconuts is about 10 m square and at this spacing about 50 per cent of incoming sunlight reaches the cocoa. The trees are planted in 2 lines between each row of coconuts at about 3 m apart (Fig. 9.2). Fertilisers for coconuts and cocoa would be about the same as that for cocoa with a banana nurse crop in the 2nd or 3rd year (Tables 9.2, 9.3 and 9.4).

Weed control Cover legumes are not generally used with cocoa; under a banana nurse crop clear weeding with herbicides should be undertaken. Traditionally sodium arsenate(V) has been much used in plantation cocoa. In native cocoa, weeding is confined to occasional slashing and hand-hoeing. However, hand-weeding is deleterious to cocoa because the superficial root system may be damaged. Satisfactory weed control for a period of several months can be achieved with a contact herbicide such as sodium chlorate(V) and diuron at the following rates:

sodium chlorate(V)	10 kg per hectare;
diuron	2 kg per hectare.

Paraquat or diquat should be used to replace sodium chlorate(V) in wet weather at a rate of 1 litre per hectare.

In bearing cocoa, the pods (husks) should be used as a mulch around the trees as they contain considerable quantities of plant nutrients and also help to suppress weeds.

Pruning The object of pruning is to produce a correctly-shaped tree. Pruning of the main stem is usually done at the first or second chupon. After formation of the first series of fan branches a new chupon will generally form just below this and continue the trunk-growth. When the height of the trunk is considered satisfactory all subsequent chupons are pruned off. Three or 4 strong side branches should be allowed to grow to give a spreading habit which closes with the canopy of adjacent trees. Two side branches (Fig. 9.8) are undesirable because they become too heavy and tend to split.

fig. 9.8 Pruning to two side branches should be avoided because these may become too heavy and split apart.

With cuttings, the tree should be allowed to develop 3 or 4 strong trunks that spread slightly. Alternatively, if chupon formation takes place readily, a single chupon should be allowed to grow on each tree giving it the same structure as a seedling tree.

Harvesting and processing

In humid tropical areas, harvesting is usually an all-year-round operation. Pods turn yellow or orange when fully ripe. Some cocoas have green pods when unripe and others have a reddish coloration. Harvesting is done with a sharp instrument to give a clean cut which minimises disease infection. Cutting should be close to the tree just above the thickened cushion but the cushion itself should not be cut. Usually pods are gathered into heaps near a roadside where they are split and the beans and pulp removed. A wooden spoon may be used for this purpose. The beans and pulp may be placed in polythene bags for transport to the fermenting bins and, as time permits, the heaps of discarded husks may be spread around nearby trees as a mulch.

Preparation of beans The first step in preparation of cocoa beans is fermentation which removes the pulp around each bean. Fermentation of quantities of about 1 000 kg or more is done in hardwood boxes of size not less than 1 m × 1·5 m × 1·2 m, located in a shed. Since the beans must be transferred from one box to another after about 48 hours, a minimum of 3 boxes is needed for any batch. For convenience, if the boxes are arranged one above the other (for example on a slope), transference is simplified since the mass of beans and pulp may simply be fed from one box to the next, by gravity, down a wooden slide. Usually 3 fermentations, each of 48 hours, will complete the process, but fermenting is an exacting task that requires experience, since it can affect the quality of the beans. The rate of heat build-up in the mass is critical to quality. Smaller quantities, of 100–300 kg, may be fermented in smaller boxes of hardwood, but should not be transferred, since too much heat will be lost. Quantities less than about 1 000 kg should be stirred during treatment to cause mixing and even fermentation.

Drying of the fermented beans must be done carefully to avoid loss of quality. Sun-drying is said to give the best quality and in a humid climate this requires some provision for covering the beans when rain occurs. Concrete floors are usually used for this purpose. Mats should be laid on the concrete to prevent contamination with grit and to facilitate gathering up the beans when rain

fig. 9.9 A concrete drying floor can be used to dry a great variety of crops in addition to cocoa. The crop may be spread directly on the concrete, or on mats. The corrugated-iron covers are used in rainy periods

threatens. Corrugated iron covers of the type illustrated in Fig. 9.9 are the simplest to construct. Drying takes about a week to 10 days. Various types of artificial driers are also used for cocoa.

Protection

Cocoa suffers from a number of serious diseases. The fungus *Phytophthora palmivora* causes several symptoms including black-pod, seedling wilt and leaf fall, particularly under shaded conditions. Pod rotting (black-pod) is serious in West Africa with 'Amelonado' cocoa. It may be controlled by spraying with Bordeaux mixture, by reducing shade and by removal of infected pods. Resistant varieties should be grown in regions infected with *P. palmivora*.

In South America a disease known as 'witches broom' occurs which produces broom-like branching of the shoots with undeveloped leaves and swollen, branched flower cushions. It is caused by the fungus *Marasmius* sp. The Upper Amazon cocoas are very resistant to this disease.

Cocoa die-back is another disease which is serious in some regions, particularly the Pacific. It has recently been shown to be caused by a fungus (*Onchobasidium* sp.). The branch tips die back and leaf fall occurs. It is controlled by planting resistant clones. Boron fertilisation at a rate of 10 kg per hectare may also help to control die-back.

A very serious virus disease, which has caused widespread losses in Africa is 'swollen shoot'. It causes swelling of fan and chupon branches and red banding of the leaf veins followed by chlorosis. It is spread by insects and best controlled by phytosanitary methods and strict quarantine in introduction of cocoa to new areas. The Amazon cocoas possess resistance to the disease. 'Amelonados' are particularly susceptible.

Of insect pests, capsids have caused the most losses in cocoa, particularly in West Africa. The insects inject a poisonous substance into the cocoa at almost any site which causes rotting. Capsids enter a cocoa area through breaks in the canopy. They can be controlled with insecticidal spray at appropriate times.

Coffee (*Coffea* spp.)

Family Rubiaceae
Other names café (French, Portuguese, Spanish)

There are several species and varieties of coffee which are important in tropical countries. The main ones are *Coffea arabica, C. canephora* (also known as *C. robusta*) and *C. liberica*.

C. arabica or arabica coffee is generally considered to be the best quality and commands the highest price. There are two known varieties, *arabica* (or *typica*) and *bourbon*. Arabica coffee is best suited to highland sites within the Tropics and is also grown at relatively high latitudes in Brazil though it may be grown in the lowland Tropics. It is very important in East Africa. The type of climate required for arabica coffee is generally similar to that required for tea. Robusta coffee can be grown in warmer locations than arabica and is grown in the lowland humid Tropics. Another type of coffee more suited for the humid tropical lowlands is liberica. This variety is relatively unimportant, but is grown in West Africa and Southeast Asia.

Botany

All coffee species are small trees and, when left to grow, will reach a height of about 10 m. In cultivation they are pruned to a smaller size. The general form of arabica and robusta coffee trees is shown in Figs 9.10 and 9.11 respectively. Liberica coffee is similar in shape to *arabica* or *bourbon* as shown in Fig. 9.12(a).

The shape of the tree is also dependent on the method of pruning and training the bush. In arabica coffee one or two main vertical shoots are allowed to grow as indicated

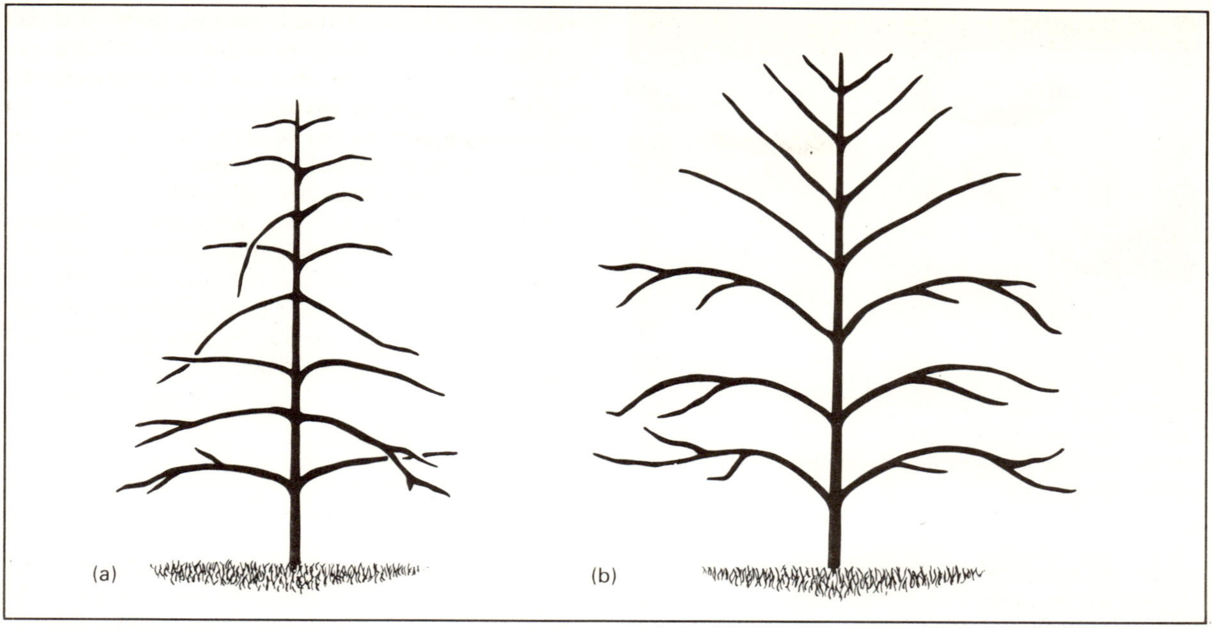

fig. 9.10 The growth characters of *Coffea arabica* (a) var. *arabica*, note the horizontal growth of the branches which tend to hang downwards (b) var. *bourbon*, note the more upright branching habit and the flatness of the top. (*Source:* **Haarer, A. E.** (1963). *Coffee Growing.* OUP: London.)

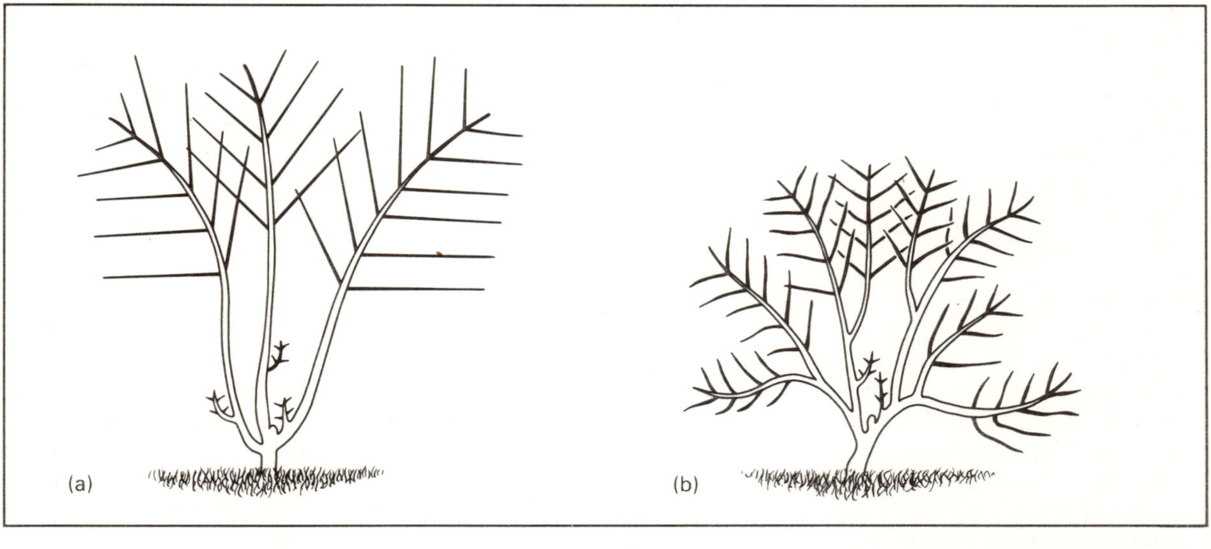

fig. 9.11 The growth characters of *Coffea canephora* (a) var. *robusta*, trained to grow into a multiple stemmed tree rather than the normal high tree; note the stiffer, more upright growth (b) var. *nganda*, assisted to grow into a multiple-stemmed tree rather than its usual sprawling habit. (*Source:* as fig. 9.10.)

fig. 9.12 (a) Liberica coffee growing in Malaya

(b) The tree is pruned by topping.
The primary, secondary and tertiary side branches produce flowers and bear fruit

fig. 9.13 Liberica coffee in humid tropical conditions produces numerous leafy suckers which are pruned off

fig. 9.14 A bearing branch of liberica coffee

on p. 65. Robusta coffee bushes are often formed by the 'Agobiado' method as shown in Fig. 9.15. This is similar to the pegging technique recommended for tea bush formation. However, robusta is also pruned by the single-stem method chiefly in Southeast Asia. Liberica coffee bushes are usually topped at a height of 2 m as shown in Fig. 9.16 and in Fig. 9.12(b).

Leaves of coffee are simple and produced in pairs (see Figs 9.12(b), 9.13 and 9.14). Branching is also opposite. The root system consists of a mass of superficial feeding roots and, in deep soils, the tap root and other deep roots may penetrate to several metres.

Flowers are often seasonal. They are white and scented. Arabicas are mostly self-pollinated but robustas and libericas are largely out-pollinated. In humid climates, flowering is more continuous but slight fluctuations occur through the year. In West Malaya for example peak crops coincide with the two relatively dry periods in December–February and May–June.

In most regions where arabica and robusta coffee is grown, one main crop is obtained per year. The berries are picked when they are red and fully ripe. Yields of arabica and robusta coffee vary considerably. The highest yields under good conditions of husbandry and good physical conditions are of the order of 2 000 kg of dried beans per hectare, average yields are less than 1 000 kg per hectare. In Brazil the average yield is about 1 100 kg per hectare.

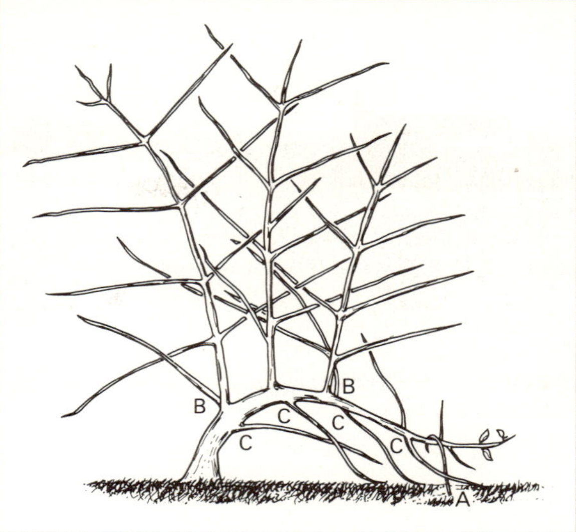

fig. 9.15 The Agobiado system of bush formation showing how the main leader-shoot of the multiple-stem system has been bent over and pegged down at A. The lateral shoots, CCCC are cut off. (*Source:* **Haarer, A. E.** (1963). *Coffee Growing.* OUP: London.)

Climate and soils

The different coffee varieties require different climatic conditions. Of the three species of coffee grown in the humid Tropics, arabicas require the cooler climate of high elevation sites and a mean temperature of around 20–22°C is about optimal for this species. A minimum elevation of about 1 300 m is required in equatorial regions down to about 500 m at a latitude of 15°. All coffee species are sensitive to frost. Higher temperatures cause early bearing and the crop will not last long, thus arabica coffee should not be grown in the tropical lowlands. Robusta coffee can be grown at higher temperatures than arabica and the optimal mean temperature is between 25–28°C. Robusta is very sensitive to cold and, while arabicas can withstand cold and even slight frost, robustas will be destroyed. Liberica coffee is adapted to warmer conditions and can be grown throughout the equatorial lowlands when rainfall is adequate.

All coffee varieties require a rainfall of about 1 800 mm for optimal yield. In some important coffee regions of Africa annual rainfall is lower; in East Africa a mulch of cut grasses is used to conserve soil moisture so that coffee may be grown at a minimal rainfall of about 1 000 mm per year. In general, however, rainfall should be in the range 1 800–3 000 mm. Arabica and robusta coffees bear well with one annual rainfall peak but with an adequate amount in most months and a short dry season of 1 or 2 months. Liberica coffee is adapted to grow under uniform humid conditions but will tolerate a short dry season.

Soils for arabica and robusta coffees should not be too acidic (pH 5·5–7·0). Liberica coffee grows in more acid soils but benefits from the application of lime when the

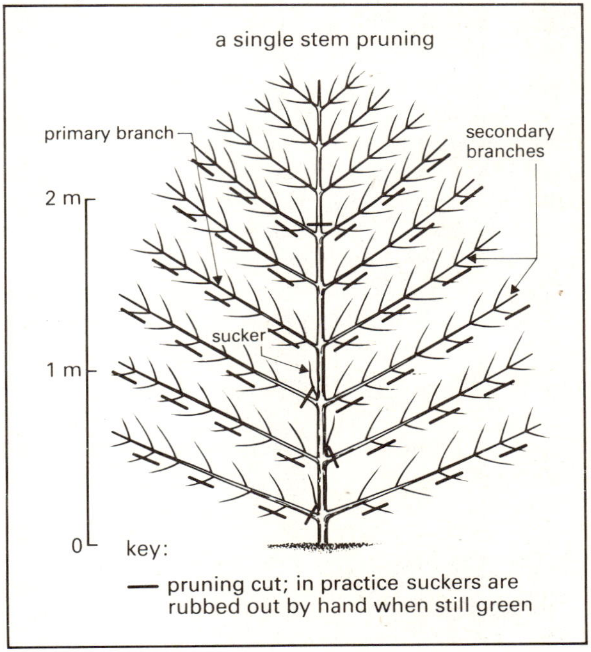

fig. 9.16 Topping and pruning liberica coffee. (*Source:* Department of Agriculture: Malaysia)

acidity is below about pH 4·5. Soil should also be deep and friable with good drainage, but with a moderate clay content and water storage capacity, particularly in regions of lower rainfall. Generally superior soils have been used for coffee growing. Liberica coffee grown in lowland regions requires field drains of at least 1 m depth. Saline soils should not be used for any of the coffee varieties.

Coffee is sensitive to strong winds and wind breaks of trees should be established in windy locations.

Cultivation and management

Nursery Coffee seed should be obtained from a reliable source. However, arabica coffee is mostly self-pollinated so that superior seedlings may be obtained by taking seed from suitable high-yielding trees.

The best type of nursery is the polythene bag nursery using perforated bags of size about 15 × 20 cm laid flat, filled with top soil or a suitable mixture. Manure or well-rotted compost may be mixed in with sub-soil and a proportion of sand added to give adequate drainage.

About 2 kg of ground rock phosphate should be added to each cubic metre of nursery soil. Pre-germination of seed in a raised bed is desirable to obtain healthy strong-growing plants for the bags. Fully ripe berries are hand pulped and air dried in open shade for a few days prior to planting. They should be sown in the pre-germination bed at a depth of 2 cm. Polythene-bag seedlings should be spaced about 30 cm apart under overhead shade providing about 50 per cent total light. They will be ready to plant to the field from about 5 months from germination and removal of shade should be carried out gradually with complete removal for about 30 days prior to field planting.

Planting and fertilisation Shade is not now recommended for coffee growing, except in extreme conditions, but temporary shade, using bananas, may be recommended as for cocoa. Because of the small size of the bushes a relatively high standard of field clearing is desirable.

Field planting should be done in a suitable wet season but flood periods should be avoided. In regions with adequate rainfall intercropping with groundnuts or other legumes may be practised if a temporary shade crop (e.g. bananas) is not planted. A temporary shade of palm fronds should also be placed at each planting point at the time of planting.

A field spacing of about 2·2 m is suitable for arabicas and 3 m for robustas and libericas. A small planting hole should be prepared at each point and if possible some top soil incorporated. In each planting hole a dressing of about 100 g of ground rock phosphate has been recommended. Coffee removes about 30 kg of nitrogen, 10 kg of phosphorus (P_2O_5) and 50 kg of potassium (K_2O) for each 1 000 kg made coffee harvested. Much of the potassium is present in the pericarp or pulp and this should be returned to the soil. Bag mixture fertilisers can be recommended for coffee having an N:P:K:Mg ratio of about 1:1:1:0·5, applied at a rate of 200 kg per hectare in the 1st year rising to 500 kg or more from the 4th year, on coffee of good genetic origin. The fertiliser should be given in 4–6 applications per year, avoiding dry seasons and heavy rain periods.

Table 9.5 indicates the required levels of straight fertilisers. It should be noted that sodium nitrate(V) (Chile saltpetre) is unsuitable as a source of nitrogen for coffee because of the high sodium content. Rock phosphate may be used on more acidic soils but on soils above pH 6·0, superphosphate should be used.

The recommendations for fertiliser use are not intended to cover all eventualities. The amount and type of fertiliser used will vary with the type of soil and with the yield level of the coffee.

Table 9.5 Fertiliser recommendations for coffee (rates in kg per hectare)

	Urea	Ammonium sulphate(VI)	Potassium sulphate(VI)	Magnesium(II) sulphate(VI)	Phosphate
Year 1, newly cleared jungle or replanted soil	50	50			100
Year 2	100	50	50	50	100
Year 3	100	50	100	50	100
Year 4, maturity	150	50	150	100	100

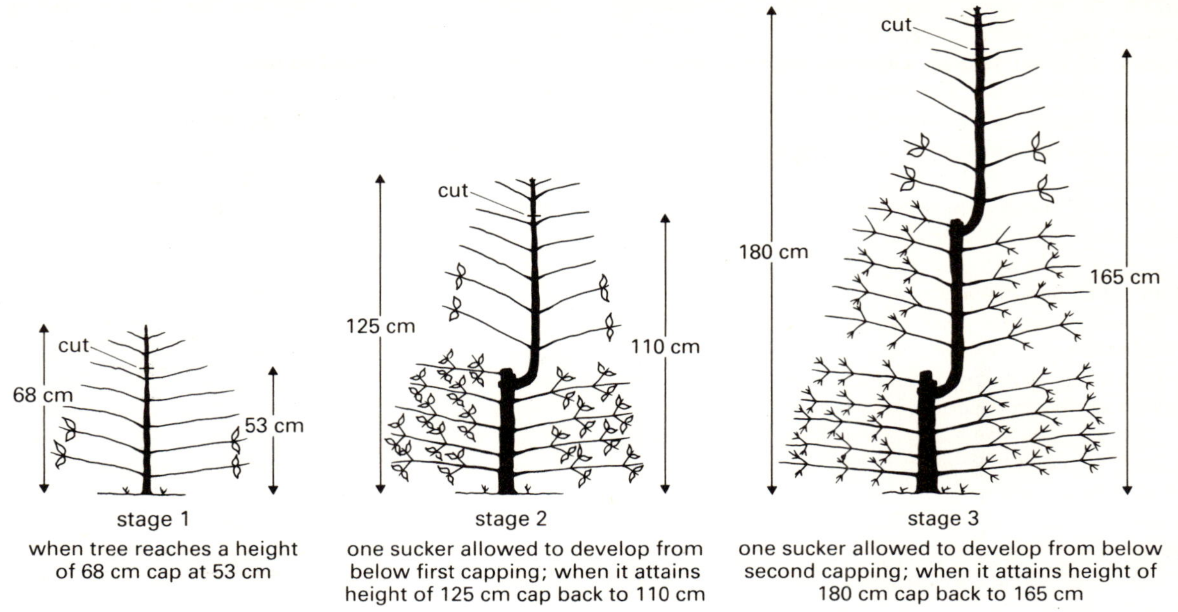

fig. 9.17 The single-stem method of pruning arabica coffee. (*Source: Handbook of arabica coffee.* Tanganyika.)

It is possible that high-yielding coffee varieties grown under good climatic conditions may respond economically to higher levels of fertilisers than those indicated in Table 9.5. In mature coffee fertiliser rates should be determined from leaf analysis of experimental plots.

Pruning Pruning is one of the most important operations in coffee to maintain high yields. Figs 9.17 and 9.18 show two ways of pruning coffee during establishment to give a satisfactory bush. Fig. 9.16 shows the way of forming the bush in liberica coffee as recommended by the Department of Agriculture, Malaysia. The method illustrated in Figs 9.16 and 9.17 is known as single-stem pruning and results in the formation of a bush with a single stout upright stem. In Fig. 9.17 (the single-stem method for pruning arabica coffee grown at cool locations) the bush is topped three times. In Fig. 9.16 for liberica, a single topping is carried out at about 2 m.

In single-stem pruning the crop is borne on lateral branches. The main problem with single-stem pruning is that eventually the tree grows too tall and spindly with an umbrella shape. In cold locations, however, arabica coffee may produce most of the flowers and crop on old wood and this method of pruning has proved successful in high latitude areas. In multiple-stem pruning, as shown in Fig. 9.18, old stems are periodically removed and the problem of the bush becoming too tall does not occur. The single-stem bush may be converted to multiple stem when it becomes too tall as shown in Fig. 9.19.

Another method of bush formation in coffee is illustrated in Fig. 9.15. This is often used for robusta coffee.

In all types of pruning for bush formation, some lateral branches should be pruned (Fig. 9.19). Downward-pointing branches and inward-growing branches should be pruned along with any weak or diseased branches. In addition suckers should be removed from the main stem and in humid locations, excessive suckers arising from the main lateral branches near the main stem (Fig. 9.13) should be removed. Pruning should be carried out soon after each peak of harvesting, leaving only healthy shoots growing outwards which will bear the next crop.

Weed control and mulching Weed control in coffee

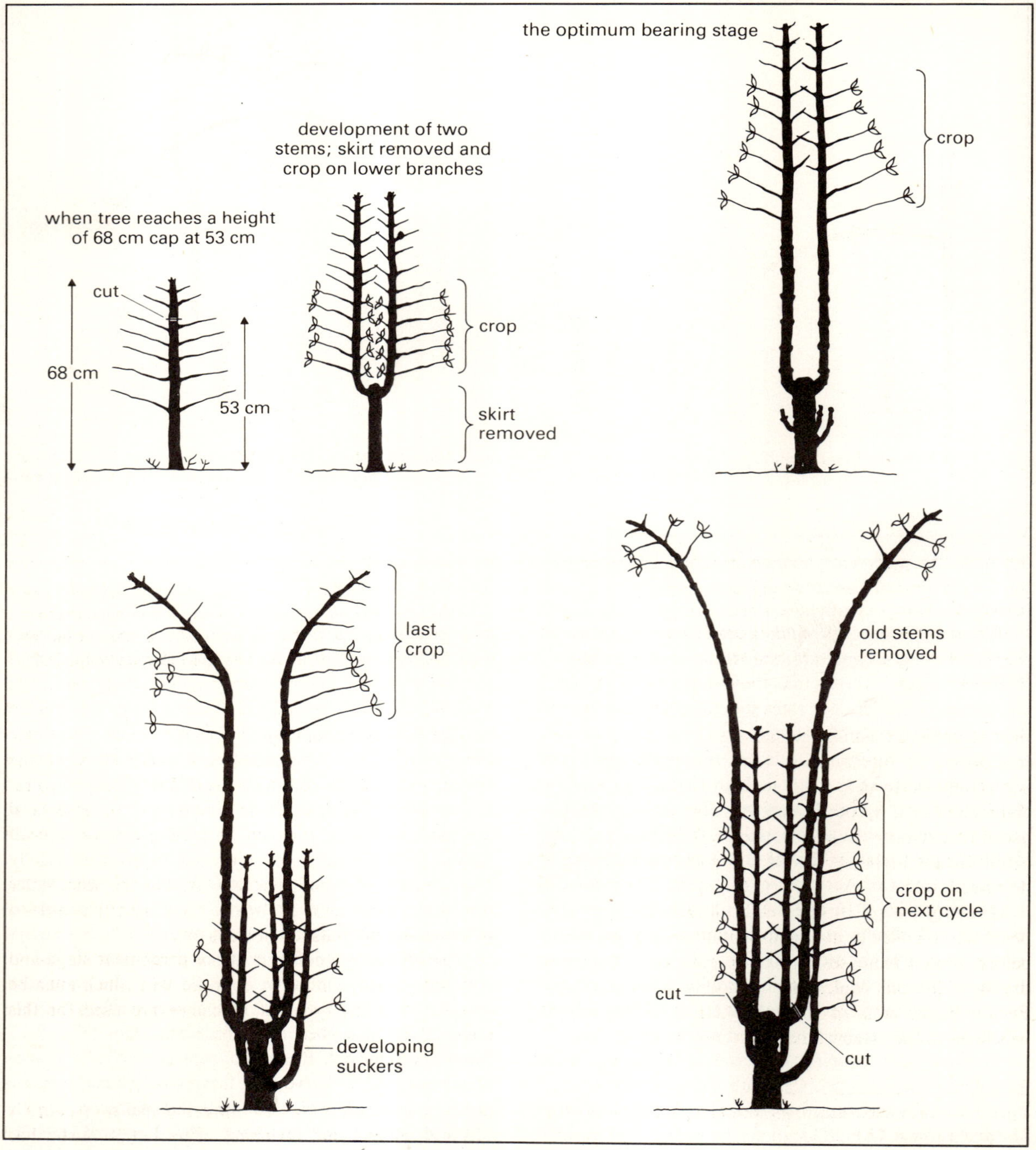
when tree reaches a height of 68 cm cap at 53 cm
cut
68 cm
53 cm
development of two stems; skirt removed and crop on lower branches
crop
skirt removed
the optimum bearing stage
crop
last crop
developing suckers
old stems removed
crop on next cycle
cut
cut

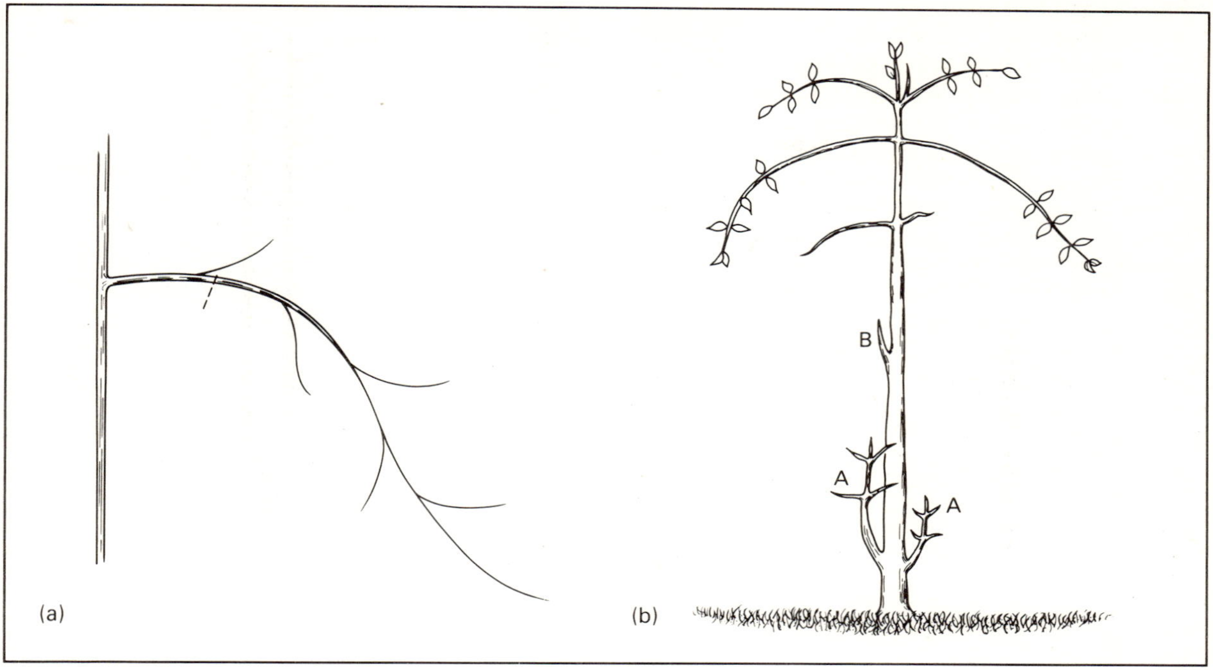

fig. 9.19 (a) Pruning back of a long, drooping lateral (b) An old tree which has lost its lower and middle branches but has a few branches growing out in the shape of an umbrella at the top. The suckers AA at the base have previously been encouraged to grow by nicking the bark above dormant buds; suckers appearing in the region of B have been pinched out at once. When the suckers at AA are well established the tree is ready to be stumped back above them. (*Source:* **Haarer, A. E.** (1963). *Coffee growing.* OUP: London.)

may be achieved with herbicides used in circles and inter-row areas. The combination of a contact herbicide and soil-acting herbicide gives extended control. Sodium chlorate(V) at 25 kg per hectare or paraquat at 1 litre per hectare are suitable contact herbicides. Diuron is suitable as a soil-surface herbicide in coffee at rates up to 2 kg per hectare.

Mulching the soil surface with grasses has been very beneficial to coffee in marginal rainfall areas as explained before. Mulch increases rainfall acceptance and reduces drying of the soil. Mulch also reduces weed competition. Even in humid areas of Southeast Asia mulch has proved beneficial to the establishment of robusta coffee.

fig. 9.18 The multiple-stem method of pruning arabica coffee. (*Source:* as fig. 9.17.)

Harvesting and processing

Picking coffee is a labour-intensive operation since the berries ripen quickly and must be picked promptly to retain quality. The berries are conveyed to a central preparation area and the pulp removed in various types of machines. After this the expressed beans are usually fermented to get rid of mucilage around the seed. After fermentation the seeds are washed and usually sun-dried by spreading on mats, open trays etc.

After drying, the coffee is at the parchment stage and each bean is surrounded by a brittle skin which must be removed. Various types of machines are used for this purpose.

Protection

Coffee rust caused by the fungus *Hemileia vastatrix* affects arabica and liberica coffee but robusta has resistance to the disease. Crosses with robusta varieties also sometimes show resistance; these could become commercially important in the future. The fungus causes lesions on the lower surface of the leaves which have a reddish coloration and give off spores. The disease is controlled by strict quarantine procedures and the use of resistant varieties. The disease may also be controlled by copper sprays such as Bordeaux mixture.

Coffee berry disease (CBD) caused by the fungus *Colletotrichum coffeanum* is serious in arabica coffee. It is more serious under moist cool conditions. It causes brown rot of the berries often starting at the base and stalk. Robusta varieties show resistance. It can be controlled by copper sprays.

American leaf spot caused by *Mycena citricolor* occurs mainly in the American Tropics. It is also controlled by copper sprays.

Various insect pests attack coffee. Green scale (*Coccus viridis*), leaf miners (*Leucoptera* spp.), the coffee-berry borer (*Hypothenemus tampei*) and others are sometimes serious. Various insecticides are used in control.

fig. 9.20 Tea planted on steep slopes (up to 45°). Run-off waterway in centre is planted with Guatemala grass

Tea (*Camellia sinensis*)

Family Theaceae

Other names thé (French); té (Spanish); chá da India (Portuguese)

There are two recognised varieties or groups of varieties referred to as *assamica* or Assam and *sinensis* or China, from their origins in Assam and China. Within each variety there are numerous selections from particular locations called 'jats'. Since tea is now chiefly propagated by vegetative means, jats often have distinct qualities characteristic of a particular region, district or estate. Most teas originate by the propagation of single tea bushes or trees having desirable properties of yield and ability to produce large populations. The fact that tea often does not adapt well to other locations with regard to quality, has led to the proliferation of large numbers of locally-adapted types.

China and Assam teas differ in morphological features and quality. As far as the western tea trade is concerned, Assams have been regarded as producing the best fermented teas. In recent years however, jats of China origin or hybrids with Assams have also been found that produce high-quality fermented teas.

Morphologically, Assam and China teas differ in the type of leaf. This affects productivity and response to fertiliser and shading.

Historically, the cultivation of China tea has been confined to China and Japan where it is prepared into dried tea without fermentation. This type of tea formed the basis of the China tea trade with the west which died out in the last century with the development of tea estates in India, Sri Lanka and other countries. These grew mainly Assams which were processed into fermented teas superior in flavour to traditional China teas.

Although tea is not tropical in origin, it is often regarded as a tropical crop since it forms an important economic crop in a number of tropical countries. It is grown at high elevations and harvested by hand plucking. It is important in Sri Lanka, India, Indonesia, Malawi, Kenya, Taiwan and many other tropical and sub-tropical countries. In recent years the development of mechanical plucking has led to the formation of tea industries in the USSR, USA and Australia.

Botany

When left unpruned, tea grows into a small tree about

10 m high with a conical shape. In cultivation it is pruned to form a hedge of convenient height for hand plucking or machine harvesting. Much of the tea produced in tropical countries is grown on steep hillsides at high elevations (Fig. 9.20).

Leaves are produced alternately on branches which originate in the leaf axils lower in the canopy. The leaves have a finely serrated margin. Assam jats generally have light-coloured leaves, horizontal in orientation. China jats have darker, smaller more vertical leaves. All tea leaves are somewhat glabrous and shiny and they contain thickened stone cells (scleroids). These are of importance in the tea-trade because their presence in made tea is used to determine its purity.

With the tea crops, leaf growth is of central importance because the leaves provide the yield. The traditional Assam jat with large horizontal leaves is typically low-yielding because the canopy does not permit good light penetration (p. 73). Yield is not increased by very high levels of nitrogen fertiliser. Furthermore, horizontal leaves are more exposed to the sun, and heat up rapidly; for this reason traditional Assam jats require shade for optimal growth. Jats with smaller, more erect leaves permit better light penetration, greater leaf production and response to high levels of fertilisers, particularly nitrogen. A yield of 1 000 kg per hectare made tea per year would be considered good for older Assam teas under shade trees. Modern tea industries are based on higher-yielding jats of both improved *assamica*, *assamica/sinensis* hybrids and *sinensis* types which are grown from cuttings and may average around 2 000–3 000 kg per hectare of made tea at maturity with adequate nitrogen fertilisation. Some exceptional varieties such as TRI 2030 from Sri Lanka can produce 5 000 kg or more of made tea per hectare.

Tea leaves are produced in flushes at regular frequency which are not strongly related to climatic changes, though the quantity of leaf produced at each flush is greatly affected by rainfall and temperature as well as by nutrition. Where tea is flushing it produces leaves of the type shown in Fig. 9.21(a). These few terminal leaves (2 leaves and a bud) are the ones that are plucked. It is very important for continued production not to pluck the mature leaves below this; these are called maintenance leaves. During a flush, several series of such leaves will be produced and this flushing is prolonged by good fertilisation, particularly of nitrogen. After a flush, a state of dormancy is entered in which smaller and smaller leaves are produced. The dormant bud is called a 'banjhi'. With nitrogen deficiency an extended series of reduced leaves will be produced. A typical banjhi shoot is shown in Fig. 9.21(b). Much of the skill of management in tea growing is concerned with the plucking

fig. 9.21(a) Healthy tea in flush. The terminal two leaves and a bud are plucked in hand harvested tea. Lower leaves are maintenance leaves important for photosynthesis and continued flushing.
(b) A bud in the phase of 'banjhi' or dormancy. During this phase the leaves become very small

of areas in flush at the right time, before the shoots get too large.

Once tea is planted, it is usually maintained for many years. There are some tea estates which have been going for more than 100 years without replanting. In estate practice, blank spaces in the field are usually replanted with higher-yielding newer varieties so that many older plantings consist of more than one variety.

True tap roots form only on bushes grown from seed, a practice now out of date except for breeding purposes, but extensive root systems form on old bushes whether grown from seed or cutting at cooler sites (as at high elevations in the Tropics) and a large food reserve accumulates in the root system. Fig. 9.22 shows the massive root system of a tea bush at 40 years. The food reserves in the root are important in regrowth after pruning. At warm sites in the lowland Tropics root growth and reserves are low and the life of the tea bush will be greatly reduced.

Tea flowers are produced in the axils of lower leaves. When mature they produce large seeds about 1·5 cm in diameter.

Climate and soils

Tea is grown over a wide range of climates from the humid tropical lowlands to regions of high latitude in Russia and Japan. In the Tropics cultivation occurs mainly at high elevations. Fig. 9.23 gives some indication of the extremely wide range of successful tea cultivation and indicates the regions of traditional cultivation of China teas. In the Tropics, lowland tea generally produces higher yields with adequate rainfall but quality is lower and root growth and longevity is much reduced. Temperature is of importance and it is found that leaf expansion only takes place above about 21°C, with optimal temperatures between 25°C and 30°C. Above 30°C however, leaf functioning is reduced. In high latitudes therefore, tea production is seasonal. Flushing is also affected by daylength. In the winter or in the short photoperiod of high latitude regions, dormancy can result.

Tea needs adequate evenly-distributed rainfall for high production without irrigation. The minimum annual rainfall, if evenly distributed is considered to be between 1 000–1 400 mm; this figure will be modified by humidity, soil storage capacity and the occurrence of mist. There is no upper limit to rainfall provided sunshine is not greatly reduced. Tea can grow healthily on soil moisture reserves so that deep root development is often important to production. One of the main reasons for the success of high-elevation sites in the Tropics for tea cultivation is undoubtedly the occurrence of adequate rainfall. In these sites, mist is important both as a source of water and in raising the humidity, since tea is adversely affected by humidities below about 80 per cent. Another reason for the success of high elevation and high rainfall sites is the occurrence of leached acidic soils.

Tea grows best on soils of acidic reaction and also soils that are sandy and freely draining. Hence the importance of root depth since these soils have rather low capacities for water storage. Soil pH above 6·0 reduces tea growth and the optimal pH is probably in the range pH 5·0–5·5. However, good yields can be obtained on soils of pH 4·5 and even pH 4·0. Consequently liming is seldom, if ever,

fig. 9.22 Massive root systems of tea (about 40 years old) grown at an elevation of 1 700 m in the Tropics

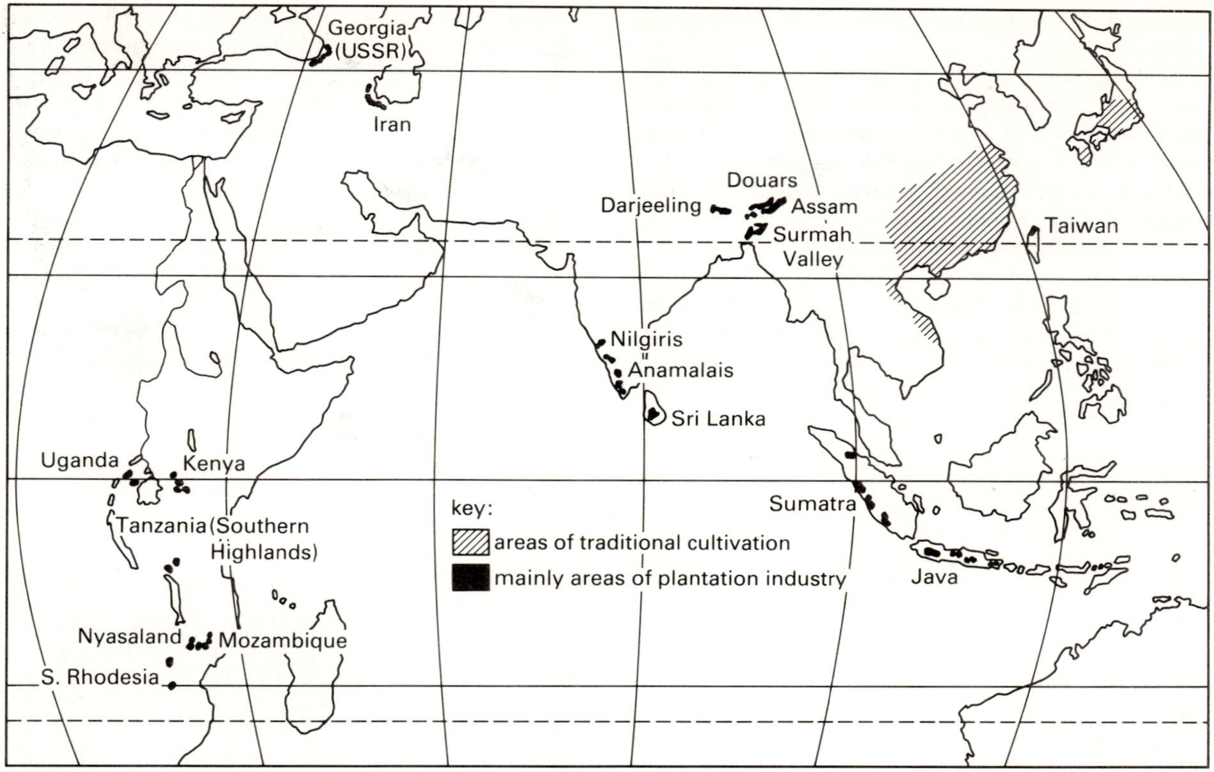

fig. 9.23 Main tea-growing areas. (*Source:* **Eden, T.** (1965). *Tea.* Longman: London.)

practised in tea growing. Furthermore, in clearing new land for planting by cutting and burning, care must be taken to spread the ash so that pockets of high alkalinity from the ash do not occur. Alkalinity may be reduced by putting sulphur (powder) in the planting hole at a rate of 100 g per hole for soils of pH 6·0 and 500 g per hole for soils of pH 6·5. Alkaline soils should not be used for tea cultivation.

Tea is not killed by frost but set-back in growth occurs. Hail sometimes causes major damage in mountainous sites, e.g. Kenya. Tea is sensitive to strong winds and the planting of wind-breaks along the ridges of hills has been shown to increase yields.

Cultivation and management

Land preparation Special mention is made of this because of the frequent cultivation of tea on very steep land. Tea is best planted on inherently stable soils in steep locations. The tea illustrated in Fig. 9.20 is on a stable granitic soil and contour planting has not been followed, nevertheless the particular area is stable and erosion free. On less stable soils careful planting on narrow, graded terraces is necessary, the grade depending on the type of rainfall to control infiltration and run-off. Excessive infiltration in certain soils leads to landslips. Run-off should be channelled into appropriately protected waterways. Guatemala grass (*Tripsacum laxum*) has been used to protect waterways in high elevations in tropical regions but on some soils concrete and stonework may be necessary to direct run-off water. During establishment, cover crops are also recommended to protect the soil but on very steep slopes cultivation of excessively unstable soils should not be attempted. For hillside planting of tea, narrow bench terraces averaging about 1–1·5 m apart are best suited to spacing requirements and are convenient for plucking and other operations.

fig. 9.24 Rooted single-leaf cuttings under a polythene tent (about 8 weeks from planting)

fig. 9.25 A group of single-leaf cutting beds protected by plastic covers to both reduce light and increase humidity

Nursery Traditionally seed was used for planting but this has largely been replaced by vegetative propagation which gives plantings of uniform quality. Any bushes with desirable properties of yield, quality etc. may be used for propagation and these should be multiplied according to the amount of propagation material required. Sometimes a tea estate will develop its own lines or jats and this is a continuing process rather than a once-only operation. The Tea Research Institute of Sri Lanka produces exceptional varieties suited to high-elevation sites in equatorial regions. In addition to the desired properties of yield and quality, a clone for vegetative propagation must also have the property of rooting easily and establishing well from cuttings, preferably without the use of rooting hormone.

Tea bushes selected for propagation are taken out of plucking and allowed to grow freely to produce branches of the size illustrated in Fig. 5.6 on p. 28. From these, single-leaf cuttings are obtained as indicated using a sharp knife to give a clean cut. Callus formation and rooting will not take place from a ragged cut. The single-leaf cuttings are laid out in the nursery bed as shown in Figs 5.6, 9.24 and 9.25. A sandy soil should be used for rooting tea cuttings and organic matter should not be added. Rock phosphate fertiliser should be mixed in at a rate of 500 g per cubic metre. Complete-mixture fertiliser should not be used until the cuttings are well rooted and then used sparingly. About 500 g of nitrophoska blue special per 1 000 seedlings should be adequate over the nursery period. Sometimes fumigation of nursery soil against nematodes is required.

fig. 9.26 Single-leaf cuttings of tea at a later nursery phase. Cuttings like those in fig. 9.24 are transplanted to polythene bags or sleeves of 12–15 cm diameter. Overhead shade is progressively reduced to planting out stage

Single-leaf cuttings are established under polythene tents supported on a hooped framework to raise the humidity (Fig. 9.25).

After establishment, cuttings are transplanted to single polythene containers filled with sandy soil, as shown in Fig. 9.26. Often polythene tubes rather than bags are used, of about 15 cm width when laid flat. Overhead

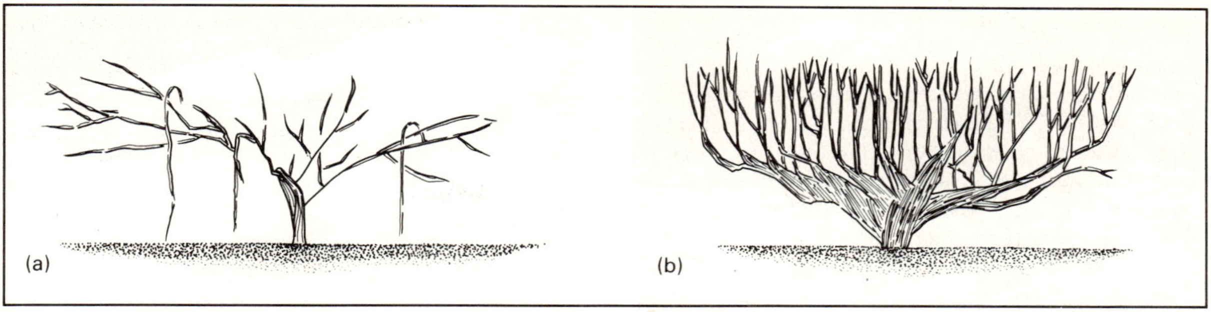

fig. 9.27 (a) Pegging of tea branches (b) Frame formation after pinching off laterals
Final plucking height should be about 1 m. (*Source:* **Eden, T.** (1965). *Tea.* Longman: London.)

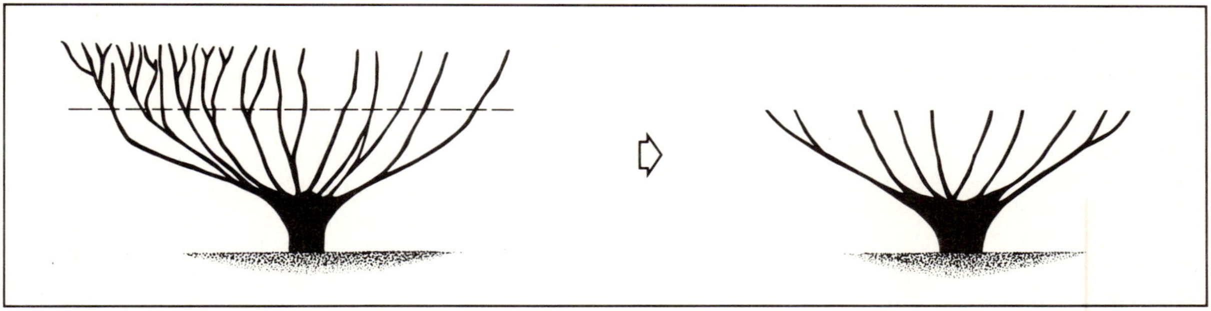

fig. 9.28 Pruning of tea to open the canopy and lower the plucking table

shade should be initially about 70 per cent and this should be progressively removed before planting out at about 6 months. Sometimes the initial shoots are pruned back as secondary shoots grow more vigorously.

Planting out, bush formation High-yielding teas should be planted at densities up to 20 000 per hectare. With Assam jats the wider, traditional spacing, of 5 000–8 000 per hectare should be used. Seedlings are planted in appropriately-sized holes after stripping off the polythene containers. Planting out should be done in a moderately moist season when soil water reserves are high. To encourage side-shoot growth the seedling is cut back to about 15 cm after field establishment. The most convenient method of forming the hedge canopy is by pegging. After seedlings have established in the field the side branches are pegged, usually using wire hooks, as shown in Fig. 9.27(a). Up to 8 side shoots may be pegged evenly around the original shoot. After about 6 months the pegged branches become fixed in this position. Vertical shoot production occurs on pegged branches and these are then pinched off to encourage a dense stand of shoots in the bush as shown in Fig. 9.27(b). Other ways of forming the bush by progressive pruning are described in the literature.

Fertilisers Nitrogen is the most important fertiliser for tea and should be supplied either in the form of ammonium sulphate(VI) or urea. Because of the possibility of sulphur deficiency in tea it is advisable to use both forms and, as with other crops, multiple applications of fertiliser are better utilised and reduce losses. Calcium(II) ammonium nitrate(V) and sodium nitrate(V) are not suitable as sources of nitrogen in tea because they increase the alkalinity of the soil. In the acid soils usually used for tea growing, ground rock phosphate is a satisfactory source of phosphorus. Micro-nutrient deficiencies of zinc, copper, magnesium and boron may occur in tea.

Table 9.6 gives general yearly fertiliser recommendations based on experience in major tea growing regions. The nitrogen dose should be split into at least 6 applications avoiding excessively wet or dry seasons and the fertiliser spread around the bushes. Often nitrogen is applied after each plucking round. Phosphorus fertilisers can be applied in 2 doses per year. Potassium fertilisers may be applied in 1 dose per year.

It is possible that very high-yielding teas will respond to as much as 800 kg urea per hectare. Tea often benefits from zinc and copper sprays. When it is necessary to spray with copper fungicides to control disease, extra copper applications are not necessary. However, zinc(II) sulphate(VI) at 2 kg per hectare should be sprayed 3 or 4 times per year. Foliar diagnosis has been used to determine fertiliser needs more precisely in mature tea and where such a service is available it should be used.

Irrigation In regions where dry seasons of more than 2 months occur tea responds to irrigation. This is usually supplied by sprinkler systems.

Weed control Satisfactory weed control in tea can be achieved by spraying between the rows below the canopy with paraquat or diquat and diuron at the following rates:

paraquat or diquat 0·5–1 litre per hectare;
diuron 2 kg per 100 litres per hectare.

Paraquat should be used where grass weeds predominate and diquat for other weeds. This herbicidal treatment should give weed protection for about 3 months. Sodium chlorate(V) should not be used to control weeds in tea.

Pruning Pruning is carried out for two reasons, firstly, to lower the canopy or plucking table when it becomes too high. Secondly, to open up the canopy when the top becomes too dense after repeated plucking thus allowing light to penetrate into the lower canopy; this encourages the growth of new branches. Tea grown in the lowland Tropics should be very sparingly pruned, if at all, because of the fact that food reserves of the plants will be low and recovery from pruning slow. At high elevations in the Tropics, periodical pruning to improve the form of the bush for penetration of light, brings about improvement in yield. However, pruning is a very specialised task and expert advice should always be sought. Fig. 9.28 illustrates the basic idea of pruning to open the canopy and lower the plucking table. This type of pruning retains most of the main branches but pruning may be more severe than this leaving only about half of the lower branches. Pruning may also be taken as far as cutting back to the main stump. This is only done to correct extremely bad bush formation, or sometimes in disease control.

Table 9.6 Suggested levels of fertiliser for different types of tea (rates in kg per hectare per year)

(a) Assam jats, under planted shade trees

	Ammonium sulphate(VI)	Urea	Nitrophoska blue	Ground rock phosphate	Potassium chloride
Year 1	50	50	50	100	25
Year 2	50	100	50	50	25
Year 3	75	150	50	50	50
Mature	100	200	50	50	50

(b) High-yielding varieties without shade

	Ammonium sulphate(VI)	Urea	Nitrophoska blue	Ground rock phosphate	Potassium chloride
Year 1	50	100	50	100	50
Year 2	100	200	50	100	50
Year 3	100	300	50	100	100
Mature	200	400	50	100	100

Since the height of the plucking table increases with time, one way of reducing this is to carry out 'top-pruning' at the time of plucking, i.e. all shoots projecting above the plucking level are removed. This reduces the frequency with which more severe pruning needs to be done.

Harvesting and pruning

Plucking The frequency of plucking depends on flushing and growth rates; it will vary from 1 to 2 weeks in the Tropics, during the periods of flushing. In quality tea production the 2 leaves and a bud illustrated in Fig. 9.21 is aimed for. If plucking is delayed the shoots will become too tall and woody; if too soon, full development of the flush to the ideal plucking stage will not have occurred and the plucking yield will be low.

Types of tea There are two commercial tea products, green tea or unfermented tea used mainly in Asian countries, and fermented tea or black tea, which forms the bulk of the tea trade in western countries. The manufacture of either from harvested fresh tea is a highly specialised task calling for a great deal of experience. The difference between green tea and black tea is that the latter is 'fermented' by enzymes which are present in the tea leaves and released on crushing (rolling), while in the former, fermentation is prevented by heating.

Protection

Like most crops, tea is affected by a host of diseases but the main one that has caused epidemic losses and set-backs to the industry is 'blister-blight' caused by *Exobasidium vexans*. The fungus first causes small translucent spots which expand and become concave depressions, looking like blisters on the under side of the leaf. The blisters turn white and release spores capable of infecting new leaves. Only young leaves are affected, mainly in wet and humid weather. Control is best achieved by using resistant varieties and by confining pruning to drier seasons. Copper sprays such as Bordeaux mixture also control the disease and should be used, when necessary, immediately after a plucking round, both to give protection to the next leaf and to give the maximum period before re-plucking for the dilution of the copper which is regarded as an impurity in tea.

Various root diseases affect tea, all caused by fungi of one kind or another. Most root diseases arise from the stumps of old jungle trees or shade trees which have died, which are the source of infection. Removing buried stumps and diseased tea bushes and replanting is the only practicable means of control.

Mites and scale insects form the major tea pests, they are controlled with systemic insecticides that break down fast enough to avoid contamination of the tea.

Minor crops

Kava (*Piper methysticum*)

Family Piperaceae
Other names yangona (Fijian)

Used as a beverage in Polynesia and the Pacific. The beverage is made from the roots which are peeled, pulped and then fermented.

Botany and propagation

It is a bushy shrub up to 3 m tall with rounded leaves. Roots become knotty and greyish-green in colour.

Propagated from seeds or stem cuttings.

Climate and soil

Kava is chiefly grown in Polynesia and other Pacific islands where it first originated. It grows well in a hot and humid climate. It can grow quite well in most well-drained soils.

Cultivation and management

The shrubs are planted at spacings of up to 1 m square. The roots are harvested about 2–4 years after planting. Fertilisation with compound fertilisers or animal manures is needed.

Protection

Weeds should be kept under control. The crop is quite free from serious pests and diseases.

Maté (*Ilex paraguariensis*)

Family Aquifoliaceae

Maté is an important beverage drink in South America where more of it is consumed than coffee or tea. The drink contains about 2 per cent caffeine (the active ingredient of coffee) and is made like tea from dried leaves. It is a mild stimulant with slight laxative properties.

Botany and propagation

It is a small tree growing up to about 7 m in height but when cultivated it is usually pruned to form a low erect bush with many stems and branches. The leaves are oval with serrated margins. Flowers are whitish, held in small bunches and fruits are round and reddish, containing many seeds. Seeds lose viability easily and should be used for planting immediately after collection.

Propagation is normally from seeds.

Climate and soil

The crop is indigenous to tropical South America and can be grown easily under humid tropical conditions. Soils should be as fertile as possible, well drained and rich in organic matter.

Harvesting and processing

The leaves are collected at intervals. Whole shoots are cut or pulled from the bushes and passed through a fire, this dries the leaves which are then beaten or ground to small pieces.

10 Drugs, stimulants and insecticide

Betel (*Areca catechu*)

Family Palmae
Other names betel nut; areca, areca nut (English in India), pinang (Malayan)

The nut is used as a masticatory stimulant mixed with lime, and betel leaf (*Piper betle*) and also sometimes with gambir or cloves. In Papua New Guinea the spadix of an arum lily is often used to replace betel leaf.

Botany and propagation

A solitary palm grown for its seeds. It has stems about 15 cm thick and grows up to 20 m tall with a crown of 10–20 leaves (see Figs 10.1 and 10.2). Both male and female flowers are enclosed in a spathe. Fruits are about 5 cm long, orange when ripe, with fibrous husk and a single seed.

Betel is grown from seeds sown in nursery beds.

Climate and soil

It is adapted to humid tropical climates and grows on a wide variety of soils in India, Sri Lanka, Southeast Asian countries, Papua New Guinea and the Pacific Islands.

Cultivation and management

The betel nut palm is cultivated to a limited extent.

fig. 10.1 Three betel palms with inflorescences containing male and female flowers. Note the bunch of young fruits near the leaf-base

fig. 10.2 Crown and young flower-bunch of a betel palm, male flowers predominate

Seedlings are transplanted to the field at the 5 leaf stage. Application of NPK fertiliser at rates of 1 kg per plant per year is recommended.

Protection

Mites are a serious problem in India; they may be controlled by spraying with sulphur. Fruit and bud rot are controlled by spraying with Bordeaux mixture.

Betel vine (*Piper betle*)

Family Piperaceae

Other names pan (Hindustani); sireh (Malayan); bulahucl (Sinhalese)

The leaves are used as a masticatory together with betel nut (see p. 76).

Botany and propagation

An evergreen climber with oval but pointed, light green alternate leaves. Flowers are brown, unisexual and formed into cylindrical male and female inflorescences.

It is propagated by cuttings about 60 cm long, planted near stakes.

fig. 10.3 Betel vines growing up wooden stakes

Climate and soil

The betel vine is native to Southeast Asia. It is grown all over tropical Asia, East Africa, Madagascar and the West Indies. Grown from sea level to elevations of 1 000 m. Rainfall of more than 1 700 mm is necessary. Loamy, fertile soil gives greatest productivity.

Cultivation and management

A number of varieties are known. Cuttings are planted a few centimetres apart and allowed to twine up stakes about 2 m tall (Fig. 10.3). Lime and organic manure should be given as often as possible. Leaves are picked from 18 months after planting. Vines must be picked 4 times a year in rotation. The upper leaves are of better quality. The cut leaves are kept covered for 2–3 days to bleach them.

Protection

Weeds should be removed regularly. The fungus *Phytophthora palmivora* causes foot-rot; this may be controlled by applying Bordeaux mixture to the lower stems and by improved drainage. The vines are seldom attacked by pests.

(a)

fig. 10.4 (a) Cinchona trees (b) A branch of a cinchona tree bearing leaves, flowers and fruit

(b)

Cinchona (*Cinchona* spp.)

Family Rubiaceae
Other names quinine tree; cinchone (French); cincona (Spanish)

Several alkaloids may be extracted from the bark; the major one is quinine which is a drug used for preventing and treating malaria. There are now many other superior antimalarial drugs and quinine is not as widely used as before World War II. Quinine is also useful as a tonic, as an antiseptic and as an ingredient in lotions protecting against sunburn and insects.

Botany and propagation

Quinine can be extracted from a number of *Cinchona* spp.; the economic ones are *C. calisaya, C. ledgeriana, C. officinalis* and *C. succirubra. C. ledgeriana* is the most widely grown; it is the one grown in Java which today accounts for 90 per cent of the world's production of quinine. The cinchonas of commerce are tall trees with leaves in opposite pairs. Yellow or pink flowers appear in bunches at the ends of the branches after 3–4 years. The fruit is a capsule 1–3 cm long; the seeds are small, flat and winged (Fig. 10.4(a) and (b)).

Propagation is mainly by seed, first sown in a seed-bed nursery. The seeds take 2–3 weeks to germinate. Cinchona may also be grafted or grown from branch cuttings.

Climate and soils
It is native to South America and needs a rainfall of 1 500 cm or more yearly. Steep mountain slopes and volcanic soils are preferred. Soil should be fertile and rich in organic matter.

Cultivation and management
Work in Java and elsewhere on *C. ledgeriana* has produced many clones, e.g. 'Ledger', 'GA 14' and 'KP 105'. Yields from improved clones are up to 100 kg per hectare of quinine compared with 25–30 kg per hectare from wild trees.

Seedlings 4–5 months old are transplanted to a second nursery of plastic bags. They are planted in the field after 1–2 years when they are 30–60 cm high. A natural cover crop is usually encouraged but legumes, e.g. *Leucaena leucocephala* and *Crotolaria usaramoensis* may also be planted. Liming and inorganic fertilisers are usually beneficial. Young trees are pruned of excess branches to obtain a single stem; the bark may be taken from these prunings.

Harvesting and processing
Trees are harvested from the 10th year onwards; by this time the bark has the highest quinine content. The bark from the trunk, branches and roots is stripped from uprooted trees by beating with sticks. The bark is dried from 70 to 10 per cent moisture content.

Protection
Damping off of nursery seedlings due to a fungus may be controlled by seed dressing with mercurial fungicide and careful watering. Field plants are attacked by root diseases; to keep them in check, remove rotting tree stumps and infected plants and use a collar dressing, e.g. PCNB. Branch canker may be kept in check by using organic fungicides. Circle weeding should be practised in the first few years.

Coca (*Erythroxylum coca* and *E. novogranatense*

Family Erythroxylaceae
Other names hayuelo, spadic, ipadú, hayo (Portuguese in Brazil)

The narcotic, cocaine is extracted from the leaves; it is useful as a local anaesthetic. The leaves are used for chewing in some South American countries.

Botany and propagation
Coca plants are small bushes 1–2 m tall with many branches, spear-shaped leaves and reddish brown bark. The flowers are small and white, appearing in bunches from the leaf axils (Fig. 10.5). They form round orange fruits 2–8 mm in diameter.

The plants are usually grown from seeds although germination is difficult. Propagation from cuttings is sometimes practised.

Climate and soil
Both plants are native to South America. *E. coca* prefers higher altitudes and is normally grown in the Andes. *E. novogranatense* grows in places up to elevations of

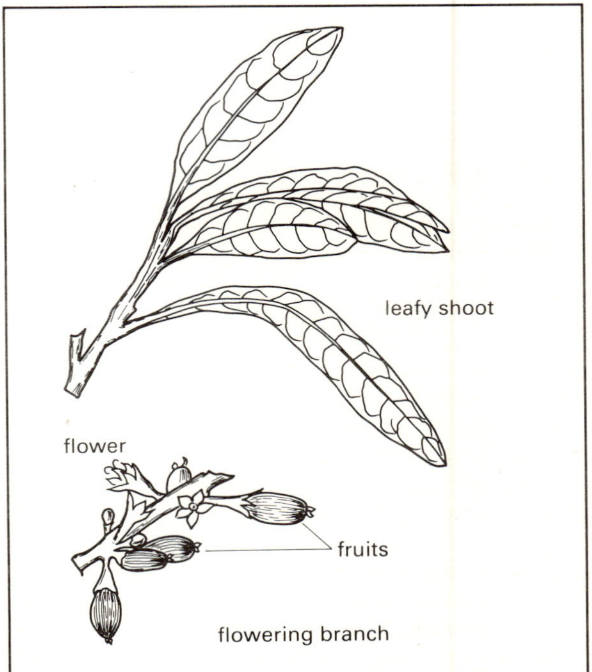

fig. 10.5 *Erythroxylum coca*

2 000 m and is more adapted to the humid Tropics. Cultivation of coca is strictly controlled in most countries as cocaine is an addictive narcotic leading to physical deterioration.

Cultivation and management

Year-old seedlings 20–25 cm tall are planted 1·5–2 m apart in the field. Leaves are first picked 1–3 years from planting. Ripe leaves are stiff and brittle and are picked 3–4 times a year. The dried leaves are exported for cocaine extraction. Yearly yields are up to 2 tonnes of dried leaves per hectare. Replanting is done after about 20 years. Regular dressings of inorganic fertilisers are usually needed; the rates depend on soil conditions.

Protection

The crop is relatively free from serious pests and diseases.

Derris (*Derris elliptica*)

Family Leguminosae (Papilionaceae)
Other names tuba root; akar tuba (Malayan)

Roots contain rotenoids toxic to animals and insects; the plant is used to poison and catch fish.

Botany and propagation

Derris is a woody climber growing up to 16 m long. The leaves have an odd-number (9–13) of oval leaflets. Pink flowers in bunches arise from leaf axils. The seed pods are flat and brownish and grow to a length of 3–5 cm.

It is propagated from cuttings about 30 cm long which are rooted in nursery beds for about 6 weeks.

Climate and soil

Derris is widely distributed in the Tropics. It is grown in areas with 1 750–2 500 mm rainfall and mean temperatures of about 85°C. It grows best on rich loams but it may be found on many types of soil.

Cultivation and management

Varieties giving up to 8–13 per cent of rotenone have been developed, e.g. 'Changi 3' and 'Sarawak Creeping'.

Plants are spaced at 90 × 75 cm. They are not usually fertilised. Roots are harvested after 2 years when they have highest rotenone content and dried to 10 per cent moisture for storage.

Protection

Derris is not affected by any serious pest or disease.

Ipecacuanha (*Cephaelis ipecacuanha*)

Family Rubiaceae
Other names ipecac (English); poaia (Portuguese); ipecacuanha verdadeive (Portuguese in Brazil); ipecacuana (Spanish)

Ipecacuanha produces a mixture of alkaloids, chiefly emetine, extracted from the roots of the plant. It is useful as a medicine for curing amoebic dysentery.

Botany and propagation

It is a small plant with creeping stems and twisted roots (Fig. 10.6).

Propagation is usually from plant divisions or stem cuttings which are first rooted in a nursery.

Climate and soil

Native to humid tropical Brazil, it is chiefly grown in India. Soils should be well drained and rich in organic matter.

Cultivation and management

Rooted divisions or cuttings are planted on raised beds at a square spacing of 30–45 cm. Roots are harvested 2–3 years after transplanting. Yields of 500–600 kg of dried

fig. 10.6 Young ipecacuanha plants with roots; commercial roots are much longer

roots per hectare have been obtained. Roots contain about 2 per cent alkaloids and about 1 per cent emetine.

Protection

Weeds should be kept under control. The crop is relatively free from serious pests and diseases.

Kola (*Cola* spp.)

Family Sterculiaceae
Other names cola

Kola is the collective name for the seed of four trees, *Cola acuminata* (Abata kola), *C. anomala* (Bamenda kola), *C. verticillata* (Owé kola) and *C. nitida* (Gbanja kola).

The seeds are used as a masticatory or for a beverage.

Botany and propagation

The trees may reach 10–20 m in height. The leaves are spear-shaped; the flowers are small, white or creamy, with male and bisexual types on the same trees. Green knobbly fruits split to show seeds with pink or red flesh (Fig. 10.7).

Kola is best planted from seed sown in polythene bags. Stem cuttings and grafts have also been successful.

Climate and soil

Kola is grown extensively in the humid area of West Africa where the rainfall is under 1 250–1 750 mm per annum. Soils should be fertile, well drained and rich in organic matter.

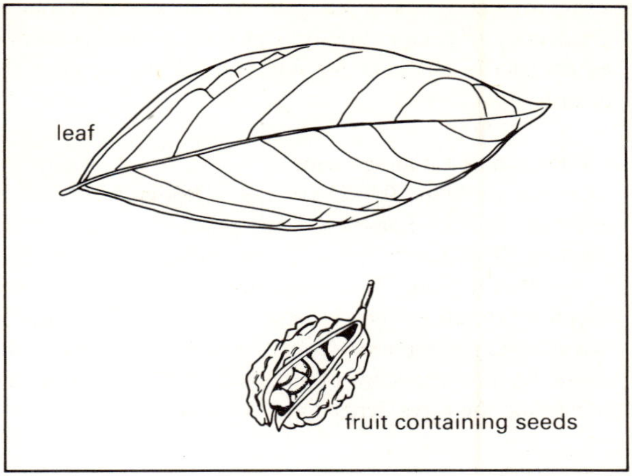

fig. 10.7 *Cola nitida*

Cultivation and management

The seedlings are planted 7–9 m apart, with intercropping in early years. Fruiting starts in the 7th year; full production is obtained at 20 years; lifespan is 70–100 years. Fruits are picked and split; seeds fermented and stored in baskets. Yields of 500–2 000 kg per hectare of nuts may be expected.

Protection

Weeds should be slashed whenever necessary. The tree is attacked by the fungus *Fomes lignosus* which causes white root rot and by thread blight (*Marasmius scandens*). *Marasmius* may be controlled with fungicides. Diseased trees must be removed. A serious pest is kola weevil (*Balanogastris kolae*), which bores into nuts.

11 Fibres

Abaca (*Musa textilis*)

Family Musaceae
Other names manila hemp; abacá, chanure de manille (French); abaca, canhamo de manilla (Portuguese); abaca manila (Spanish)

Abaca is the fibre obtained from the pseudo-stem of the plant, which is related to the banana. About 20 varieties are of commercial importance. Abaca is important in the Philippines where 95 per cent of the world's crop is grown; it is also grown throughout Southeast Asia, India and Central America.

Botany

Abaca has a branching underground stem or rhizome with many small roots. Aerial stems or pseudo-stems shoot at different periods so that an old plant may bear up to 20 or more pseudo-stems of various ages. Abaca stalks are more slender and the plant has smaller and narrower leaves than the banana (Fig. 11.1). The fruits are not edible and contain several black seeds.

fig. 11.1 Abaca plants

Climate and soil

Abaca grows best in areas where the rainfall is between 2 500–4 000 mm and seasonal in incidence, and in deep free-draining soil with a high level of organic matter. In the Philippines the best crops are grown on volcanic soils.

fig. 11.2 Abaca fibres drying

Cultivation and management

Abaca may be propagated in 2 ways:

1 by suckers, which arise at the base of the parent plant or corms (heads) as with bananas;

2 by seeds.

Generally whole corms or sections, depending on the availability of planting material are preferred to planting with seeds. Abaca planted from seeds takes 1–2 years longer to mature. When planting out in the field temporary partial shading is provided to protect the abaca plants from excessive heat.

Three methods of planting abaca are used in the Philippines.

1 **Square method** The corms are planted in a square-planting arrangement at 2·5 m apart and about 30–40 cm deep. The soil covers the corm to form 'hills' or mounds, this practice protects the corms from becoming waterlogged. This method of planting will allow 1 600 hills per hectare.

2 **Avenue method** In this method the corms are planted as above in sets of two rows at 2·5 m spacing but 4 m apart from the next set of 2 rows. Cash crops are generally grown between rows. This method of planting will allow 1 240 hills per hectare.

3 **Modified square method or triangular method** The

corms are planted at 2·5 m spacing. This method of planting will allow 1 800 hills per hectare.

Continuous harvesting of abaca will deplete the soil of nutrients and fertiliser has to be applied to maintain good yield. In the Philippines an application of nitrogen at 60 kg per hectare per year can increase the yield by 50 per cent (about 68 kg per hectare of urea or 136 kg per hectare of ammonium sulphate(VI) may be used per year).

Harvesting and processing
Abaca can be harvested 1·5–2 years after planting. The leaves are cut just before flowering and the pseudostem is cut close to the ground. The fibre is produced in the outer and inner part of the pseudostem, the inner fibre being of lower quality. The fibres are extracted either mechanically or by hand and hung to dry (Fig. 11.2). The quality of fibre varies depending on the agronomic practices, time of cutting, position of the fibre in the pseudostem, method of extraction, etc. After the first harvest about 3 stalks may be harvested from each hill every 5 months. From the 4th year 2–5 tonnes of fibre per hectare can be obtained per year and this yield is generally maintained to the 8th year. Abaca is normally replanted after about 13 years.

Annual world production of abaca is about 85 000 tonnes. Abaca is manufactured into the best grade of commercial ropes and cables, especially for marine use, as it is resistant to rotting in salt and fresh water. Abaca is also made into pulp for use in manufacturing high quality paper, tea bags, upholstery and filler used in acoustical insulation.

Protection
Abaca is often quite free of diseases but abaca mosaic disease can be a serious problem. The virus that causes this disease is believed to be transmitted by aphids. Abaca is affected by the same diseases and pests as bananas.

Congo jute (*Urena lobata*)

Family Malvaceae
Other names caesar weed (English); aramina (Portuguese in Brazil); carillo (Spanish in Central America)

Congo jute is a superior jute substitute producing a fine pale fibre. It is normally used as an admixture in spinning jute. In Malaysia it is used as herbal medicine.

fig. 11.3 Congo jute plants in flower, in Malayasia

Botany and propagation
Congo jute is a shrub with a short tap root. It grows up to 4–5 m tall and has many widespread branches. The leaves alternate and may be simple, notched, or lobed. The red flowers arise from the leaf bases (Fig. 11.3). The fruit is round with stiff hooked hairs containing 5 seeds.

The seeds are treated with concentrated sulphuric(VI) acid for 3 hours and soaked in water for 1–2 days. The fibres are obtained from the bark of the stem.

Climate and soil
Congo jute is widely grown in the Tropics but chiefly in Zaïre, also in Brazil and Madagascar. The shrub needs better conditions than kenaf. Soils should be rich, loose and well drained.

Cultivation and management
Plants should be spaced at 5 cm in rows 60–80 cm apart. Planting at a seed rate of 60–80 kg per hectare is recommended. High rates of complete fertilisers or manure are needed.

Harvesting is done when flowers appear, at 4–6 months and height 2–3 m. The stems are soaked in water (retting) for 5–14 days to remove non-fibrous material. Yields of 700–2 250 kg per hectare of fibre may be obtained. Rotation with other crops often proves beneficial.

Protection
Weeds must be removed up to 1 month. Varieties resistant to fungal 'cancer' disease (*Macrophomina urenae*) should be used. The fungus *Stagonospora urenae* causes

stem rot; these may be controlled by spraying with fungicide.

Pests include root-knot nematodes and stem borers.

Jute (*Corchorus capsularis* and *C. olitorius*)

Family Tiliaceae
Other names jute (French); juta (Portuguese); yute (Spanish)

The fibre, second in commercial importance to cotton, is fine but weaker than hemp. It is used chiefly for the manufacture of sacking. A smaller variety of *C. olitorius* is used as a vegetable.

Botany and propagation

These plants are herbaceous annuals with straight thin stems; they grow to 3–5 m tall, branching near the top. Jute fibres are extracted from the stems. The light green and serrated leaves grow alternately. Yellow flowers arise from leaf axils. The fruit of *C. capsularis* are round and ridged; the fruit of *C. olitorius* are long, like small lady's fingers (*Hibiscus esculentis*).

Both species are grown from seeds.

fig. 11.4 Young jute plants

Climate and soil

Jute originated in Asia and Africa and grows best in deep alluvial soils in a hot humid climate (temperature about 25–30°C, rainfall about 1 500 mm per annum). Soils must be well drained as the young plants are sensitive to waterlogging.

Cultivation and management

Early and late maturing varieties are obtainable, mainly from India. Soil is worked to fine tilth and seed broadcast at a rate of 6–10 kg per hectare; germination occurs in 3–4 days. Thinning may be done to 9 cm spacing when plants reach 15 cm tall (Fig. 11.4). Rich soils are not fertilised. On poor soils some complete fertiliser is applied. The crop is harvested when 50 per cent of the plants bear young fruits at about 4 months after planting. Late harvesting reduces fibre quality. Jute fibre is cut and retted as for kenaf; yields of 1 500–2 500 kg of dry fibre per hectare may be obtained. Seeds are produced at 7–10 months at a yield of about 280 kg of seeds per hectare.

Protection

No major pests are known but the plants are sometimes attacked by yellow mites (*Hemitarsonema latus*). Stem rot (*Macrophomina phaseoli*) is the most serious disease and soft rot is a soil-borne disease attacking the collar region; these are controlled by improved drainage. Early weeding is necessary.

Kapok (*Ceiba pentandra*)

Family Bombaceae
Other names white silk-cotton tree

The silky floss or kapok from the inside of the fruit is useful for stuffing pillows, life-jackets, etc. and also for electrical and heat insulation. It may be spun into yarn and cloth. The young pods are sometimes eaten as a vegetable. The seeds contain 20–25 per cent oil useful for cooking, lubrication, soap-making, etc. The soft and light timber is used in carpentry.

Botany and propagation

It is a tree growing up to 30 m tall. The trunk is broad with small buttresses at its base. The bark is greyish and in some varieties is covered with rounded spines. The branches are broadly spreading and almost horizontal.

Leaves consist of up to 15 leaflets each up to 15 cm long. It periodically drops all or some of its leaves and usually flowers when leaves have dropped (Fig. 11.5). The white or pink flowers are small and grow in bunches up to 20 cm long. The fruit is a capsule 10–30 cm long, brown when ripe and pointed at both ends. It contains many small black seeds embedded in silky floss which is usually white but may be greyish or light yellow (Fig. 11.6).

Propagation is from seeds sown in a nursery and transplanted 8–10 months later; direct field sowing is sometimes practised. It may also be grown from branch cuttings each 1·5–2 m long.

Climate and soil

Kapok is native to tropical America, it requires plenty of rainfall during early growth and a drier period for flowering and fruiting. It does best at elevations below 500 m in the humid Tropics and needs protection from strong winds. Soils should be deep and well drained.

fig. 11.6 Fruits of kapok

fig. 11.5 A mature kapok tree, in fruit; note the defoliation

Cultivation and management

Trees should be spaced about 6 m apart in the field. They start to bear after 3–4 years and continue for more than 60 years. Full bearing is after 7–10 years when mean yields of 300–400 pods per tree are produced per year. Pods contain about 17 per cent floss and 100 pods give up to 1 kg of floss. Yields may be up to 5 kg of floss per tree per year.

Harvesting and processing

Pods are usually plucked by hand when ripe and before they open. The pods are broken and the floss removed and dried. The seeds are then removed by beating with sticks or by machine.

Protection

The kapok tree is relatively free from pests and diseases. Weeds should be kept under control by circle weeding.

Kenaf (*Hibiscus cannabinus*)

Family Malvaceae

Other names mesta, deccan hemp, Guinea hemp, bimli or bimlipatum, jute (general, English); apocino, kenaf (French); caruru azêdu, cararu da Guiné (Portuguese)

A jute substitute, often used as an additive to jute.

Botany and propagation

Kenaf is an annual with straight thin prickly stems reaching 4 m in height. The leaves are oval or lobed. The flowers (Fig. 11.7) are red or yellow with a bright red centre yielding round and pointed fruits. High yields of fibre result from maximum stem growth before flowering. Flowers appear when the day shortens to 12·5 hours or less.

The plant is grown from seeds.

Climate and soil

Kenaf originated in Africa and is grown throughout the Tropics. India is the major producer. Temperatures in the range 15–25°C are suitable and rainfall of at least 600 mm annually is required. Well-drained sandy loams, rich in humus have proved to be the most suitable soils.

Cultivation and management

Many varieties differing in form of plant parts and response to day length are available. Top soil must be worked to a fine tilth. Seeds are planted at 25–35 kg per hectare at spacing of 20 cm by 5–8 cm. Wider spacing is necessary for seed production. Adequate complete fertilisers must be applied. For best fibre quality the stems should be harvested in 3–5 months at the 10-flower stage (Fig. 11.8). The leafless stems are soaked in water (retting) for 5–14 days to remove non-fibrous tissues. Fibre, washed and sundried makes up 16 per cent of stem. Yields of 1 000–6 000 kg per hectare of fibre may be obtained.

fig. 11.7 Kenaf plants in flower

Protection

Weeding is necessary until the crop reaches 20–30 cm tall.

Diseases include dry rot, leaf spot, leaf blight, stem rot and anthracnose. These are caused by fungi and can be controlled by fungicide sprays and the use of resistant varieties.

Pests include root-knot nematodes and stem borers.

fig. 11.8 Kenaf plants ready for harvest

Ramie (*Boehmeria nivea*)

Family Urtricaceae

Other names rhea, China grass (general, English); ramié (French); ramio (Spanish)

A high quality fibre of great strength and resistance to rotting. Used for sail cloth, Chinese linen, high quality rope, etc. The young plant is also used as an animal feed.

Botany and propagation

Ramie is an erect perennial herb 1–3 m tall. The stems bear few branches; leaves are round or heart-shaped. The male or female flowers may be greenish yellow, pink or red (Fig. 11.9). Rhizomes are formed underground and the crop is usually propagated from these rootstocks or from seeds; 3–4 crops are obtainable from each planting.

Climate and soil

A fairly warm moist climate with even rainfall is needed.

fig. 11.9 Ramie plants in flower

fig. 11.10 Preparation of ramie fibres

Somewhat acid, sandy soil rich in humus is preferred. The crop is chiefly grown in southern China.

Cultivation and management

There are two main groups of varieties: *nivea* from China and Japan; *tenacissima* from Malaysia with smaller and greener leaves. Rhizome pieces, stem cuttings or suckers are planted at spacing of 70–120 cm × 30–50 cm. Animal manure at about 7·5 tonnes per hectare is applied in the planting holes. Heavy dressings of complete fertilisers are needed after every harvest and plant residue is used as mulch. The first harvest is at about 10 months and 120–180 cm tall and after that at 2–6 monthly intervals. The stems are cut when in flower. Fibre is extracted and the gum scraped off. Yields of 800–2 000 kg of fibre per hectare may be obtained (Fig. 11.10).

Protection

Black leaf-eating caterpillar *Cocytodes coerula* attacks the crop and is controlled by insecticide spray. White fungus disease caused by *Rosellinia necatrix* kills roots; it is necessary to remove the diseased plants. The fungus *Rhizoctonia solani* causes die-back; control is by fungicidal spray and removal of diseased plants.

Roselle hemp (*Hibiscus sabdariffa*)

Family Malvaceae
Other names bonga, ceiba de carc (Spanish in Colombia); sumaing, painera (Portuguese in Brazil)

There are two types: *H. altissima* grown for fibre is a jute substitute and *H. sabdariffa* grown for its edible flower bracts.

Botany and propagation

Roselle hemp is an erect, annual shrub growing 30–80 cm tall; the leaves are alternate and are divided into 3 or 5 lobes, with toothed margins (Fig. 11.11). Single yellow flowers arise from the leaf axil; flowering occurs during short days. It is usually grown from seeds but sometimes from cuttings.

Climate and soil

Native to India and West Africa, roselle hemp is grown all over the Tropics. It requires a warm, humid climate, with even rainfall of 1500–2 000 mm yearly and 250 mm per month during the growing season. It grows up to 800 m elevation on moderately fertile and well-drained soils.

Cultivation and management

Soil must be prepared to a fine tilth. Spacing is 120 cm × 90 cm when grown for fruit and 60 cm × 5 cm for fibre. Fruits are picked 2–3 weeks after flowering giving yields of about 11 tonnes per hectare. Complete fertilisers are needed. The fibre is harvested at 7–8 months when the flowers start to appear. Retting is practised as for kenaf. Fibre yields are 600–2 000 kg per hectare.

Protection

Pests and diseases are the same as for kenaf.

fig. 11.11 Young roselle hemp plants

12 Fruits and nuts

Major crops

Banana (*Musa* spp.)

Family Musaceae
Other names See text for local varietal names.

The commercial bananas are mutants of wild *Musa* species which produce seedless and therefore edible fruits. The two wild species considered to give rise to the commercial bananas are *M. acuminata* and *M. balbisiana* and there are diploids, triploids and tetraploids with chromosome numbers of 22, 33 and 44. The diploids are usually smaller plants than the triploids and tetraploids and generally have smaller bunches and fruits. Tetraploids are the largest but there are not many of these; a number, however, have been produced by artificial breeding. One natural tetraploid is 'Klue teparod' mentioned below. Sometimes bananas are given the general name *M. paradisiaca.*

In the Pacific, a different group of bananas occur known as the Australimusa bananas; these are not economically important on a world basis but are collected and eaten locally. Of these the 'Tahiti' bananas are the best known. The 'Tahiti' or F'ei bananas produce erect bunches pointing straight upwards unlike the commercial bananas in which the bunch hangs downwards (see Fig. 12.2). The Australimusa group also contains the fibre crop known as abaça; once the most important material for high quality rope (manila hemp) but now largely replaced by synthetic rope fibre, chiefly nylon. Abaca is considered as a separate crop on p. 81.

There are hundreds of banana clones which are bud mutants of wild types and their hybrids which have been multiplied by man over the ages, but almost all (with the exception of the Pacific Australimusas) are thought to have the basic chromosomes of either *acuminata* or *balbisiana*. Naturally, those mutants which have seedless, and therefore edible, fruit have been the ones selected and grown in gardens and farms. Of the hundreds of clones only a few are important commercially on a world wide basis. These are mentioned below, and the suspected contributions of the chromosomes of *acuminata* and *balbisiana* are indicated by the letters A and B. Thus a diploid derived only from *acuminata* would be a AA and triploid from *acuminata* and *balbisiana* might be AAB or ABB, and so on.

Important banana clones

'Sucrier' (AA) Other names include lady's finger, pisang mas (Indonesia and Malaysia), lue kai (Thailand), surya kadali (India), banana ouro (Brazil), honey (West Indies).

This is the only diploid banana of commercial importance, it is valued for its highly flavoured fruits. The plant and the bunches are small (Figs 12.1 and 12.2) and the yield low. Furthermore, the fruits are thin-skinned and are easily bruised on handling and transporting. They are therefore only important locally in tropical countries. 'Sucrier' is resistant to Panama disease but very susceptible to leaf spot (see p. 98).

There are many other AA clones and they are particularly abundant in New Guinea. In Kenya the banana called 'Gatumia' is probably AA as are 'Mjenga', 'Sikuzani' and 'Paka' (Zanzibar) and 'pisang lilin' (Malaysia). 'Pisang lilin' was used to breed one of the main tetraploids (AAAA) 'Bodles Altafort' (see p. 92).

'Gros Michel' (AAA) Other names include pisang ambon (Malaysia, Indonesia), klue hom tong (Thailand), disu (New Guinea), bluefields (Hawaii), raimbaud, makangaia (West Africa).

This was the first important banana of international commerce and was the main banana of the American and West African banana trades, but it has been replaced by others because it is susceptible to Panama disease. It is still important throughout tropical Asia. It is a large plant as shown in Fig. 12.1, and produces high yields of good quality bottle-nosed fruit on even, cylindrical bunches (Fig. 12.2). The skin is fairly thick and the bunches, because of their even shape, are easy to pack for transport.

'Dwarf Cavendish' (AAA) Other names include Chinese and Canary banana, Johnstone, klue hom kom (Thailand), malindi (East Africa), camayenne (West Africa), caturra (Brazil).

This is a small hardy, high-yielding clone now widely planted and resistant to Panama disease. However, it is very susceptible to leaf spot. The fruit are of good quality with a fairly thick skin suitable for handling and transport but the bunches are uneven in shape and the bananas must be removed and packed as 'hands'. Unlike 'Gros Michel', the fruits are blunt ended.

A similar but taller clone is the 'Giant Cavendish'.

'Robusta' (AAA) Other names include valery (Congo), pisang buai (Malaysia, Indonesia), saina (New Guinea), mtwika (Zanzibar), poyo (Guadelupe), veimama (Fiji), fa'i palagi (Samoa).

'Robusta' is a tall vigorous clone producing a heavy bunch and a high yield; it is the main trade banana in parts of the West Indies, Central America, Fiji and Samoa. Like 'Cavendish' it is resistant to Panama disease but very susceptible to leaf spot.

A similar clone is 'Lacatan' which is also resistant to Panama disease. 'Lacatan' has greenish fruit when ripe and is also an important trade banana.

'Mysore' (AAB) Other names include pisang keling (Malaysia), klue kai ferang (Thailand), kikonda (Zanzibar), etc.

A very large and vigorous clone which is the main banana of India; although not widely grown elsewhere, this is a most promising clone to replace 'Gros Michel' because of its high resistance to Panama disease. It is also hardy, and tolerant of poor soils and periodic drought; it

(a) (b)

fig. 12.1 The largest and smallest banana clones (a) 'Gros Michel', and (b) 'Sucrier'

is resistant to leaf spot. It produces a uniform bunch of bright yellow, well-flavoured fruits.

Plantains (AAB) There are many plantain varieties. They are mostly vigorous plants resistant to Panama disease and leaf spot but susceptible to stem borers (see later). The fruits are large and starchy and mostly used for cooking throughout the Tropics. Some of the plantains are strongly curved and called 'horn' plantains (see Fig. 12.2). Straight varieties have the general name of French plantain.

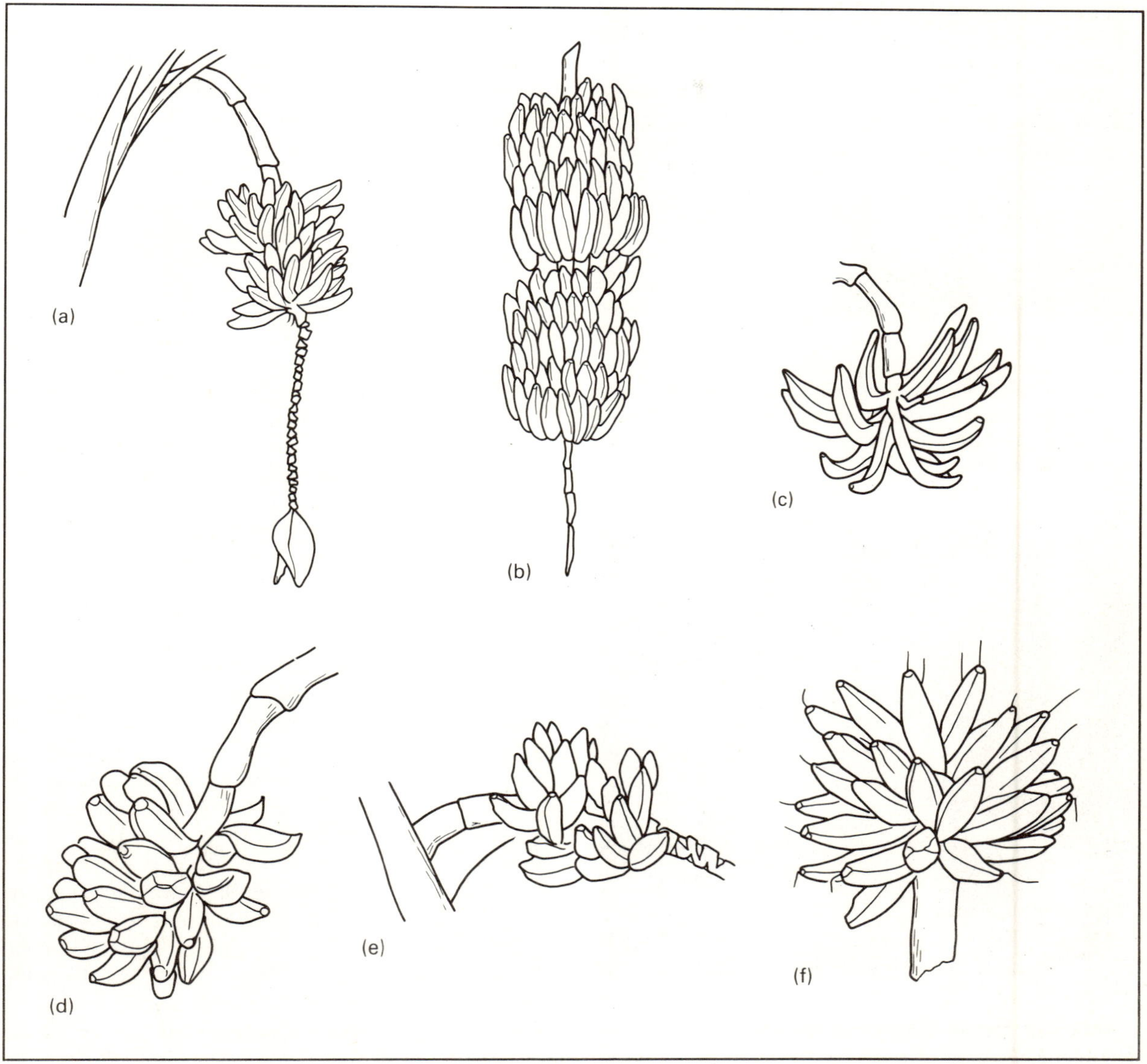

fig. 12.2 Bunches of various clones of bananas (a) 'Bluggoe' (b) 'Gros Michel' (c) 'Horn Plantain' (d) 'Klue teparod' (e) 'Sucrier' (f) F'ei banana (a)–(e) are true (Eumusa) bananas; (f) is an Australimusa banana.

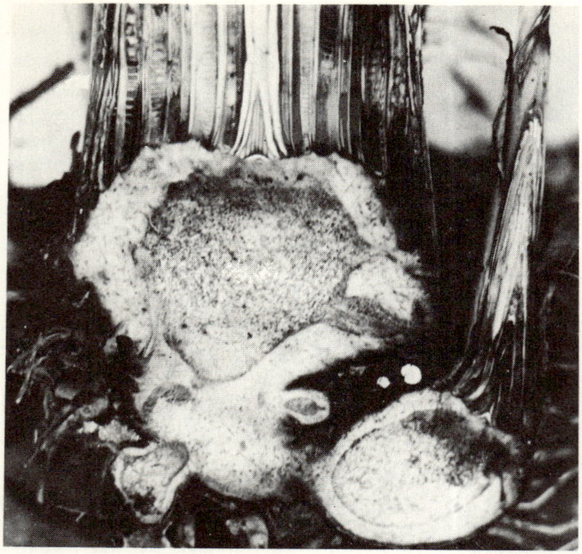

(a)

fig. 12.3 (a) Vertical section through base of banana plant in the vegetative stage; all the leaves are produced from the growing point or apex (b) Cross-section through pseudostem showing interfolded leaf bases. The true stem or flowering stalk is the solid white centre

(b)

'Bluggoe' (ABB) Other names pata (Samoa), saba (Philippines), jamani (Fiji), nalla (South India) and mkojosi, punda and other names (East Africa).

It is a vigorous clone producing bunches of thick almost straight fruits (Fig. 12.2). It has resistance to Panama disease and leaf spot, and is important in Samoa, Fiji, Philippines, South India, the West Indies and parts of East Africa.

'Klue teparod' (ABBB) Other names include pisang siam (Malaya), pya-ye-san (Burma).

This is a tetraploid banana and is the only one of importance known to occur in nature. The plant is large and robust and immune to Panama disease and leaf spot. It produces bunches of large blunt fruit characteristic of *balbisiana* (Fig. 12.2). The fruit is of low quality but is used in Thailand for cooking purposes. It is not an important banana in commerce and is mentioned here only because it is a tetraploid.

Other tetraploids with AAAA chromosomes have been bred. One of some commercial importance is 'Bodles Altafort' which was bred from 'Gros Michel' and 'Pisang lilin'.

There are many other banana clones of local importance throughout tropical countries. In East Africa for example the 'Lujugira-Mutika' group of bananas are important in upland areas. They are bananas of AAA make-up and several clones exist. The red bananas known by various names are also of AAA origin, and the pleasant-flavoured thin-skinned banana known generally as 'silk' (pisang rastali in Malaysia and Indonesia) which is common in tropical countries is of ABB composition.

Botany

The basic structure of all banana clones is the same. The so-called vegetative stem (pseudostem) of the plant is actually made up of interfolded leaf sheaths which grow from the corm (Fig. 12.3). The true stem appears only when the inflorescence is formed. This true stem grows up inside the pseudostem until it emerges and bears the bunch. The inflorescence is formed only after a certain number of leaves have grown and there is a high enough rate of photosynthesis to form the bunch. Nitrogen fertilisation (see below) is important in promoting leaf development, early bunch production and high yield. Bananas produce large simple leaves and generally between 20 and 50 are formed before the inflorescence is produced (Fig. 12.1). The youngest leaves form at the centre of the plant and older ones are displaced outwards to form the crown of leaves. When a new leaf emerges it is rolled to form a thin 'spear' and then progressively opens

(a)

(b)

fig. 12.4 Opening of a leaf (a) a newly-emerged leaf at the spear stage (b) the leaf half-unfolded

as it emerges (Fig. 12.4). When the inflorescence is formed, no more leaves develop on the shoot and after the bunch is ripe the shoot dies. The plant continues growth by producing side shoots.

Bananas are planted from side shoots or pieces of the corm (see below p. 95) and under tropical conditions take from 8–12 months to produce the bunch.

The bunch is formed from the inflorescence which is made up of successive layers of flowers each surrounded by a bract in the young stage. The bracts curl back one after another revealing the layers of flowers which later grow into 'hands' of bananas. The first-formed flowers (that is, the ones nearest the base of the stalk) are always female and those formed towards the top of the inflorescence are male. However, in commercial bananas the fruit generally develops without fertilisation and without seed as mentioned before. The male flowers which become exposed, dry up and fall in the mature bunch, but at the end of the bunch stalk a male bud remains, containing unopened male flowers. This bud, in certain varieties, is sometimes used as a vegetable. The details of the inflorescence are shown in Fig. 12.5.

The banana bunch may hang vertically downwards or project sideways (diatropic) or hang at an intermediate angle. Generally heavy bunches such as those of 'Gros Michel' hang vertically and light bunches such as 'Sucrier' tend to be horizontal. Since the banana fruit itself bends upwards when it grows (negatively geotropic) the shape of the bunch varies according to the angle of the main stalk. In 'Gros Michel', for example, with a vertically hanging bunch, the fruits all grow upwards evenly to give a uniformly shaped bunch (Fig. 12.2). With small-bunched clones that hang sideways, the fruits on the lower side bend upwards while those on the upper side grow straight

(a)

(b)

(c)

fig. 12.5 Stages in development of the inflorescence (a) Inflorescence just emerged from centre of crown, enclosed in sheath (b) Young inflorescence with first bracts removed to show young female flowers which become the first hands of bananas in the mature bunch (c) An inflorescence at the female stage with young developing fruits; the male flowers exposed will eventually wither and fall off

up, giving a very uneven bunch shape (e.g. 'Sucrier', Fig. 12.2). When packing such bananas for transport the hands of fruit have to be removed and packed separately.

The roots of bananas grow from the corm and most occur in the top 30–50 cm of soil. The roots are about pencil thickness and no tap root is formed.

Climate and soil

Bananas grow best in the humid Tropics where bunches are produced in the shortest time; that is after 8 to 12 months depending on the clone. For optimal growth the monthly rainfall must be at least 100 mm and can be as high as 250 mm. Where dry seasons occur maturity is delayed unless the crop is irrigated. The region of South-east Asia centred on Malaysia, and the wetter parts of South and Central America are the best regions for growing bananas and for banana industries. Regions which have a dry season are less suitable.

High temperature is the other requirement for banana growth and temperatures between 25 and 30°C are optimal. Temperatures lower than this delay maturity and temperatures below freezing destroy the crop. However, many of the largest banana industries are established in drier and colder climates where the plant may take up to 18 months to produce the bunch.

Strong winds also affect the banana crop since the shallow roots and absence of a tap root, combined with the large leaf and soft pseudostem, make the plant very susceptible to wind damage. Winds above 20 kph cause damage to bananas and winds reaching 80 kph may completely destroy the crop.

The shorter types of bananas such as 'Dwarf Cavendish' and certain hardy clones such as 'Mysore' are best suited to withstand adverse conditions.

It will be apparent that the humid equatorial Tropics, outside of the cyclone belt, present altogether the best climates for bananas.

Bananas can be grown on many types of soil provided drainage is good. Hill soils are generally adequately drained but alluvial clayey soils generally require drains to about 60 cm depth at suitable intervals. Very heavy clay soils need special treatment to raise the level of organic matter and improve drainage. They will need more frequent drains perhaps between each 2 or 3 rows of plants and this will require construction of cambered beds. In heavy soils deep cultivation to break up the clay should be carried out at the time of field preparation and laying of the drains.

Less acid soils are best for bananas, especially for sensitive clones such as 'Gros Michel' that suffer from Panama disease. The best pH range is pH 5·5–6·5. Soils more acid than this will benefit from liming with dolomitic limestone. A soil of pH 4·5–5·0 requires about 2 tonnes of magnesium limestone per hectare.

Cultivation and management

Planting material and planting Bananas are not planted from seed but from various types of side shoot or from the corm, or pieces of corm which will have buds in various stages of development (Fig. 12.3). Side shoots are obtained from around old banana clumps and when many shoots are required for planting, cutting back of the large shoots before bunch formation will encourage more of these side shoots to grow. The corm is obtained from the base of old mature shoots and is used for planting by digging it up and trimming off old roots and dead material but not the small buds present on the sides. Only healthy shoots or corms should be chosen for planting. Usually the corm may be cut into 4 if it is large (about 25 cm broad), or into 2 if it is small (15 cm broad). There should be at least 1, preferably 2 buds ('eyes') per piece. There are several types of shoots: maidens, swords, peepers and water suckers. Maidens are large shoots which have not yet produced a bunch inflorescence. Swords and peepers are different stages of strong side shoots. Peepers are young with up to about 30 cm of the shoot showing above the soil and swords are larger, producing a number of thin sword-like leaves, which precede the expanded foliage leaves. A sword shoot and peeper are shown in Fig. 12.6. Water suckers are weak superficial side shoots producing large leaves (unlike swords). Of these various kinds of shoots, swords, peepers and maidens are the best for planting. Water suckers may also be used if they have a corm more than about 8 cm in diameter but will be slow to start. Swords and peepers are dug up and cut off where they join the main corm. Maidens are trimmed back leaving a short length of pseudostem above the corm, and the leaves are trimmed from water suckers before planting.

In planting, holes are dug about 30 cm deep (rather larger holes are recommended for heavy clay soils) and the planting material is planted at this depth. Pieces of corm should be planted on the slant with eyes on the upper side. Maidens should be planted upright and swords, peepers and water suckers should be planted whole with the ends protruding if they are taller than the hole.

(a)

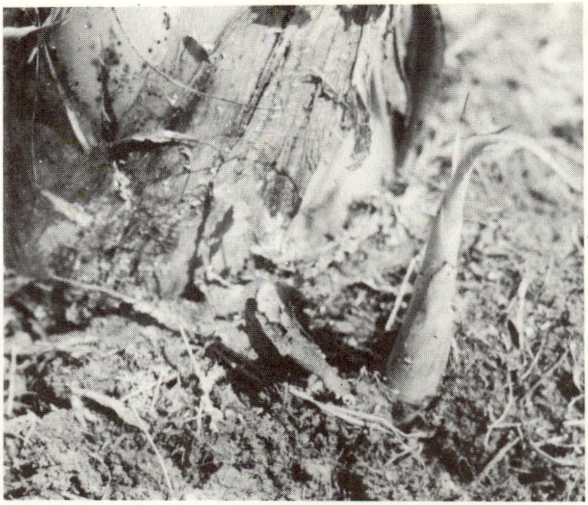

(b)

fig. 12.6 (a) A sword shoot, and (b) a peeper, each growing next to the mother pseudostem

Spacing The spacing adopted for bananas varies considerably but there has been a tendency for closer spacing in recent years. Spacing also varies with clone. The recommendations in Table 12.1 are considered to be suitable for average conditions for a crop which is to be retained for 3 or more years before clearing. If only one crop is to be grown spacing should be decreased by 30 cm.

After the first shoot has produced a bunch the number of stems increases by production of side shoots and it is a common practice to keep 2 vigorous shoots per planting point; pruning out the others, particularly the weaker ones. These shoots are called 'followers'. Thus density increases following the initial planting. In selecting followers, pruning should be done so that one small side shoot such as a developing sword and one larger shoot (e.g. a strong growing maiden) are retained so as to give a spread of growth and bunch production and not overtax the root system. Other shoots are pruned using a heavy blade (chisel) on a pole which is used to destroy the growing point.

General cultivation and weed control Mechanical cultivation should be avoided in bananas because the superficial root system is easily damaged, but earthing up or building up the soil around the pseudostem may be recommended when wind damage is likely to occur.

Normally weed control is best achieved by means of herbicides. For weed control around the stems the following herbicides may prove useful, sprayed in 250 litres of water per hectare: sodium chlorate(V) (10–20 kg) followed by diuron (3 kg). Sodium chlorate(V) should be sprayed only in clear weather since it kills weeds by desiccation. Under wet conditions paraquat (Grammoxone), at 1 litre per hectare should be used to replace sodium chlorate(V). Sodium chlorate(V) and paraquat are contact weed killers which destroy standing weeds; diuron gives lasting control by preventing new weeds from growing. This mixture should give at least 3 months weed supression. If climbing weeds are present, 1·5 litre of 2:4-D amine may be added to the sodium chlorate(V) when the banana plants have become tall. Great care must be taken not to spray the

Table 12.1 Spacing data for various bananas

	Diploid clones e.g. 'Sucrier' and small-sized triploids	Triploid clones e.g. 'Gros Michel' 'Mysore', 'Lacatan'	Tetraploid clones e.g. 'Klue teparod'
Planting distance (cm)	200	250	330
Approximate number of plants per hectare	2 500	1 600	900

banana foliage but only the soil between the plants. Simazine and atrazine may replace diuron for weed control in bananas.

After harvesting, old pseudostems should be cut at ground level, chopped up and spread over the inter-plant spaces. Pruned old leaves may also be spread on the ground. This practice helps to control weeds and also returns potassium and other nutrients to the soil.

Fertilisers Bananas require heavy applications of fertiliser for optimal yield. In particular high nitrogen and potassium fertilisation is necessary. Nitrogen is required early in the growth stage when the canopy is forming. However, under humid tropical conditions cropping is likely to be more or less continuous after establishment due to the original planting material and followers becoming out of phase. Fertilisers should be applied in fractional doses at least 4 times in a year; more frequent application of smaller doses makes for more economic fertiliser use for continuous cropping with bananas. For a crop grown once only (unusual in the Tropics) nitrogen, potassium and phosphorus should be applied in fractional doses (from about 4 weeks) over the first 4 months and a final dose of potassium chloride at flowering. Intermediate seasons without peak rains or dry spells should be used for applying fertilisers. Total rates of fertilisers per year or per crop which can be generally recommended are shown in Table 12.2.

When legume covers are established at planting, the nitrogen fertilisation may be greatly reduced or omitted over the first year. Soils of pH below 5·5 should be limed

Table 12.2 Fertiliser rates recommended for bananas (rates in kg per hectare)

	Ammonium sulphate(VI)	Urea	Potassium chloride or sulphate(VI)	Magnesium(II) sulphate(VI)	Rock phosphate	Single superphosphate
Virgin crop on newly-cleared and burned jungle or other rich soils in the first year	200	200			100 g in each planting hole	50
Following crops or crops on previously cleared land	200	300	300	80	100	50

using magnesium limestone, and on very acid soils calcium(II) ammonium nitrate(V) may be used to replace two-thirds of the ammonium sulphate(VI). On soils above pH 6·0 rock phosphate should be replaced by superphosphate and when calcium(II) ammonium nitrate(V) is used to replace ammonium sulphate(VI) single superphosphate should be used instead of rock phosphate as this also supplies sulphur.

Irrigation Irrigation is not usually practised for bananas in the humid Tropics but it is known that bananas will respond to irrigation and rainfall up to 65 mm per week.

Harvesting

Banana fruits can be harvested green and will ripen after harvesting. Even when harvested for local marketing, particularly for weekly markets, it is normal to harvest the bunch while it is still green just before yellowing occurs, because at the green stage it can stand rougher handling. Sometimes a vendor will put the hands on sale as they ripen because the lower hands generally ripen slightly sooner than those towards the tip. When bananas are harvested for transport over long distances, for example by ship, they are cut even sooner, when they are about three-quarters mature, and the fruits are not yet fully expanded. In certain clones which are grown for export trade, e.g. 'Gros Michel', 'Cavendish' and others, the fruits will increase in size for about 1 month as they ripen at the expense of stored food and water in the stalk and skin. Thin-skinned varieties such as 'Sucrier' and many others ripen much more quickly. For the purposes of marketing, green bananas are sometimes ripened artificially using ethene (ethylene) known as 'Ethrel', or ethyne (acetylene). To use ethyne, the bunches are placed in a small closed room or shed or covered with a polythene sheet and ethyne is introduced. Ethyne is generated by placing carbide in a tin half-filled with water under the polythene cover. 'Ethrel' should be used according to the manufacturer's instructions.

Protection

The most widespread pest of bananas is the stem borer *Cosmopolites sordidus* which is a beetle that lays its eggs in the pseudostem. The larvae tunnel in the pseudostem and weaken the plant and may cause death of the shoot if the growing point is eaten. The adult beetles live on old banana stumps and debris and are best controlled by good field hygiene; that is, by cutting back unwanted shoots, and old pseudostems which have flowered, to ground level, and cutting up and spreading the old shoots between the clumps as a mulch which quickly dries out. Removal of other rubbish also reduces the incidence of borers. Traps may also be made by splitting lengths of old pseudostem and placing them around the clumps. The beetles shelter under the pseudostems and may be collected by hand.

Insecticides recommended for borers are aldrin and dieldrin at 2 kg per hectare at 6-monthly intervals; they are sprayed on the sides of the pseudostems together with a sticker such as tenac.

The two most important diseases of bananas are Panama disease and leaf spot. Both have been mentioned above in regard to the resistance of various important clones. Panama disease is caused by a variety of the fungus *Fusarium oxysporum*, strains of which cause disease in many plants. Of the important trade bananas 'Gros Michel' is particularly susceptible and this disease has caused widespread losses in the West Indies and tropical America. This has led to replacement of 'Gros Michel' by other clones such as 'Lacatan' and 'Robusta'. The disease attacks the roots giving rise to wilting and collapse of the leaves. The most effective control is to grow resistant clones, although increasing soil pH by liming and improving drainage also reduces the disease. Panama disease is known to occur particularly after wind damage which ruptures root tissues.

Leaf spot is caused by another fungus, *Mycosphaerella musicola*. It attacks leaves, first causing pale yellow spots which later become dark and necrotic (that is the leaf tissue dies and decays). Older leaves first show the symptoms. The best method of control is by growing resistant strains but the disease can also be controlled by spraying with copper fungicides such as Bordeaux mixture. Spraying with mineral oils with an ultra-low-volume sprayer also controls leaf spot. In humid regions the disease is most serious and with susceptible banana varieties continued use of Bordeaux mixture will lead to copper toxicity. The fungicide dithane M-45 may be used in combination with spray-oils to replace Bordeaux. Spraying every 2 weeks is recommended under humid conditions.

In the humid Tropics bananas are also sometimes attacked by the bacterium *Pseudomonas* spp. which causes symptoms similar to Panama disease. It is controlled by improving soil drainage and its spread is prevented by dipping pruning tools in a solution of calcium(II) hypochlorite or methanal (formaldehyde).

Bananas are also attacked by nematodes, these build up in the soil over the years and may reduce yields considerably. Clean planting reduces the rate at which nematodes build up and this can be done by dipping all planting material in a 0·1 per cent solution of the nematicide DBCP for two minutes prior to planting. Annual field sterilisation using nemacur at 3 g per square metre of soil surface also reduces nematode build-up. Nematodes may also be controlled by periodically, once every 3–5 years, growing a different crop on the land and then replanting with bananas. A suitable alternate crop is sweet potatoes.

Citrus (*Citrus* spp.)

Family Rutaceae

There are a large number of citrus species which differ greatly in size and fruit characteristics. There is also a lot of disagreement among botanists about their correct names. Here, however, we will be dealing only with well-known citrus cultivars and using their well-known names. The largest citrus industries are located in the sub-tropics and warm temperate regions of the world and of these the largest industry consists of growing various types of sweet orange (*Citrus sinensis*). In sub-tropical and temperate regions lemons (*C. limon*) and mandarins or tangerines (*C. reticulata*) are also grown. Sweet oranges have only recently been grown in the Tropics and there appear to be few varieties well adapted to tropical conditions. In the Tropics, the citrus species best adapted are limes (*C. aurantifolia*), grapefruits (*C. paradisi*), pummelos (*C. grandis*) and mandarins which are grown in both tropical and sub-tropical regions.

Within all these species there are many varieties which are largely propagated by vegetative means. Various hybrids are also grown such as the tangelo, which is a cross between a grapefruit and a mandarin; and the ortanique (from Jamaica) which seems to be a cross between a sweet orange and a mandarin. Some varieties of so-called sweet oranges from Taiwan and China also appear to have genes of mandarin in them. One, for example, the 'Teo Chew' variety, looks like an orange but resembles a mandarin inside. Bud mutants (sports) form readily in most citrus species with the result that within each species there are a large number of varieties.

Botany

Citrus trees vary greatly in size. Size is also determined by the method of planting; whether from buds grafted on to rootstocks (Figs 12.8, p. 101 and 12.10, p. 104) or by cuttings or marcotts. Cuttings and marcotts both produce a smaller tree with a branched main stem, rather than a stout single trunk (Fig. 12.7). Cuttings and marcotts also produce a more superficial root system without a deep tap root. Trees grown by cleft grafting or budding the desired type of citrus on to stock seedlings are larger, they have a single main stem and a deeper root system. The size of the root system also depends on the rootstock used. Of the two most commonly used rootstocks, sour orange (*C. aurantium*) generally produces a deeper root system and rough lemon (*C. limon*) a more spreading root systemn with many surface roots.

With rootstock trees, size varies from small trees up to 5 m high, such as the limes, to medium-sized trees up to about 10 m, such as pummelos and grapefruits.

The leaves of citrus are mostly shiny glabrous, often fairly dark green, oval in shape and with winged petioles in most species and varieties. The branches are often thorny.

In the humid Tropics citrus trees may flower and fruit continuously or irregularly, or may show one or two seasonal peaks and troughs in yield. Many varieties produce special seeds (nucellar seeds) which bear the genetic characteristics of the mother tree only; of the seeds in a fruit only some will be true seeds from both male and female parents. This has the advantage of producing uniform stock seedlings.

Common rootstocks used in the Tropics are:

Rough lemon (*C. limon*) This rootstock is used in areas of the Caribbean and tropical America. It produces a superficial root system and has been recommended for shallow soils, inland soil and sandy soils. It is good for sweet oranges and mandarins and their hybrids, usually giving good yields. However fruit quality (sweetness and solid content) is generally poor on rough lemon rootstock, although this can be overcome to some extent by careful fertiliser use. Rough lemon produces 90–100 per cent of nucellar seeds.

Sour orange (*C. aurantium*) Less widely used in tropical countries. Very susceptible to tristeza virus but

(a)

fig. 12.7 Citrus trees grown from marcotts
(a) A tropical lime variety (b) a mandarin orange; note the characteristic branching and absence of a strong main trunk

(b)

resistant to gummosis and *Phytophthora* foot-rot which commonly occur in the humid Tropics; it can be used on heavier coastal soils. About 80 per cent of seeds are nucellar.

Sweet orange (*C. sinensis*) This is used to some extent as a rootstock in the Caribbean and tropical America. It generally produces lower yields than rough lemon with mandarins and oranges but the fruit is of very good quality. Up to 90 per cent seeds are nucellar. Soils should be well drained as sweet orange rootstocks are susceptible to foot-rot.

Mandarins (*C. reticulata*) These are widely used as a rootstock in Asia. They show some resistance to tristeza virus and are quite resistant to foot-rot and gummosis. They can be recommended for humid tropical regions. A mandarin variety, the calamandarin *C. reticulata* var. *austera* which has also been called *C. madurensis* is a major rootstock in the Philippines. It is very resistant to foot-rot and also has resistance to citrus scab. Mandarin rootstocks are used chiefly for mandarins and oranges. Fruit quality is quite good. Nucellar seeds vary between 10 and 100 per cent. Mandarin stocks are usually slow to develop.

Grapefruits (*C. paradisi*) These are not widely used as rootstocks but are best when used for other grapefruit and pummelo varieties. They are not very compatible with oranges. Nucellar seeds are between 60 to 90 per cent.

Citropsis This is a native citrus in the Congo region and has potential as a rootstock highly resistant to foot-rot and gummosis.

Limes (*C. aurantifolia*) Limes are not commonly used as a rootstock, except in Egypt and Israel. Citrus species on limes generally develop and mature quickly producing early fruit but are also short-lived and many are sensitive to foot-rot.

Whether rootstock trees or trees grown from marcotts or cuttings are used is often a matter of tradition. Throughout Southeast Asia marcottage is a very common method of propagation for limes, mandarin oranges and pummelos. Grapefruits are grown from cleft grafts (Fig. 12.8) since these large-fruited species require strong branch support. When graftage is used, it is necessary to use a rootstock which is compatible with the species being grafted. The above account of common rootstocks is only general since citrus species show a great deal of variation and there are hundreds of different types within each species. Local knowledge is essential

fig. 12.8 A young grapefruit tree grown from a grafting. Note the strong main trunk compared to the marcott trees in Fig. 12.7

when selecting a rootstock. A healthy combination of stock and the grafted variety (referred to as the scion) usually gives a smooth union of the stock and trunk on the tree without stock or scion overgrowths.

Fruit development Fruit development varies greatly in citrus varieties and may have major effects on yield and tree growth. Limes, in the humid Tropics, tend to come into bearing very early and may fruit so heavily that growth of the tree will be retarded. Some mandarin, grapefruit and pummelo varieties also bear very heavily. Ways of overcoming the problem of overbearing are considered on p. 105. In the other direction, some varieties bear few fruit or hardly bear at all. This may be due to a number of causes. The most common in the equatorial Tropics is the absence of a cold or dry spell, as one or the other may be necessary for flowering. Oranges, in the humid equatorial Tropics, as well as lemons and some mandarins, must be grown at high elevations to flower well. The growing of true oranges is best developed in the American Tropics. In the drier Tropics, which are mostly above 10° latitude, some orange growing is successful at sea level.

Fruit development may be poor because of self-incompatibility and seedlessness. Generally seedless varieties or varieties with few seeds fruit poorly. In this case cross-pollination by growing of mixed varieties may be the answer.

Climate and soil

Different citrus species require different types of climate. In general, however, they do not stand severe frost and are therefore not grown in very cold regions. Vegetative growth of most citrus species virtually stops at temperatures below about 10°C.

In the humid Tropics oranges do not produce good quality fruit. Some varieties of orange are grown in the Tropics but generally quality deteriorates rapidly with increasing humidity. This is probably due both to too much leaf growth and to a low incidence of sunshine which causes fruit to develop too fast making them acid and watery. In a continuously humid climate flowering and fruiting of oranges is also poor since most orange varieties need a cool spell or a dry spell to flower. In equatorial regions the oranges generally increase in quality (sweetness and colour) from a wetter to a drier zone or from sea level to higher elevations when sunshine is not reduced. However, there are varietal differences. For example, Valencia oranges can be grown successfully at sea level in Colombia but navel oranges require an elevation of about 1 000 m. Mandarins are similar to oranges in their climatic requirements but generally are slightly better adapted to humid climates. Pummelos, grapefruits and limes do well in humid climates and these species of citrus should be grown in the wetter regions of the Tropics as indicated in the diagram overpage.

For good growth citrus trees require light well-drained deep soil. On such soil trees will grow strongly and also stand periods of moisture stress. Varieties and rootstocks susceptible to foot-rot especially require well-drained soil and absence of flooding. Some varieties can stand heavier soil than others however, and generally limes, pummelos and mandarins can be grown on heavier soil than oranges. In regard to rootstocks, rough lemon is better for shallow soils because it produces an abundance of surface roots.

Most citrus species do not stand very acid soils.

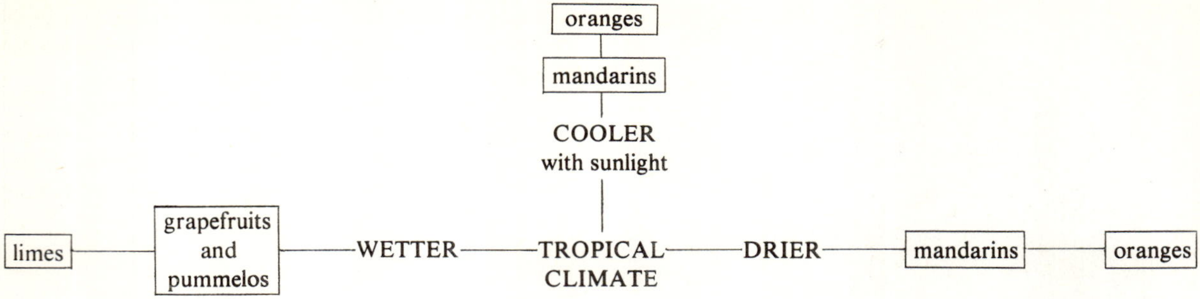

Magnesium limestone will generally benefit citrus when the soil pH is below 5·0. Alkaline soils and saline soils are unsuitable for citrus.

Cultivation and management

Propagation and nursery There are three different ways of propagating citrus; by graftage, marcottage (or cuttings) and from seeds. Generally the first two must be recommended because seed of most good citrus varieties contain out-pollinated genes and produces variable trees. Uniform, even plantings can only be obtained by marcottage or graftage.

Marcottage and cuttings These are recommended for limes and may also be used for mandarins. A marcott is made by tying earth around a suitable branch shoot. Sometimes the bark is also ringed at this point as shown in Fig. 12.9. A strip or square of polythene should be used to tie the soil to the branch. These operations encourage roots to grow from the branch into the ball of soil and after a variable period the branch together with the ball of soil and roots is cut off and planted.

Producing cuttings is faster. The shoots cut off should be firm (not too green and soft) but not too old and hard either. They may be pruned of leaves, treated with rooting hormone and planted in polythene bags of size about 20 × 30 cm laid flat, filled with well-draining soil. The cuttings should be given light overhead shade (about 50 per cent) and watered regularly until they have rooted.

Graftage There are two methods of grafting. First stock seeds of the desired variety are planted in polythene bags. When these reach a thickness of about 10 mm they are ready for graftage. This takes 6–12 months in tropical conditions. For cleft grafting a branch of the desired variety, such as is used for cuttings, is cut and its end trimmed to a V-shape. The stock seedling is cut off at a height of about 10 cm where the stem thickness is about 10 mm and a V-shaped notch is carefully cut in the end into which the above cutting is inserted. The join is taped up with polythene. A type of cleft grafting is also used to re-graft new material on old stocks which have been cut back as shown in Fig. 12.9.

For bud graftage, the method illustrated in Fig. 12.10, p. 104, is recommended. This is called T-patch budding. The bud grows on the stock after a few weeks and the the stock-seedling is cut back to a point just above the bud patch.

In all types of graftage the grafted shoot is called the scion. The scion is allowed to grow to a height of 50–70 cm in the polythene bag; during this time any shoots coming from the stock must be pinched off. The scion shoot should be tied to a stick if it is weak to ensure straight growth, for this will become the trunk of the tree. After this the tip of the shoot is cut back to allow about 6 strong side shoots to develop; this will form the main frame of the tree. Seedlings may be planted in the field as soon as the side shoots are growing strongly. Tree spacing depends on tree size, some indication is given in Table 12.3.

Fertilisation Citrus species generally respond well to fertilisation, especially in areas where moisture stress does not limit yield, and with varieties adapted to grow under such conditions. Nitrogen is the most important fertiliser in both developmental and bearing stages of the tree. Excessive potassium should be avoided. In newly cleared and burnt land potassium fertiliser may be omitted for the first few years. Table 12.4 gives recommended rates for fertilisers.

It is possible that yields may be increased by rates up to twice these values. In soils below pH 5·0, magnesium

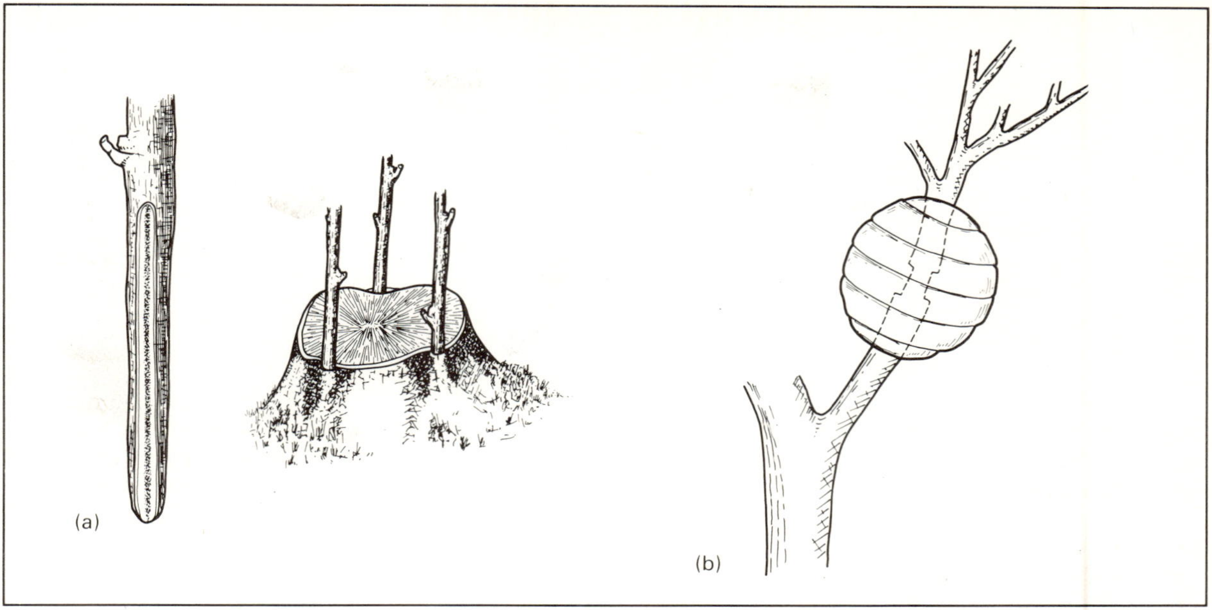

fig. 12.9 (a) Bark grafting of old trees with new scion wood (b) Marcottage of citrus

Table 12.3 Suggested planting distances for different citrus species and varieties

	Type	Distance in metres	Approximate number of trees per hectare
Orange	small varieties	6	280
	large varieties	10	100
Grapefruit	small varieties	9	120
	large varieties	12	70
	grafted	9	120
Mandarins	marcotts	4·5 × 6	280–500
	small varieties	10	100
Pummelos	large varieties	12	70
Limes	grafted	3–4·5	500–1 000

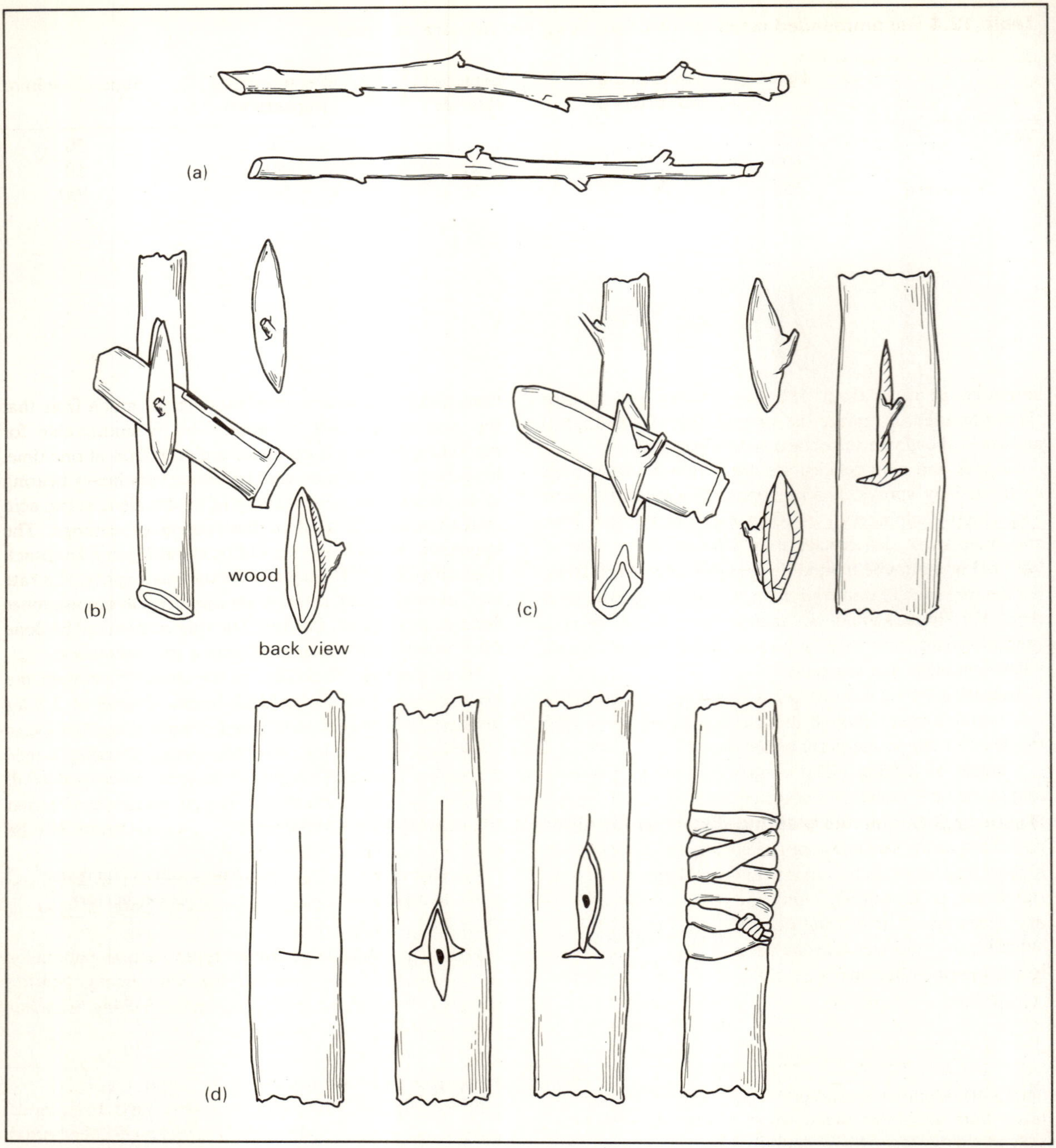
(a)
(b)
wood
back view
(c)
(d)

Table 12.4 Recommended rates for fertiliser in kg per hectare per year

	Urea	Ammonium sulphate(VI)	Potassium chloride	Magnesium(II) sulphate(VI)	Superphosphate
Year 1	100	50		50	50
3	100	50	25	50	50
5 onward	150	100	50	100	100

limestone at rates from 500 to 1 000 kg per hectare should be used to replace the magnesium(II) sulphate(VI) and rock phosphate to replace superphosphate.

Copper and zinc deficiencies are common and should be treated by sprays of 5 kg copper(II) sulphate(VI) and 5 kg zinc(II) sulphate(VI) in 10 kg lime per hectare. Iron and manganese deficiencies may also occur in soils of high pH and may be treated by spraying at rates of 10 kg per hectare each of iron(II) or manganese(II) sulphate(VI) in lime. Fertilisers should be applied over the 6 months preceding fruit maturity or in 6 equally-spaced applications where fruiting is non-seasonal.

Irrigation Irrigation of citrus is not recommended in the humid Tropics though it is normal in the drier sub-tropics and temperate parts of the world.

Control of fruiting The control of fruiting is important in oranges, mandarins and pummelos. Limes in humid tropical regions fruit nearly all the time. Fruiting usually coincides with seasonal variations in temperature, at higher latitudes where annual temperature variation is significant, or in rainfall. Flowering occurs in the cool or dry spells and fruit bearing occurs several months later. Where 2 dry and wet seasons occur, there will be 2 fruit bearing periods but one is generally larger than the other.

fig. 12.10 Methods of bud-grafting in citrus. Buds may be taken from shoots of round wood or younger irregularly-shaped shoots as in (a), (b) and (d). Buds are removed from round wood cutting as shown in (c). Young angular wood buds are cut off and 'side-slipped' under the stock bark as shown in (d)

Sometimes one season may produce so much fruit that the next season will be poor. This is undesirable for marketing because it produces a glut of fruit at one time. Fruit thinning has been used to reduce very heavy bearing in one season by using a spray of naphthalene acetic acid (NAA), which is used in the rooting of cuttings. The strength of the solution should be about 2 g in a knapsack type sprayer. 2:4:D may also be used as a spray, at a rate of 1 ml commercial solution per sprayer. Do not use more because this will kill the tree. The spraying should be done after flowering at the time of young fruit formation.

Weed control Weeding, in the circle or between the rows, is best done by herbicide spray. Diuron at 2·5 kg per hectare will give good weed control if applied every 3–6 months on normal soils (not sands). Standing weeds should first be killed. Paraquat or dalapon can be used to kill grasses. 2:4-D amine at about 1 litre per hectare can be used for creeping weeds between the rows but must not be sprayed on the foliage of the trees.

Pruning Citrus trees should be pruned to keep the branches from touching each other. Once a year is adequate for pruning.

Propping With large-fruited types such as pummelos and grapefruit and other species with heavy bearing branches, it is necessary to prop up the fruiting branches with poles to prevent breakage.

Harvesting and fruit ripening

In the Tropics fruit harvesting is done by hand. Fruit tends to ripen and spoil quickly in the Tropics and in the humid Tropics usually has a poor colour being green or blotched yellow instead of bright yellow or orange. Fruit colour can be improved before marketing with 'Ethrel' or ethyne.

The fruit is placed in a basket and covered with a polythene sheet for treatment with these chemicals. Ethyne is introduced by placing a tin of water with a small piece of carbide in it under the polythene sheet.

Protection

Citrus species suffer from a number of serious virus diseases of which tristeza virus, also known as quick decline, is the most serious and widespread. It causes swelling of the rootstock. Resistant rootstocks should always be used and budwood, scion shoots and cuttings taken from disease-free trees. Because the disease is transmitted by insects, removal and burning of all diseased or suspected trees is necessary. Nurseries should be sprayed with insecticides to protect the seedlings from insects, particularly aphids. Some mandarin rootstocks tend to be resistant to tristeza. Sour orange is most susceptible to the virus. Psorosis is another virus disease which causes leaf yellowing and bark scaling. Xyloporosia (cachexia) and exocortis are virus diseases; the first causes honeycombing of the bark of the trunk which leads to tree leaning, and the second causes scaling and cracking of the bark. All three of these diseases are controlled by using healthy budwood and stock seedings and by eliminating all diseased trees.

Root-rot caused by *Phytophthora* is common in citrus especially on poorly-drained soil. Sour orange and mandarin rootstocks show some resistance. Gummosis or the exuding of gum is also caused by *Phytophthora*.

Citrus scab is caused by the fungus *Elsinoe fawceti* which is controlled by copper fungicides and the use of resistant types such as the calamandarin. This is also resistant to bacterial canker caused by *Xanthomonas* spp.

There are numerous pests of citrus in the Tropics where absence of good quarantine procedures often makes their control very difficult. Most conspicuous are the various insects which attack the fruit causing galls in the skin, boring through the flesh or sucking the fruit. Large-fruited species such as pummelos should be wrapped in paper while the fruit is developing but this is not generally practical with small-fruited species. All spoiled fruit should be buried or otherwise destroyed. Numerous insect species also attack foliage. Sometimes insecticide sprays are effective.

Many species of trunk borers also occur. Some of these attack older wood, sometimes through wounds and cracks. All infected branches should be pruned and burned and wounds painted with lime-sulphur. Sometimes the adult borers (various beetles) feed on the citrus foliage and may be controlled by spraying with a suitable insecticide.

Papaya (*Carica papaya*)

Family Caricaceae
Other names paw-paw

Papayas are among the most widely grown tropical fruits and there are few back-yards in tropical countries that do not have their papaya tree. The original home of the papaya is tropical America where other closely related species such as *C. candamarcensis, C. monoica* and *C. cauliflora* occur. Papayas are mostly grown in small gardens or as isolated trees for local consumption but in Australia, South Africa, Hawaii, and elsewhere, they are also grown in larger farms. The papaya fruit is nutritious and contains significant quantities of vitamins A and C. Papaya is sometimes grown for the production of papain and/or pectin.

Botany

The papaya tree is small, reaching about 5 m and generally produces only one unbranched, soft fibrous stem supporting a crown of large-lobed leaves on long petioles (Fig. 12.11). The roots are also soft and easily damaged. The flowers are formed in the axils of the leaves and, if female, develop to produce a succession of fruits as shown in Fig. 12.11(a). The fruit shape varies from round to long and pendulous, when they dangle down as in the tree illustrated in Fig. 12.11(a). Some trees only produce male flowers which occur on long stalks or inflorescences as shown in Fig. 12.11(b); these produce no fruit at all. Other trees (hermaphrodites) produce both male and female flowers (Fig. 12.11(c)) and these are very variable. Some, like the 'Solo' variety of papaya from Hawaii have been bred and selected to produce uniform small oval fruits about 15 cm long. Others produce variable fruit sometimes very unevenly since the trees may change sex with different seasons. When planting it is necessary to obtain a high proportion of bearing trees.

The fruits of papayas are thin-skinned and the flesh varies in colour from pale yellow to red. There is also much variation in flavour and flesh thickness. Inside the

(a)

(b)

(c)

fig. 12.11 (a) Mature papaya tree bearing fruits (b) Papaya with male inflorescences (c) An hermaphrodite papaya producing both male flowers and fruits on long stalks

hollow centre of the fruit the mucilaginous shiny black seeds occur. They may be few in the near seedless types to more than a thousand in number. In growing papayas for the market the grower should try to obtain those types which are most favoured by the consumer. Often smaller types are more favoured as fresh fruit and large types for canning. Papayas are very locally adapted and the grower should select his own seeds (see p. 108).

Climate and soil

Papayas are locally adapted but widespread, they are grown from the equatorial Tropics to temperate latitudes. Growth in the Tropics is faster and trees will flower within 6 months and produce ripe fruit within 9 months. Papayas are killed by frost. In very wet climates they can suffer from root diseases if the soil is poorly drained. Hill sites up to about 1 500 m (without frost) in the Tropics are good locations for papayas, if not too wet and cloudy.

Soil should be well drained and for this reason a fairly sandy soil is often best. It is thought that acid soils below pH 5·0 should be limed to minimise root diseases. Papayas should not be grown successively on the same soil unless fumigation is carried out.

Cultivation and management

Seed production In some tropical countries papaya seeds may be obtained from government agencies or from commercial firms but generally throughout the Tropics the farmer must obtain his own seeds. There are basically two ways of doing this:

1 Seeds are collected from high-yielding trees bearing fruits of the desired type.

2 Crossing of high-yielding trees may be undertaken. Since the papaya is often out-pollinated, better results are obtained by crossing specific trees. Crossing a heavy-bearing female tree with a superior hermaphrodite gives a high proportion of seeds which will produce female trees—which is desirable in planting up a field.

If small-fruited types are required for the market, hermaphrodite trees may be crossed.

Pollen is collected from hermaphrodite flowers and stored in clean cotton plugged test-tubes in a desiccator. Pollen release takes place on sunny days from about the mid-morning. Female flowers should be enclosed in grease-proof paper bags fixed with pins prior to and for 10 days after the petals are removed and pollen has been dusted on the stigmas. Fig. 12.12 shows a female tree with flowers at various stages. Parent trees which are known to produce good seeds should be clearly marked and kept for seed production.

Seeds are removed from the ripe fruits and may either be planted straight away or dried and stored for periods

fig. 12.12 A female papaya tree with flowers and young fruit

fig. 12.13 A papaya seedling, grown in a polythene bag, at the correct stage for transplanting. The polythene bag is stripped off with a blade

of up to a year. Sometimes the mucilaginous layer around the seeds is removed before planting.

Nursery and planting Seeds should be germinated in flat boxes of sandy soil which is well drained. The seeds should be sprinkled evenly about 2 cm apart and covered with 1 cm of soil. They germinate best if the boxes receive morning sun. Papaya seedlings suffer from damping off if over-watered; the soil should be lightly watered twice daily. Sometimes it is necessary to use steam-sterilised soil or soil fumigated with bromomethane. After about a week the germinated seedlings should be carefully raised and planted in small polythene bags (about 15 × 20 cm laid flat) in which they can be kept for a further 3–4 weeks before transplanting (Fig. 12.13).

Papayas are generally planted between 2–3 m apart. Closer spacing gives higher yields in the first year but trees become more etiolated and yields decline in the second year. Generally papayas are not grown for longer than 3 years. To obtain a high proportion of female trees, plant 3 seedlings about 25 cm apart at each point with out-pollinated seed. The purely female trees can be recognised by the absence of male flowers before 3 months in the Tropics; at this time the best-looking female tree is left and the others cut back.

Fertilisers Organic manure and inorganic bag fertilisers are both used for papayas depending on availability. For high yields, high levels of fertilisers are required. Levels of N:P:K mixture fertiliser (approx. 15:15:15 or similar composition) of up to 1·5 kg per plant per year have been recommended; the rates given in Table 12.5 may prove acceptable.

Table 12.5 N:P:K mixture (rates in g per tree per month)

Month	Rate
0–3 months	20
4–6 months	50
7 months onwards	100

When the fruit is required for canning the nitrogen content of the fertiliser should be reduced because this induces a high nitrate(V) content which affects the tins. For canning, a fertiliser mixture of composition N:P:K in the ratio 1:2:2 should be used and at a rate of not more than 50 g per tree per month.

Weed control Hand weeding should be avoided because of the risk of damaging the roots. Diuron at 2 kg per hectare and paraquat at 1 litre per hectare give good weed control when sprayed on circles around the trees in normal soils, but not in very permeable soils and peats.

Harvesting and processing

Papaya fruit should be harvested at the half ripe stage when the flesh is still hard and of a carroty texture. They ripen rapidly within one to three days and should be transported to market before they become soft. Papayas for canning are usually processed by hand because the fruit is too soft to peel by machine. However, papaya can be successfully canned in syrup at the carroty stage when the fruit is still hard and easy to handle, this gives a canned product not unlike peach.

Papain and pectin These two products are both obtained from papayas. Papain is an enzyme used to tenderise meat. It is present in the milky sap of the plant and is generally obtained by 'tapping' the green fruits. Thin cuts are made vertically down the fruits towards the tip and the milky latex which oozes out is scraped off after it hardens. This is done in the early morning. The hardened latex after drying is the product which is sold.

Pectin is used in jam and jelly making. Papayas contain up to 2 per cent which may be extracted from the by-products of canning.

Protection

Papayas suffer much from nematodes and land should not be planted with more than one crop (1–3 years) before being rotated with other unrelated crops. On badly infected land nematicides such as 'Nemogon' are recommended. Formalin (25 ml of a 4 per cent solution of methanal) poured in the planting hole has also been recommended.

In poorly drained soils, and soils previously planted with papayas *Phytophthora* and various other root rots cause serious losses to papayas.

Various leaf and stem diseases also occur and can sometimes be controlled by spraying with fungicides.

There are many virus diseases of papayas giving symptoms of mosaic, distortion, general slow growth and stunting, leaf yellowing and small (pencil point) crowns. They are spread mainly by insects but these are difficult to control. Any abnormal trees should be up-rooted immediately and burned or buried. Some trees show resistance and these should be used in producing seed as described above. The mountain papaya *C. candamarcensis* and other species also have some resistance to some virus diseases.

Pineapple (*Ananas comosus*)

Family Bromeliaceae
Other names ananas (French, Portuguese); piña (Spanish)

Pineapples originated in the American Tropics and became important in tropical agriculture in the late nineteenth century. Like sugar cane, many of the major pineapple industries are established in the sub-tropical and drier regions of the Tropics as shown in Fig. 12.14. However, pineapples do very well in the humid Tropics and the fact that most industries are established elsewhere is due to largely historical circumstances. Very successful pineapple industries within the humid Tropics exist in the southern Philippines, Malaysia and elsewhere. The Malaysian pineapple industry is unique in that organic peat soils are almost exclusively used; these are quite unlike the soils of most other major pineapple-producing regions.

Botany

The pineapple plant consists of a single main stem bearing a rosette of long thin leaves as shown in Figs 12.15 and 12.16. The leaves may be smooth-edged (as in 'Cayenne' and 'Singapore Spanish') or the margin may be serrated or spiny. The leaves curve upwards in section and form natural channels for collecting rain water and dew in the leaf axils. Roots grow into the leaf axils and this forms an important way of obtaining water under dry conditions.

Under humid tropical conditions most pineapple varieties will flower naturally after 12–15 months. At flowering a change takes place in the growing point of the plant and instead of continuing to produce vegetative leaves, the stem produces the fruit. The fruit of pineapples

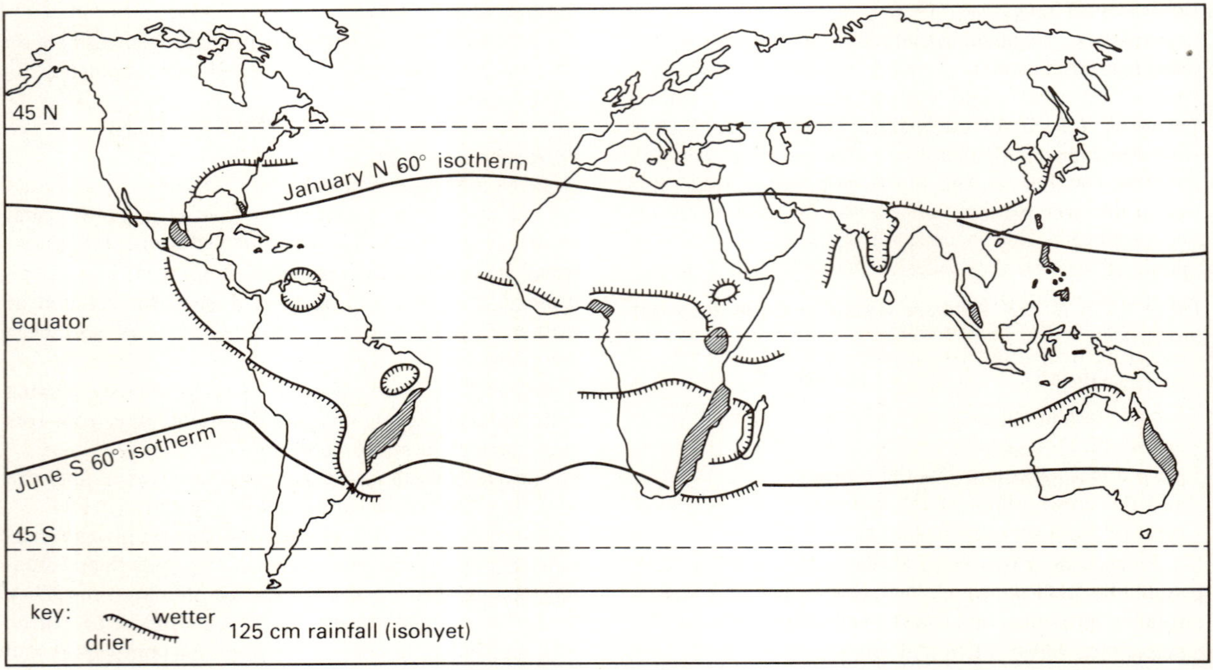

fig. 12.14 Important areas for pineapple growing

fig. 12.15 A pineapple plant with the inflorescence just beginning to form into a fruit

is actually a swollen inflorescence made up of many flowers. Figs 12.15 and 12.16 also show the flowering details. The normal fruit of the pineapple is seedless but when different varieties cross-pollinate, seeded fruit may be produced. This is undesirable from the commercial point of view so that different varieties should never be planted close together.

The pineapple plant continues vegetative growth after flowering by the growth of side shoots or suckers (Fig. 12.16). When the fruit is ripe the old shoot dies off and one or more of the suckers will form new plants as shown in Fig. 12.17. In addition to these side shoots, the pineapple produces shoots below the fruit which are called slips and the crown shoot at the apex of the fruit. All three types of shoot are used in planting.

Pineapples have very shallow and restricted root systems which relates to their origin as epiphytic plants (plants growing on trees). Generally the roots of pineapples will not penetrate deeper than about 30 cm. Pineapples also produce roots in the leaf axils to take up water which collects there.

Varieties and yields

'Cayenne' ('Smooth Cayenne') A widely planted variety for the canning industry. It has large yellow-fleshed fruits with good flavour. The leaves are spineless.

'Red Spanish' A hardy variety with fruits weighing 1·0–1·5 kg. Flesh pale and rather acid, particularly outside the equatorial zones. Grown chiefly in Cuba and Puerto Rico.

'Sugar loaf' Generally recognised as the sweetest of

fig. 12.16 Section through fruiting pineapple plant showing stem and foliage leaves, peduncle bearing fruit, suckers arising from the stem, slip below the fruit and the crown shoot

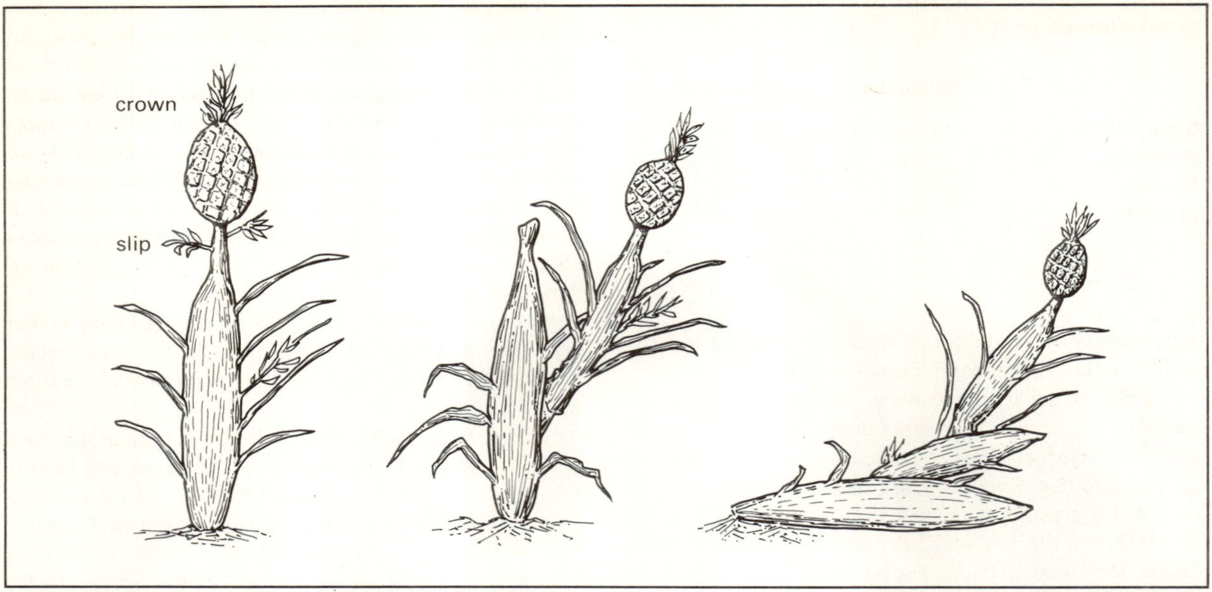

fig. 12.17 To show how new shoots are formed in pineapples.

the white-fleshed pineapples, with fruits weighing about 1 kg. Grown widely in tropical America.

'Queen' ('Natal Queen', 'Palembang', 'MacGregor') Sweet fruit with good flavour of size 0·5–1·0 kg. It produces numerous suckers and there are many strains. Best adapted to sub-tropical conditions.

'Abakka' ('Abachi', 'Abacha', 'golden Abachi') Sweet, well-flavoured yellow-fleshed fruits of size 1·5–2·5 kg. Grown in Indonesia, Florida and Venezuela. Produces characteristic basal slips on the fruits.

'Cabezona' ('Cabezone') Large fruits up to 7 kg. It is a very vigorous clone.

'Congo Red' Yellow-fleshed, medium-sized fruit of good flavour.

'Elenthera' ('Pernambuco', 'Abacaxi', 'English') Similar to Abakka but without basal slips on fruit. Double or multiple crowns are common.

'Singapore Spanish' A good quality fruit of weight about 1·0–2·5 kg. Yellow-fleshed and grown mainly for canning in Malaysia.

Fruit size, as indicated above, is used to characterise varieties. However, fruit size is very much dependent on growing conditions, particularly spacing, and the above size ranges refer to average conditions and commercial spacings.

In addition to the major varieties of pineapples there are many races within each variety. Undesirable mutants of pineapples occur from time to time and one which is of common occurrence has a ring of slips beneath the fruit. Other undesirable mutants have multiple crowns (Fig. 12.18). In planting, shoots should not be taken from such undesirable plants. With good husbandry yields in excess of 50 tonnes of fruit per hectare have been obtained in the humid Tropics over an 18-month growing period.

Climate and soil

Pineapples are often grown under sub-optimal climatic conditions. The average ambient temperature of some important regions for pineapples are shown in Table 12.6. The optimal temperature for pineapples is around 25–30°C or close to the temperatures of the humid tropical lowlands. The time taken to mature however, is not greatly affected by temperature as in some other crops (e.g. sugar cane) because the pineapple is a naturally slow-growing species. With the use of flower-stimulants

Table 12.6 Maximum and minimum temperatures in °C

	Maximum	Minimum
South Africa	41 to 47	–1 to 5
Australia	32	11
Hawaii	32	10
Malaya	30	25

pineapples may produce ripe fruits after 15–20 months under humid tropical conditions against about 22 months under conditions with a definite cold season. There are varietal differences in temperature requirements and tolerance. 'Queen' pineapple is suited to regions of low temperature (e.g. South Africa); the 'Cayenne' variety is suitable for growing in the sub-tropics or at elevations up to 1 000 m in the Tropics. 'Spanish', 'Singapore Spanish', 'Sugar loaf' and 'Abachi' are all suited to the lowland Tropics; if they are grown in cooler locations the fruit will be acidic and have a low sugar content.

fig. 12.18 Multiple and double crowns in pineapples

Pineapples are generally considered to require less water than other tropical crops and can be grown in regions with a rainfall as low as 600 mm. However, in such low-rainfall regions, dew which forms on the leaves and percolates down to the roots in the leaf axils is essential for growth and yield. Pineapples have an ability to store water for later use but their root system is small so that even with this storage ability, growth is better with regular rainfall. In Malaya improved growth has been observed with irrigation even in a rainfall region of 2 500 mm per year, evenly distributed.

Pineapple fruits are scorched by strong sunlight and are sometimes covered with a layer of straw or similar material before harvesting, when required for sale as fresh fruit.

Soils for pineapples should be on the acid side and generally liming should be avoided unless the soil is very acid. Calcareous soils are unsuitable for pineapples. Both inland latosols and alluvial soils may be used but the latter need adequate drainage to a depth of about 50 cm. Peat soils are used in Malaysia for pineapples. Special attention to micronutrient deficiencies is needed on peat, particularly for molybdenum, copper and zinc.

Cultivation and management

Planting material All types of shoots are used in planting (Fig. 12.19). However, different types of shoots grow at different rates and large shoots grow faster and mature sooner than small ones, thus planting material should be graded and only one type planted in any field. Generally the rate of growth depends on the size and conditions of the shoots, and the time to maturity will range between about 17 and 24 months, depending on the planting material. In Malaysia, with 'Singapore Spanish', crowns (tops) are found to be faster-growing followed by slips and suckers, but in Hawaii with 'Cayenne', generally suckers (side shoots) are faster than slips, and crowns are slowest. The old stem or 'butt' may also be used for planting but it rots easily and is not recommended for the humid Tropics.

Whatever material is used, it should be collected up to 4 weeks before planting and may be stored in a dry place base end upwards. Sometimes shoots are stored by placing them on the leaves of the old crop. Before storage the shoots should be dipped in Bordeaux mixture. Side shoots may also be trimmed at the base and this is thought to increase rooting.

fig. 12.19 Types of pineapple shoots for planting: (a) large side sucker (b) slips from beneath the fruit (c) multiple crown (this will produce fruit with multiple crowns)

Field preparation Generally a high degree of field preparation is used in pineapples to allow easy access to fields and planting in straight rows. On peat soils access roads are prepared by laying split logs on the soil surface to prevent subsidence. Field drains are required to a depth of about 50 cm on flat land in humid tropical locations.

Planting The double-row planting method as illustrated in Fig. 12.20 is recommended for pineapples. The shoots are spaced 30 cm apart on each pair of rows 60 cm apart with 1 m between the pairs. This gives an initial density of about 40 000 per hectare. Subsequent density for ratoon crops will increase and should be maintained at around 70 000 per hectare by pruning for optimal yield. Increasing the initial density of the first planting by using a spacing of 30 × 45 × 75 cm may increase yield of the first crops somewhat but fruit size will decrease. The shoots are planted with the base about 5 cm below the soil surface and leaves pointing vertically.

Fertilisation Pineapples do not require high levels of phosphorus and excessive levels may reduce yield. If a bag mixture is used it should be low in phosphorus, with a ratio of about 12:4:12 (N:P:K). Fertilisers should be

fig. 12.20 The double-row system of planting pineapples

Table 12.7 Fertiliser recommendations for pineapples in kg per hectare

	Ammonium sulphate(VI)	Urea	Potassium chloride	Rock phosphate
Plant crop (*c.* 20 months)	100	200	200	50
Ratoon crops	100	200	250	50

split and applied at 6 weeks, 3 months and 9 months but nitrogen should not be applied after this as it will reduce flowering and yield. After chemicals to promote flowering have been applied, no further nitrogen fertilisation should be undertaken until the bulk of the fruits have matured. Potassium fertiliser, however, may be given at the young fruit stage and may increase sugar and fruit quality. Table 12.7 gives recommendations for straight fertilisers.

In the acid soils normally used for pineapples, ground rock phosphate may be used. If the pH of the soil is 5·5 or above superphosphate is recommended.

Pineapples suffer from copper, zinc and molybdenum deficiencies on some soils. Copper(II) and zinc(II) sulphates(VI) may be applied as sprays in a lime solution at a rate of 2 kg in 30 kg lime per hectare. Molybdenum is usually supplied with the phosphate fertiliser.

Flowering control Under the conditions of uniform rainfall and temperature in the humid Tropics all-year-around harvesting can be achieved. Quite often there are peaks and troughs in maturing of fruit even under uniform conditions, often following small climatic variations. For example, in peninsular Malaysia peaks occur in May–July and November–December when there are slight dry spells. Flowering and fruit set can be controlled to give even output by the use of chemicals which promote flowering, if different fields are treated at prescribed intervals. The chemicals however, should not be applied more than 6 months before natural fruit set which means, under average humid-tropical conditions and with average planting material, treatment should be given at about 10–12 months. If given too soon, growth could be set back. Three chemicals are commonly used:

1 Naphthalene acetic acid (NAA). Use a solution of 20 parts per million and apply 50 ml of the solution in the crown of each plant.

2 Ethyne (from carbide). Add 25 g calcium carbide to 4 litres of water and pour 50 ml of the solution into the crown of each plant (the solution must be fresh). This is about the cheapest method and is fairly effective.

3 Ethene (from 'Ethrel'), used according to instructions.

De-crowning and de-slipping In some varieties of pineapples removing the crown or top increases fruit size (e.g. 'Singapore Spanish'). Some varieties produce more than two slips below the fruit, removal of these may also increase fruit size. The removal of crowns and slips is done on the immature fruit when the shoots are about 5 cm long.

Weed control Chemical weed control is recommended in pineapples because of the superficial root system. Simazine may be used at a rate of 2 kg per hectare on normal soils, but not on peats and sands; it is applied immediately after planting. Thorough field preparation and pre-planting cultivation reduces early weed growth. Herbicide may be used for spraying between the rows. Sodium chlorate(V) at 10 kg per hectare and paraquat at 1 litre per hectare are effective contact herbicides. Diuron and simazine at 2 kg per hectare are also suitable as soil-surface, long-acting herbicides in normal soils. Hormonal herbicides such as 2:4–D or MCPA must not be used in pineapples because they upset flowering.

Harvesting and processing

The fruits are usually picked by hand and carried in baskets or other containers to a roadside for collection. Pineapples are mainly processed by canning and most pineapple-producing regions have canneries which produce canned pineapple in various forms.

Protection

Nematodes are the most serious pests of pineapples. In commercial growing the infection is controlled with soil fumigation, using for example D-D nematicide. The small farmer is best advised to rotate with another crop such as sweet potatoes, sugar cane or grass after the third ratoon crop. Mealy bugs carry a pineapple virus disease. They are controlled by spraying the border rows of pineapple (those nearest the outside of the field) with an insecticide such as malathion. *Phytophthora cinnamomi,* which also attacks pepper, causes root rot in wet weather and is controlled by improving drainage. Fruit rot occurs if the fruit is left too long before picking. Fruit should also be protected from strong sunlight by covering with straw.

Minor crops

Akee (*Blighia sapida*)

Family Sapindaceae

The flesh around the akee seeds is sometimes eaten raw but more usually after cooking. The cooked flesh looks like scrambled eggs. The pink material connecting the flesh to the seeds should first be removed because it is poisonous. The flesh of unripe fruits is also poisonous.

Botany and propagation

It is a tree growing up to 20 m tall (Fig. 12.21(a)) with leaves consisting of 3–5 large leaflets. The flowers are small and greenish white. The fruit is red and yellow when ripe; it opens along rounded valves and inside are three black seeds covered with a yellowish flesh (Fig. 12.21(b)).

Akee is usually grown from seed.

Climate and soil

Akee requires a humid climate and at least 2 000 mm of rain per annum. It grows on most tropical soils; a well-drained loam, rich in organic matter is preferred. The tree is a native of West Africa; today it is important in the West Indies, particularly Jamaica. It is not very easy to cultivate in the ever-humid Tropics because of its poor adaptability.

Akee is cultivated and protected in a manner similar to rambutan, see p. 134.

(a)

(b)

fig. 12.21 (a) Mature akee tree (b) Fruit of akee showing the seeds

Avocado pear (*Persea americana*)

Family Lauraceae

Other names avocat, avocatier or poire d'avocat (French); abacate (Portuguese), aguacate, palta (Spanish, in Mexico)

The smooth, buttery and well-flavoured flesh of the fruit is eaten as a salad vegetable. The oil is used for special purposes, notably in cosmetics.

Botany and propagation

A tree reaching up to 20 m in height, native to Mexico and Central America; its shape varies from erect with stout trunks to bushy and spreading. The trunks are covered with a fleshy bark. The leaves are round or pointed, held alternately on the branches, and reddish in colour when young. Many small greenish-yellow flowers in compact bunches are held at the ends of the young branches. Fruits of tropical varieties are round, egg- or pear-shaped, 0·85–1·40 kg in weight, with thick skins; they may be pale yellow, green, reddish or purple in colour. The fruit flesh is yellow or yellowish-green and soft like butter when ripe; it encloses a large, round seed, whitish or pinkish in colour (Fig. 12.22).

fig. 12.22 Leaves, fruit and flowers of avocado

Until recently it has been grown usually from seeds. Now it can be vegetatively propagated, e.g. by veneer grafting or budding, using 2–4 month-old stocks. Leaf cuttings may be used successfully but are not economical.

Climate and soil

West Indian varieties grow better in the humid Tropics than Mexican or Guatemalan ones. Avocado needs a minimum rainfall of about 750 mm yearly and prefers alternating wet and dry seasons to even rainfall. Sandy loams that are fertile, slightly acid to neutral in reaction and well drained, give the best growth; on heavy and poorly-drained soils, incidence of foot-rot disease is high. Waterlogging should therefore be avoided.

Cultivation and management

Varieties are grouped into three major types: Mexican, West Indian and Guatemalan; West Indian varieties, e.g. 'Pollock', are best suited to the humid Tropics.

Budded or grafted plants are transplanted when 9–30 months old. As most humid-tropical varieties have a spreading habit, wide plant spacings, e.g. 10 × 12 m, are used. Pruning is often done to remove dead branches, to produce a spreading habit or to thin excess branches; tall trees are often lopped to make harvesting easier.

Fertiliser requirements differ according to the variety, age of trees, soil type, climate, cropping history, etc.; they are somewhat similar to those of citrus. A sample recommendation is 0·5, 1 and 1·5 kg of a 10:6:10 (N:P:K) fertiliser for 1-, 2-, and 3-year-old trees, respectively; trees in bearing may be given 2 kg per year of a 12:6:10 fertiliser at quarterly intervals. Organic manures are beneficial. Deficiencies of zinc and copper are fairly common; they can be remedied by foliar sprays.

Budded or grafted plants start flowering after 1–2 years in the field but flowers are usually removed until the 3rd or 4th year to allow the trees to grow. Seedlings start bearing much later, about 5–10 years from planting. To achieve best fruit set, irrigation may be necessary just before flowering and during fruiting in drier periods. The tree tends to bear fruit once in 2 years. Yields of up to 6–12 tonnes per hectare per year (100–500 fruits per tree) may be obtained.

Protection

Circle weeding or clean cultivation with mulching are often used, sometimes with cover and green manure crops. Excess weeds can be controlled by spraying a desiccant herbicide such as paraquat.

Root rot caused by the soil fungus *Phytophthora cinnamomi* is controlled by avoiding waterlogging and by planting healthy trees. *Verticillium* wilt due to a soil fungus is similarly controlled; in addition dead branches should be removed. Rots of the surface and stem end of fruits (caused by fungi *Colletotrichum gloeosporioides* and *Dothiorella* sp.) are avoided by careful orchard hygiene, pruning of dead wood, careful fruit handling and by spraying protective fungicides. *Cercospora* leaf spots can be controlled by copper fungicide sprays.

The main pests of avocado are fruit-fly, fruit-spotting bug, scale insects and red-shouldered leaf beetle. They may be controlled by regular sprays of insecticides, e.g. 0·1 per cent DDT or trichlorphon.

Bilimbi (*Averrhoa bilimbi*)

Family Oxalidaceae
Other names cucumber tree; tree-sorrel

The fruits are like small cucumbers; they are very acid and are used for pickles, preserves and curries.

Botany and propagation

The bilimbi tree grows up to 15 m tall with 11–21 leaflets and dark red flowers arising from the trunk (Fig. 12.23).

Methods of propagation, cultivation, climate and soil requirements, resemble those for carambola, see p. 121.

fig. 12.23 A young bilimbi tree, fruits grow from the trunk and mature branches

Brazil nut (*Bertholletia excelsa*)

Family Lecythidaceae
Other names pavory nut, pava nut (English); neuz de para (Spanish)

The seeds or nuts are consumed as food. They contain about 65 per cent oil which is often extracted.

Botany and propagation

A large tree reaching about 35–40 m in height with large leaves 25–50 cm long, it has large, light-yellow flowers held in bunches (Fig. 12.24). Ripe fruits are brown and woody, about 10–15 cm in diameter; within the thick shell are 12–24 angular brown seeds or nuts (Fig. 12.25). The fruits take more than a year to ripen.

The brazil nut is usually grown from seed. The seeds

fig. 12.24 A mature brazil nut tree

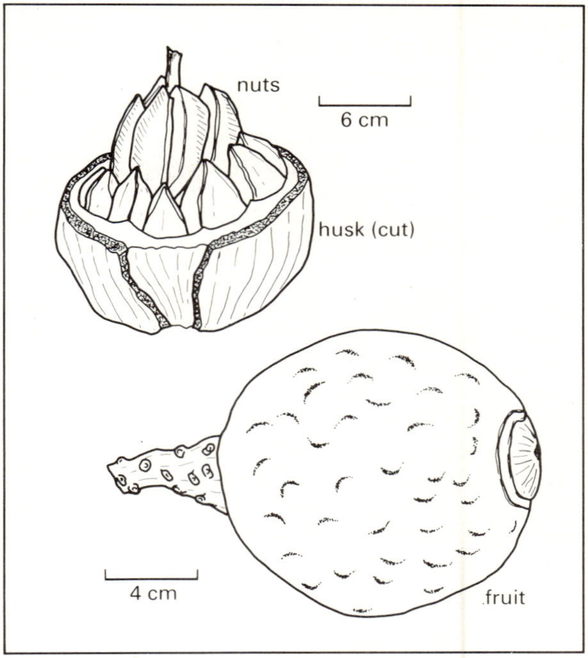

fig. 12.25 *Bertholletia excelsa*

are germinated in a seed-bed or in containers. Seeds take about 14 months to germinate.

Climate and soil

The tree grows wild in the Amazon basin of South America. It grows best in a well-drained rich loamy soil.

Cultivation and management

Seedlings are transplanted when about 30 cm tall at a plant spacing of about 12–15 m square. In plantations, regular compound fertilisers are needed, depending on the soil type. Nuts are first harvested after about 10 years. A good tree can yield about 300 fruits at a time.

Protection

Regular weeding is necessary during establishment. The tree is relatively free from pests and diseases.

Breadfruit (*Artocarpus altilis*)

Family Moraceae
Other names uto (Fijian); arbe à pain (French); fruta pào (Portuguese, in Brazil); fruta de pan (Spanish)

An important starch staple in Polynesia and the West Indies, it is eaten boiled, fried or made into various dishes.

Botany and propagation

The breadfruit tree grows to about 14 m tall with striking large-lobed leaves (see Figs 12.26 and 12.27). The fruit may be 20–30 cm in diameter, more or less round in shape with a thick, tough yet waxy skin varying in colour from yellowish-green to brown. Inside a moist pale yellow or whitish pulp surrounds a central core. Fruits with seeds are called breadnuts.

It is propagated by marcottage, buddings or cuttings.

Climate and soil

Adapted to hot moist climates, the breadfruit tree grows up to elevations of 1 000 m, requiring an annual rainfall

fig. 12.26 A young breadfruit tree

fig. 12.27 Two ripe breadfruit

of 1 500 to 2 500 mm and temperatures of 25–30°C. It may be grown on a variety of deep and well-drained soils.

Cultivation and management

Seedless and seeded varieties exist in Polynesia. Planting is at a spacing of 10–13 m apart. Fruiting starts from 3–6 years. The fruits should be plucked while firm, before they are fully ripe. Yields of about 700 fruits per tree per year, each 1–5 kg in weight, may be obtained.

Protection

Rarely attacked by major pests and diseases.

Bullock's heart (*Annona reticulata*)

Family Annonaceae

Other names custard apple, ramphal, sweetsop (English); anone, coeur de boeuf (French); araticumape (Portuguese); coracão de boi (Portuguese, in Brazil); anona, anona colorado, corazon (Spanish)

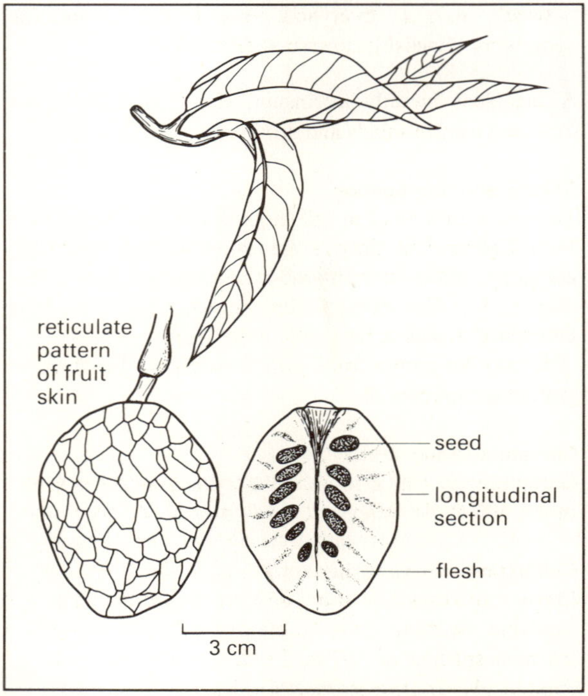

fig. 12.28 *Annona reticulata*

This tree is grown for its fruit which has a white pulpy flesh. Its quality is considered somewhat inferior to that of soursop and sugar apple.

Botany and propagation

Bullock's heart grows as a shrubby tree up to 6–8 m tall. It has spear-shaped leaves and small greenish yellow flowers. Fruits are round, 10–12 cm across, yellowish-brown in colour and with faint angular markings. There are numerous dark brown seeds in the flesh of the fruit (Fig. 12.28). Trees bear about 20–40 fruits per year, each about 300 g in weight.

Cultivation and protection are similar to those for soursop, see p. 139.

Carambola (*Averrhoa carambola*)

Family Oxalidaceae
Other names averrhoa, star-fruit, coromandel gooseberry (English); carambole (French)

A succulent, sweetly astringent, star-shaped fruit eaten fresh and used in salads and drinks.

fig. 12.29 A young carambola tree in bearing

Botany and propagation

The tree grows 5–12 m tall (Fig. 12.29). The leaves have 7–11 leaflets. The flowers are rose-coloured. The fruits are juicy, edible, star-shaped in section and light yellow (Fig. 12.30). The seeds are dark brown, about 1 cm long, embedded in pulp in the centre of the fruit.

It may be grown from seed in containers, or budded, grafted or marcotted.

fig. 12.30 Carambola fruits

Climate and soil

Carambola is native to humid tropical Asia. It is grown below 500 m elevation and on deep, well-drained soils.

Cultivation and management

Many cultivars of sour and sweet carambola are available. Seedlings may be planted out when 30–50 cm tall at a spacing of 5–7 m. Fertilisers are needed to get good yields. Fruiting starts in 4–5 years and the tree fruits throughout the year.

Protection

Regular circle weeding is necessary. Fruit-flies are the main pest. They are controlled by bagging fruits, removing infested fruits and raking the soil to kill pupae.

Cashew (*Anacardium occidentale*)

Family Anacardiaceae

Other names cachou (French); caja (Portuguese); merci (Spanish, in Venezuela)

The kidney-shaped nut is used as a dessert nut and for confectionery. The nut is embedded in a juicy 'apple' which is fermented to produce a drink.

Botany and propagation

The tree has small spreading branches and grows 7–14 m tall (Fig. 12.31). The flowers are produced on the branch ends; they are red in colour and form mature nuts in 2–3 months (Fig. 12.32). The greyish-brown kidney-shaped nut with a hard shell is embedded in the end of a bright yellow or red juicy cashew apple.

It is usually grown from seed sown in containers. Seeds are sun-dried for 2–3 days to improve germination. Direct field planting is sometimes practised. Propagation may also be done by air-layering, grafting or in-arching.

Climate and soil

Cashew is hardy and drought-resistant; it grows well in most tropical soils and climates. It grows optimally at elevations of 1 300 m, with a rainfall of 1 750 mm per year and on sandy soils.

Cultivation and management

Many varieties produce small, medium and large nuts. Yields are about 80–200 kg per hectare yearly. Trees are spaced at 7–10 m apart; planting holes are given dung, ash and rock phosphate (120 g per hole). Complete fertilisers should be applied 2 to 3 times a year. Intercropping is practised in the first 2 years. Pruning is necessary for the first 3 years to shape the tree. The tree starts bearing in the third year; maximum yields are obtained at about 10 years and lifespan exceeds 20 years.

fig. 12.31 A large mature cashew tree

fig. 12.32 Apple and nut of cashew

Mature nuts are picked, dried for 2–3 days, split, dried and roasted. Cashew apples can be eaten as a fruit or vegetable.

Protection

Cashew is attacked by leaf-eating caterpillars, e.g. *Attacus atlas, Euthalia aconthea* and *Orthaga incarusalis*; these are controlled by the use of insecticides. The stem-boring *Rhytidodera simulans* is controlled by pruning and treatment with dichlorobenzene.

Champedak (*Artocarpus champeden*)

Family Moraceae
Other names lemasa (Tagalog, in the Philippines)

The champedak is very similar to jackfruit, see p. 127. The tree is smaller than jackfruit with hairy leaves and branches (Figs. 12.33 and 12.34); fruits have a stronger smell. Fruiting begins after 4–5 years and each tree produces 50–100 fruits per year, each weighing 15–20 kg.

fig. 12.33 Young flowering shoot of champedak

fig. 12.34 Mature champedak fruits

The tree is native to Southeast Asia and is not grown elsewhere in the Tropics. Many varieties are grown in Malaysia and Indonesia.

Cherimoya (*Annona cherimolia*)

Family Annonaceae
Other names anone, cherimolier (French); cherimolia (Portuguese)

Altogether similar to sugar apple, see p. 139. The fruits are considered to be the sweetest and best-flavoured of all fruits of the genus *Annona*.

Durian (*Durio zibethinus*)

Family Bombacaceae

Other names durione (French)

The fruit is famous in Southeast Asia for its flavour and distinctive smell. The flesh around the seeds is eaten fresh or used as flavouring.

Botany and propagation

Durian is a tall tree reaching up to 40 m in height; budded trees are shorter, up to 12 m (Fig. 12.35). It bears large, olive-green and thorny-looking fruits containing many seeds covered with soft flesh (Fig. 12.36). Some clones, e.g. from Thailand, are seedless. The leaves are small and bronze-green.

Propagation is traditionally from seed, but selected clones may be propagated by patch budding in which 4–5 month old seedlings grown in bags are budded with scions from known cultivars. Buddings may be successful in 20–40 days; they should be hardened for a month before field planting.

Climate and soil

Durian is thought to originate in Malaysia; it is now grown all over Southeast Asia. It needs a warm humid climate, rich soil and elevations below 800 m.

Cultivation and management

Many clones have been developed, e.g. D 2, D 7, D 8, D 10, D 16 and D 24 from Malaysia. Many clones from Thailand have thick flesh with good flavour. Clones differ in fruit quality and yield. Plants are spaced at 13 m square. Holes 60 cm square and 60 cm deep are dug and soil is mixed with 20 kg of well-rotted manure and 0·25 kg of a complete fertiliser. A complete fertiliser is applied in 4 doses per year at the rate of 0·75 kg per tree in the first year, rising to 7 kg per tree in the 8th year.

fig. 12.35 A young durian tree

fig. 12.36 Durian fruits on a mature tree

Trees are pruned to obtain 1·5–2 m of clear trunk. Budded durians start bearing 2 crops per year in 5–7 years and can continue for over 50 years, but yields decline after about the 15th year. Usually fallen fruits are collected to obtain the flesh. In Thailand, fruits are picked to increase their shelf life.

Protection

Circle weeding must be done regularly by hand or using herbicides, e.g. 2 kg per hectare of diuron. The fungus *Phytophthora palmivora* causes the only serious disease known as patch canker. The durian fruit is also attacked by a fruit borer.

Granadilla (*Passiflora quadrangularis*)

Family Passifloraceae
Other names barbadine (French); maracujá melao (Portuguese, in Brazil)

The flesh as well as the pulp of the fruit is edible. The green, unripe fruit may be cooked and eaten as a vegetable.

Granadilla is very similar to the passion fruit (see p. 133) except that fruits are larger (20–30 cm long) and fleshy (Fig. 12.37). It only grows in hot and humid lowland climates. The fruit is often attacked by rats, bats and birds.

fig. 12.37 Granadilla fruit and leaves

Guava (*Psidium guajava*)

Family Myrtaceae
Other names goyave (French); goiaba (Portuguese); guayaba (Spanish)

The fruit is eaten fresh and canned. It is used for making jellies, jams and juice. The most important commercial products are the jellies. It contains 2–5 times as much vitamin C as fresh oranges.

Botany and propagation

The guava tree is native to tropical America where it grows wild but today it is found all over the Tropics and also the sub-tropics. It is an important minor fruit crop, often planted as a backyard tree. In Queensland and the Pacific Islands it is sometimes considered to be a weed.

fig. 12.38 Guava budlings ready for planting in the field

The tree is variable in size, 3–10 m in height. Its thin trunk which has scaly bark, branches fairly low. Branches are often pruned in orchards to get a regularly shaped tree. The small branches are square in section and carry two opposite rows of oval leaves. The flowers appear singly from the leaf bases at the ends of branches and are white in colour. After pollination, the flowers develop into round or pear-shaped fruits in which a variable number of small seeds are surrounded by white or pink flesh.

In the past, guava has commonly been propagated by seed. This has resulted in wide variations from tree to tree in the shape, size, flavour and flesh colour of fruits. The wild guavas are usually small, with thick skins and thin flesh; they are full of seeds and stone cells. The improved types have large fruits, thin skins, thick meaty pulp with few seeds and fewer stone cells. The best guavas have a mild and pleasant taste; others produce highly sour and acrid fruits. At present a number of cultivars are available in different countries with the wider use of vegetative means of propagating the plant. In Florida, for example, a white-fleshed variety produces high yields; three better known red-fleshed varieties are 'Red Indian', 'Rolfs' and 'Ruby'. In Malaysia a few cultivars have been developed or introduced, e.g. 'Taiwan Pear', 'Maha 3', 'Gu 4', 'Gu 5', 'Hong Kong Pink' and a seedless cultivar which tends to have irregular shape. In India a good number of varieties have also been produced, e.g. 'Safeda' and 'Chittidar' which are round, white-fleshed and sweet, 'Hafsi', which is red-fleshed and 'Beybugee' which is seedless.

fig. 12.39 A branch of mature guava tree in bearing

Climate and soil

The tree is hardy and able to grow with little attention. It grows in the Tropics up to 1 500 m and is quite adaptable to a wide variety of climatic and soil conditions. It is somewhat drought-resistant and tolerant of high temperature; it can also survive flooding and can grow even on waterlogged soils. It cannot, however, stand low temperatures and frost. For best results, the guava should be given adequate water. Commercially, the crop is grown below 1 000 m and in places with 1 000–3 000 mm of rainfall per year. Too much rain during fruiting can, however, cause fruits to crack and spoil. The crop can grow on poor soils with high acidity.

Cultivation and management

When seeds are used for planting, they should first be treated with a fungicide to reduce damping off disease and sown in nurseries in a sandy loam soil to a depth of about 1 cm. The seedlings are later transplanted into containers. Better results are obtained by direct sowing of seeds in containers, as this reduces transplanting shock. The seedlings are given sufficient moisture all the time and in 5–7 months they are ready for planting. In some places, seedlings are not planted until about 1 year old and 30–45 cm tall.

Vegetative propagation of clones is to be highly recommended. A successful method is bud grafting by the patch budding method. Other methods of bud grafting, marcotting and cuttings of stem or root may be used with success for producing clonal materials. Root suckers may also be used for planting; more suckers develop when the roots are cut about 60–90 cm around the trunk. In the bud grafting method, seedlings and budwood should be 10–20 mm in diameter; budwood is prepared by removing leaves and branches 10–15 days before bud grafting to allow the buds to enlarge. Buds from selected trees are grafted on seedling rootstocks (Fig. 12.38).

The usual plant spacing is about 7 m square. The crop is often irrigated and organic manures or inorganic fertilisers supplied. The amount and type of fertiliser used varies with soil conditions. Generally 150 g per tree of a 8:8:8, N:P:K fertiliser is given 3–4 times in the first year. This amount should be increased to 220 g and 450 g per tree in the second and third years, respectively. Mature trees should be given 150–200 kg of nitrogen per hectare split into 3–5 doses over the year. The trees are sometimes pruned to maintain the required shape but this may not be required. Suckers from the bases of trees and low hanging branches should be removed. Under good management, guava trees fruit about 2 years from planting; the yields tend to increase until the 8th year and fruiting may continue for 30 years or more. The trees flower throughout the year in most places and are open-pollinated; flowers become fruits about 5 months after appearing (Fig. 12.39). Fruits may be picked before they are ripe for sale in distant markets. Mature trees each produce 400–500 fruits per year, weighing a total of 60–80 kg. Grafted trees of improved types yield 1 000–2 000 fruits per hectare per year, weighing a total of 150–300 kg.

Protection

The fungus *Clitocybe tabescens* causes roots and crown to rot and often kills the tree. Control is by removal of infected trees. Another fungus *Glomerella cingulata* causes blackening and rotting of the young fruits whereas the mature fruits rot due to the fungus *Colletotrichum gloeosporioides*. The alga *Cephaleuros virescens* causes fruits and leaves to rot especially in wet areas and may be controlled to some extent by suitable fungicide sprays.

Fruit-flies of different species (*Dacus dorsalis* in Hawaii, *Ceratitis capitata* in the Mediterranean and *Anastrepha* spp. in the West Indies) are very serious pests of guava. The larvae of the flies spoil ripe fruits by boring through them. Insecticides like malathion should be sprayed every fortnight to give satisfactory control. The fruits are also attacked by mealy bugs, scale insects, thrips and the larva of the moth *Argyresthia eugeniella*. Root-knot nematodes also cause some damage but can be controlled to some extent by heavy fertilisation, irrigation and spraying nutrients on the leaves.

Jackfruit (*Artocarpus heterophyllus*)

Family Moraceae

Other names jack (French); buah nangka (Malayan); jaca (Spanish, Portuguese); langka (Tagalog, in the Philippines)

A large pungent fruit, the pulp and seeds of which are relished in tropical countries. May be eaten fresh, cooked (when under-ripe or ripe) or preserved in syrup.

Botany and propagation

The tree reaches 20 m in height bearing one of the largest of edible fruits (Fig. 12.40). The trunk is relatively short and straight, bearing thick branches. The leaves are dark green and shiny. Flowers are borne in male and female flower-heads on mature branches. A multiple fruit develops from the female flower-head; inside are many seeds covered with golden yellow flesh between unfertilised flowers (Fig. 12.41).

It may be propagated from seeds grown in containers and by budding on to champedak (p. 123) rootstocks.

fig. 12.40 Jackfruits hanging from woody branches

Climate and soil

Jackfruit is native to India and grown all over the Tropics. It grows well in lowlands in wet and humid climates. Many different types of soil are suitable but rich and well-drained soils are preferred.

Cultivation and management

Many types varying in fruit quality, shape and size are available. Some are known cultivars, e.g. 'Singapore Jack'. Seeds germinate in 3–8 weeks; seedlings are transplanted at 4–leaf stage. Budding is usually done at 8–11 months on champedak seedlings. Planting is at a spacing of 13–15 m apart. Little pruning is needed except to shape the tree. The crop is sometimes manured; orchard trees should be given complete fertilisers. Fruiting of seed-grown trees starts in about the 8th year. Buddings fruit in 1·5–3 years. Fruits mature about 8 months after flowering; they should be protected by some form of wrapping. Up to 250 fruits, each weighing 10–30 kg, may be obtained from the largest trees. Each fruit contains 100–500 seeds.

fig. 12.41 A jackfruit tree in bearing (the fruit-bags are protection against fruit-flies)

Protection

Stem borers and brown weevils attack young stems and fruits; control is by hand picking or insecticide sprays. Fruit rot and leaf spots due to fungi can be controlled by spraying fungicides.

Langsat and duku (*Lansium domesticum*)

Family Meliaceae

Other names lansium (French); ayer ayer (Malayan); arbol de lanze (Portuguese); lansone (Tagalog, in the Philippines)

A jungle fruit with sticky sap and flesh in sweetish jelly-textured white segments. Almost unknown outside Southeast Asia.

Botany and propagation

Medium-sized trees in the range 7–15 m high; langsat has a narrow straggly crown and duku a dense round crown (Fig. 12.42). Bunches of round yellow fruits 3–5 cm in diameter are produced once or twice a year (Figs 12.43 and 12.44).

It is usually planted from seed but can be budded (Fig. 12.45).

Climate and soil

A strictly humid tropical crop, grown principally in Southeast Asia and below about 600 m elevation. Rich, deep well-drained soil is required.

Cultivation and management

Trees commence yielding after 5–8 years. They should be spaced at 8–10 m. Compound fertilisers containing nitrogen, phosphorus, potassium and magnesium at the rate of 2 kg per tree may be given every 3 months.

Protection

Circle weeding by hand around the trees is recommended although it seems likely that herbicides such as diuron may be safely used.

fig. 12.42 A mature duku tree

fig. 12.44 Flowers of langsat or duku

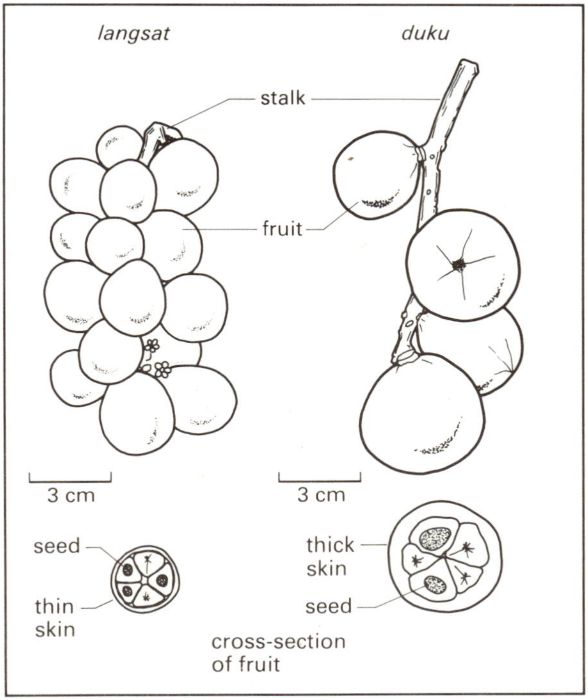

fig. 12.43 *Lansium domesticum*

fig. 12.45 Budlings of langsat or duku ready for transplanting in the field

Macadamia nut (*Macadamia ternifolia* syn. *M. integrifolia* and *M. tetraphylla*)

Family Proteaceae
Other names Queensland nut

The seed of this tree is eaten as a delicious 'nut', either raw or after roasting. It is a confectionery nut of growing importance in commerce.

Botany and propagation

A small tree, growing up to 15 m tall, native to the rain-forests of Queensland and New South Wales in Australia. Leaves, 3–4, are held in whorls at each node, spear-shaped and pale green or reddish. Flowers are creamy-white or pinkish, in terminal bunches. Fruits have a fleshy husk and a single round, smooth or rough and sometimes pointed seed, with a hard brown shell, about 2 cm thick (Fig. 12.46).

Propagation is by cleft or side grafting or by patch budding named varieties on to seedling rootstocks. *M. tetraphylla* rootstocks are considered about the best because of their resistance to trunk canker. Seeds are germinated in boxes or seedbeds, containing soil; germination takes 3–6 weeks. Germinated seedlings are planted in plastic bags. Grafting or budding is done when the stock is 12–18 months old.

Climate and soil

The macadamia is a rain-forest tree. Although better suited to the sub-tropics, it also grows in the humid Tropics. It needs a lot of moisture and yields are usually poor in a dry season unless irrigation is provided. Shelter from strong winds is desirable to avoid breakage of branches and poor yields.

The tree is not very selective of soil type, but it does best on deep, friable and fertile loams. A soil pH of 6·0 to 6·5 is recommended.

Cultivation and management

Commercial cultivation is most advanced in Hawaii, although practised in other places, e.g. Queensland. Well-known commercial varieties are chiefly from Hawaiian selections, e.g. 'Keauhou (246)', 'Kakea (508)', 'Ikaika (33)' and 'Keaau (660)' which vary in tree habit, yield and kernel recovery. Some Australian varieties are 'Hinde (H2)', 'Rickard (B5)', 'Tinana (B6)', 'Schimke' and 'Own Choice'.

Land is prepared by ploughing and ripping. Further cultivations may be necessary to loosen the soil and kill weeds. Lime should be added to acid soils.

fig. 12.46 Fruit and leaves of the macadamia nut

Trees are transplanted at a spacing of 9–10 m square, depending on tree habit (upright or spreading). Mulching is often practised to conserve moisture and control weeds. A 10:4:7, N:P:K fertiliser mixture is applied at the rate of about 0·5 kg per tree (in 3–4 doses yearly) for the first 10 years after which rates may be increased depending on the tree vigour and yield. Zinc and copper deficiencies are often observed; they may be corrected by foliar sprays.

Young trees should be watered regularly until well established. Mature trees benefit from irrigation during dry seasons.

Grafted or budded trees may start to crop lightly from the third year after transplanting. Yields increase with time; a 10-year tree yields 10–18 kg of nuts per year. Yields can reach a maximum of 40 kg per tree in 15 years.

Harvesting and processing

Mature nuts that have fallen are collected from the ground, at 1–3 week intervals, dehusked as soon as possible and laid on racks to dry for at least 3 weeks. Nuts are sold nut-in-shell for use as a dessert or for confectionery. As a confectionery nut, they are dried to 1·5 per cent moisture content, cracked, shells removed, roasted in oil, further dried and sprinkled with salt before packing for sale.

Protection

Clean or circle weeding is usually practised, manually or by the use of herbicides, e.g. paraquat. A sod culture of grass is often practised; the grass should be cut regularly.

Common flower pests include the macadamia flower caterpillar, hairy lime blue butterfly, macadamia lace bug and citrus aphid; they are controlled by spraying 0·05 per cent trichlorphon or endosulfan or 0·025 per cent demeton–S–methyl. Nut pests such as the fruit-spotting bug, banana spotting bug, macadamia nut borer and yellow peach moth may be controlled by sprays of trichlorphon, endosulphan or 0·1 per cent carbaryl. The leaves are attacked by the macadamia felted coccid, leaf miner, twig girdler, Molepta beetle, orange fruit borer and macadamia cup moth. Control is achieved by spraying with trichlorphon, endosulfan, carbaryl or 0·05 per cent diazinon or methidathion.

The only major disease is macadamia trunk canker, caused by the fungus *Phytophthora cinnamomi*. It is controlled by prevention of wounds on trunks, tree surgery followed by painting with a copper fungicide and by spraying the trunks with Bordeaux mixture. *Armillaria* root rot can be avoided by removing roots and stumps during replanting. Blossom blight and root anthracnose are minor diseases.

(a)

fig. 12.47 (a) A mango tree in bearing (b) Flowers of mango

(b)

Mango (*Mangifera indica*)

Family Anacardiaceae
Other names arbre de mango, manguier (French)

Botany and propagation

The mango is a large tree growing to 10–25 m tall (Fig. 12.47(a)). The leaves are alternate, spear-shaped up to 30 cm long and often brightly-coloured when young. Male bisexual flowers, reddish-pink or white, are carried in bunches at the ends of branches (Fig. 12.47(b)). Only 1–35 per cent of flowers are bisexual of which only 0·1–0·25 per cent set fruit. Cross-pollination occurs at night. Fruits mature in 2–4 months.

Propagation is usually by in-arching, shield budding or

grafting. It may be grown from seeds but fruits from seedling trees are usually inferior.

Climate and soil

Mango is grown in the Tropics from 0–1 300 m, with an annual rainfall of 3 500–2 500 mm. A marked dry season is needed for pollination and fruiting, hence it does not do so well in the humid Tropics. However, in recent years varieties capable of yielding well in humid zones have been developed, e.g. 'Haru Manis' (Indonesia), 'Apple' (USA), 'Malgoa' and 'Malda' (India). It can grow on poorer types of soils provided they are not waterlogged, too acid or shallow. Very fertile soils produce poor crops due to excess vegetative growth.

Cultivation and management

Thousands of varieties are established today, including in addition to the above, 'Alphonso' from India, 'Haden' from Florida, 'Cambodiana' from Indo-China, 'Peach' from Queensland and 'Julie' and 'Bombay' from the West Indies. Varieties are chosen to suit their local conditions. Spacing should be 6–12 m apart depending on variety. Planting holes are given organic manure and wood ash. Budded trees fruit in 4–5 years; seedlings take longer. Yields increase up to the 20th year and are up to 400–600 fruits per year per tree. Ripe fruits are picked before they soften. Most fertilisers are needed in the first 4–5 years.

Protection

Circle weeding should be done regularly. The most serious disease is anthracnose due to the fungus *Colletotrichum gloeosporioides* which causes leaves, twigs, flowers and fruits to rot. Powdery mildew causes flowers and fruits to drop. These diseases are controlled by fungicides. The most serious pest is mango-hopper, controlled by spraying with DDT. Fruit-flies spoil ripe fruits; they can be controlled by destroying fallen fruits.

Mangosteen (*Garcinia mangostana*)

Family Guittiferae
Other names mangoustan (French); mangostão (Portuguese); mangostan (Spanish)

A delicious, sherbet-flavoured tropical fruit. The purple-crimson skin of the fruit produces a dark-staining, very fast dye.

fig. 12.48 A large mature mangosteen tree

fig. 12.49 Leaves and young fruits of mangosteen

Botany and propagation

The mangosteen is a slow-growing tree, 7–15 m tall with yellow latex (Fig. 12.48). The leaves are pink when young, spear-shaped and leathery (Fig. 12.49). The pink or red flowers are single or paired; male and bisexual flowers appear on the same tree. The ripe fruits are purple-crimson and round; they grow to 8 cm across with 5–8 white fleshy segments inside. 'Seeds' found in some segments are not true seeds but consist of vegetative material.

The trees are grown from these 'seeds', sown in polythene bags soon after extraction.

Climate and soil

A hot and humid climate with even rainfall is needed. Mangosteen grows at 0–500 m altitude and on many types of soil. Clay or loam soils rich in organic matter are preferred.

Cultivation and management

Seedlings which are very uniform, are transplanted to the field when 2–3 years old. Animal manure and fertilisers should be placed in the planting holes, spaced 10 m apart. The young trees should be shaded and will bear fruits at 8–15 years. Trees treated with fertiliser usually bear earlier. Yields of 200–2 000 fruits per tree per year may be obtained.

Protection

Regular circle weeding is necessary. The crop is relatively free from pests and diseases. A physiological disease produces flesh filled with sap from bruised rind.

Passion fruit (*Passiflora edulis*)

Family Passifloraceae
Other names granadilla, sweetcup, simitoo (general, English); pomme liane (French); maracujá (Portuguese)

Botany and propagation

The passion fruit is a strong semi-woody climber with tendrils. The young leaves are oval, becoming trilobed when older. Large showy flowers appear from leaf axils with a purplish corona at the base of the ovary. Fruits are oval, and the seeds covered with yellow juicy pulp (Fig. 12.50).

fig. 12.50 Yellow passion fruit growing on a wire trellis in Malaysia

It may be planted from seeds, sown in containers and transplanted at 3–4 months. Mature rooted cuttings or layering may be used for propagation.

Climate and soil

The passion fruit is native to tropical America, but grown throughout the Tropics and sub-tropics. The purple edulis form grows best in highlands but the yellow form tolerates lowlands. Rainfall of 750–1 250 mm is necessary; excess rain causes poor fruiting. The crop requires a rich, well-drained, not too acid soil.

Cultivation and management

Two forms are grown. The edulis form has purple fruits and is sweeter. The flavicarpa form produces yellow fruits (also called Jamaica honey-suckle, watermelon and yellow granadilla). Plants should be spaced 3–7 m × 2–3 m apart and trained on trellises 2·5 m high. Only two branches are allowed to grow by removing other shoots. Complete fertilisers should be applied at 1·4 kg per tree, quarterly. Fruits are harvested from the second year onwards. Dropped fruits are picked every 2–3 days; edulis forms yield about 12·5 tonnes per hectare of fruits yearly; yellow types give up to 5 times more yield in humid regions.

Protection

Woodiness disease of leaves is due to a virus spread by

aphids. The fungus *Alternaria passiflorae* causes leaf spots and fall. Serious pests are the fruit-flies *Dacus dorsalis*, *D. cucurbitae* and *Ceratitis capitata*. Red spider mite and nematode root galls are less widespread problems.

Pulasan (*Nephelium mutabile*)

Family Sapindaceae

Pulasan is a closely related and very similar fruit tree to the rambutan. The fruits are similar to rambutan in general appearance except that the hairs are very stout and thick, the skin is thick and the flesh is softer and jelly-like (Fig. 12.51). It is not grown to the same extent as rambutan.

Rambutan (*Nephelium lappaceum*)

Family Sapindaceae

Other names hairy lychee (general, English); ramboutan, litchi chevula (French); ramustan (Spanish)

Botany and propagation

The rambutan is a bushy and wide-crowned tree achieving a height of 13 m when grown from seeds or buds; marcotted trees are much smaller. The trunks are thick and short with low, spreading branches. Leaves consist of 2–4 pairs of alternate leaflets. The flowers are small, greenish and in bunches at the ends of branches. Male and female flowers are borne separately on the same tree. Fruits have long hairs, green at first (Fig. 12.52) and turning crimson,

fig. 12.51 Young, green pulasan fruits

fig. 12.52 A young rambutan tree in bearing

red or yellow when ripe. Inside the skin is a seed, covered by the sweet white or translucent flesh for which it is grown. It can be propagated from seeds but then produces fruits which often have sour flesh that do not peel easily. Normally it is propagated by budding chosen clones on to 5–6 month old seedlings in polythene bags. It may be marcotted.

Climate and soil

Rambutan is strictly a tropical tree, thought to orginate in Malaysia, from where it spread to other parts of the Tropics. It needs a humid climate with at least 2 000 mm of rain annually. It grows on most tropical soils, the most suitable being a well-drained loam, rich in organic matter.

Cultivation and management

Common clones are 'Seematjan', 'Seejonja', 'Seetangkoowah', 'Seelengkeng' and 'Seekonto' from Indonesia and R 3, R 4, R 7 and R 99 from Malaysia. Budded plants are transplanted in the field at a spacing of about 13 m. Planting holes should be given well-rotted manure and 0·25 kg of a complete fertiliser. On acid soils, 1·5–3·0 tonnes per hectare of dolomite should be added. Complete fertilisers are given at the rate of 1 kg per tree in the first year, rising to about 7 kg per tree in the 8th year and thereafter. For marcotts, a 6–7 m square spacing and one-quarter of the above fertiliser rates per tree are recommended. Trees are sometimes topped to produce low branches for easier picking. Seedling trees take 5–6 years to fruit but clonal buddings and marcotts start bearing after about 2 years. Trees usually fruit twice a year giving about 1 000–1 500 fruits per tree in the first 5–7 years; older trees can give 5 000–6 000 fruits per tree per year. Fruits are usually picked and sold in bunches.

Protection

Circle weeding should be carried out using herbicides (e.g. 2 kg per hectare of diuron) or manually. Seedling blight due to the fungus *Phomopsis* sp. can be checked by spraying with copper or other fungicides and by removing and burning infected seedlings. Root fungal diseases caused by *Fomes lignosus* and *Ganoderma pseudoferreum* are controlled by removing and burning all affected trees. Powdery mildew (due to *Oidium nephelii*) causing fruit fall can be controlled by applying sulphur. Rambutan is relatively free from pests except for some fruit-flies and mealy bugs which sometimes attack the fruits. Insecticides such as 'Labacyd', 'Tamuron', etc. can be used for control and should be sprayed on at the young fruit stage.

Rose apple (*Eugenia jambes* or *Syzygium jambes*)

Family Myrtaceae
Other names jambu mawar (Malayan); pomarosa (Spanish)

The rose-scented fruits are eaten raw or made into preserves.

Botany and propagation

It is a small tree growing to about 6–10 m tall with narrow pointed leaves 15–20 cm long. The young leaves are brownish-red and very attractive. The flowers are greenish-white in colour. Fruits are round and about 4 cm in diameter with whitish or pinkish, rose-scented flesh surrounding a pale brown seed (Fig. 12.53).

Propagation is by seed, marcotting or grafting.

Climate and soil

The tree grows well in the humid Tropics, being native to India and Southeast Asia. It prefers a deep loamy soil but can perform quite well on poorer soils, although it cannot tolerate a high water-table.

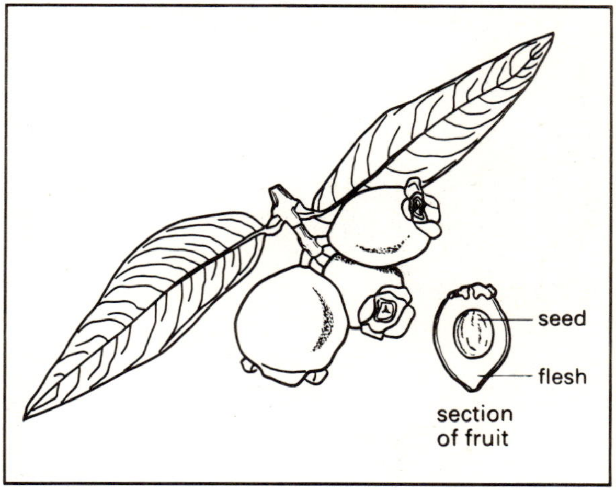

fig. 12.53 *Eugenia jambes*

Cultivation and management

Spacing of trees in the field is 6–8 m square. Much organic matter in the planting holes and regular fertilisation is usually needed to produce good crops.

Protection

Circle weeding is needed. The crop is often attacked by leaf-eating insects but usually not seriously; insecticides may be used if necessary.

Java rose apple (*Eugenia javanica* or *Syzygium javanica*)

Family Myrtaceae

Other names jambosa, wax jambo, Java apple, macopa, semarang, jambu ayer rhio

The fresh fruits are eaten raw, usually with some salt or sauce.

Botany and propagation

It is a tree about 14 m tall, fruiting when quite small. It is similar to the watery rose apple (*E. aquea*) but different in the waxy type of fruit and the presence of a short leaf stalk (Fig. 12.54).

It is usually propagated by marcotting or budding.

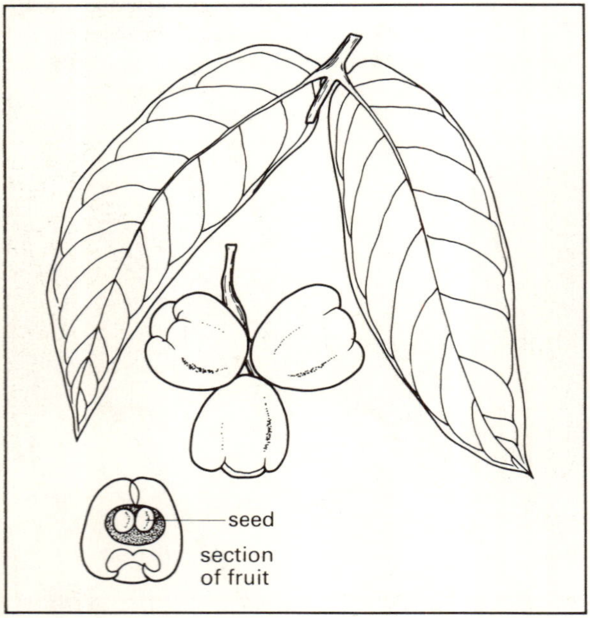

fig. 12.54 *Eugenia javanica*

Climate and soil

Native to Southeast Asia, it is not grown outside humid tropical Asia. It needs a deep, fertile soil.

Cultivation and management

Spacing in the field should be 8–10 m square. Adequate quantities of well-rotted animal manure should be given in the planting holes and compound fertilisers should be supplied. Well kept mature trees can yield very large crops.

Protection

As for rose apple.

Malay rose apple (*Eugenia malaccensis* or *Syzygium malaccensis*)

Family Myrtaceae

Other names jamalae, jamboissier rouge, pomme de Tahiti (French); ohia (Hawaiian); jambu merah (Malayan); jambe de Malacea, pomoreja de Malaca (Spanish)

fig. 12.55 Fruits of Malay rose apple

The pleasantly scented fruit is eaten raw.

Botany and propagation

It is a heavy looking tree with large pointed leaves up to 45 cm long. It grows up to about 20 m in height. Flowers are deep pink or red, appearing in bunches of 2–8 from mature branches. Fruits are somewhat apple-shaped and rosy red when ripe. The flesh is white and thick and contains one round seed (Fig. 12.55).

Propagation is usually by seeds and budding.

Climate and soil

It is native to Southeast Asia and requires a well-distributed rainfall above 1 200 mm. It grows best where the soil stays moist but drainage is good.

Cultivation and management

There are many cultivars differing in fruit colour, shape and taste. A spacing of 10 m is generally used in orchards. Organic matter in the planting holes and compound fertilisers are usually necessary. It generally blooms once a year; fruits ripen in 2–3 months.

Protection

As for rose apple.

Watery rose apple (*Eugenia aquea* or *Syzygium aqueum*)

Family Myrtaceae
Other names jambu ayer (Malayan)

The fruit is eaten fresh as a dessert or in salads. They are very juicy and are good thirst quenchers. Ripe fruits may also be made into a syrup or beverage.

Botany and propagation

Watery rose apple is a tree about 5–10 m tall (Fig. 12.56). Leaves are in opposite pairs, pointed and 12–25 cm long. White or pale cream flowers appear in clusters of 3–5. Fruits are somewhat conical and pale green, whitish, pink or red; each fruit contains 1–2 brown, loosely-attached seeds (Fig. 12.57). Seedless cultivars are available and preferred.

fig. 12.56 A watery rose apple tree

fig. 12.57 Light pink fruits of watery rose apple

It is grown usually by marcottage or budding.

Climate and soil
Native to south India, it is grown mainly in Southeast Asia and is of some economic importance in Indonesia. It grows well in the humid Tropics on most soils that are well drained.

Cultivation and management
There are now many cultivars varying in fruit characters, e.g. colour, shape and taste. Trees are spaced 6–8 m apart in the field; planting holes should have ample organic matter and inorganic fertilisers are usually needed. The tree fruits when quite small; fruits are produced twice a year.

Protection
As for rose apple.

Sapodilla (*Manilkara achras* syn. *Achras zapota*)

Family Sapotaceae
Other names naseberry, sapota (general, English); sapotille (French); chika (Malayan); sapoti (Portuguese); chicu (Talalog, in the Philippines); nispero (Spanish, in Colombia)

Cultivated as a dessert fruit and for the latex which is used to make chewing gum.

fig. 12.58 Sapodilla tree with fruits

Botany and propagation
Sapodilla is a tree reaching 20 m tall with a low-branching trunk and dense oval leaves (Fig. 12.58). The flowers are white and appear singly from leaf bases. The fruits are oval, 5–10 cm long, with yellowish-brown flesh sweet and gritty to taste; containing 0–12 black hard seeds. White latex is found in all plant parts.

It can be grown from seed sown in containers. Vegetative propagation, e.g. marcotting, budding and grafting, is now commonly practised.

Climate and soil
Native to Central America, it is grown all over the humid Tropics. Young plants need well-distributed rainfall. Older trees can stand some drought and wind. Altitudes of 0–2 500 m are suitable. Soil should be rich, well drained and loamy.

Cultivation and management
Many varieties have been developed, e.g. 'Prolific' and 'Russell' from Florida and 'Betawi', 'Koolon', 'Apel Benar' and 'Apel Leelin' from Indonesia; they vary in tree shape and fruit quality. It is planted at a spacing of 7–15 m square and is responsive to complete fertilisers given quarterly. Fruiting starts at 3–5 years and continues for about 30 years. Yields of 2 500–3 000 fruits per year from large trees may be obtained. Fruits are picked while hard;

they are edible after 2 weeks. Latex tapped from trunk (every 2–3 years) contains 20–40 per cent gum.

Protection

Rust fungus (*Scopella sapotae*) and leaf spot (*Septoria* spp.) cause leaf fall; they are controlled by a monthly spray of 'Ferbam'. Fruit-flies and leaf caterpillars infesting the crop can be controlled by appropriate insecticide sprays (e.g. malathion, 'Lebacyd', 'Tamuron').

Soursop (*Annona muricata*)

Family Annonaceae
Other names sappadille (French); durian belanda (Malayan); graviola, jaca do Pará (Portuguese, in Brazil); guanábana (Spanish)

fig. 12.59 A young soursop tree in bearing

The pulpy white, rather astringent, flesh of the fruit is eaten fresh or made into an ice cream or drinks.

Botany and propagation

A small straggly fruit tree, the soursop grows up to 8 m high. It is native to tropical America. It flowers and fruits throughout the year (Fig. 12.59). The fruits are oval or irregular, 15–30 cm long, with sparse soft green curved spines. The flesh is pulpy white, stringy and sour containing shiny black seeds.

Propagation is usually by seed but it can be easily grafted.

Climate and soil

Soursop is widely grown in the Tropics on a great range of soils with good drainage up to about 1 000 m. A warm humid climate is required.

Cultivation and management

The tree starts bearing 3–5 years from planting. No known varieties are available but seed-grown trees are fairly uniform throughout the Tropics. Trees are spaced 5 m square. Fruits are produced throughout the year but often there is a period of heavier cropping. Individual trees may yield up to 30 fruits each of various weights. Spacing should be about 5 m. Compound N:P:K:Mg fertiliser at rates of up to 100 g per tree applied every 3 months is recommended.

Protection

Circle weeding by hand is recommended although it seems likely that herbicides such as diuron may be safely used.

Sugar apple (*Annona squamosa*)

Family Annonaceae
Other names sweet sop, sugar apple, custard apple, anona (general, English); ata, pinha (Portuguese); anón (Spanish)

The fruit is eaten fresh or made into drinks. It is very similar to soursop but the fruits are heart-shaped,

fig. 12.60 Sugar apple fruits on a young tree

7–10 cm long, covered with round fleshy knobs and very sweet (Fig. 12.60). The trees yield up to 100 smaller fruits. The fruits are sometimes damaged by the atlas moth (*Attacus atlas*).

Surinam cherry (*Eugenia uniflora* or *E. michelii*)

Family Myrtaceae
Other names Brazilian cherry, cayenne cherry (general, English); cerise de Cayenne (French); pitanga (Portuguese); ceraza de Cayena (Spanish)

The fruit is eaten raw or after cooking. It makes excellent jelly, pickle or chutney. A wine can be made from the ripe fruits and the young leaves have been used as tea. Plants are also used for hedges.

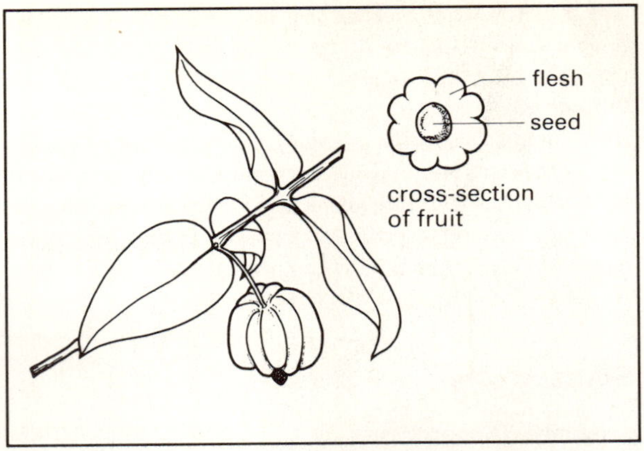

fig. 12.61 *Eugenia uniflora*

Botany and propagation

It is a woody shrub reaching about 5 m in height. Leaves are held in opposite pairs. Flowers are small and white and fruits are also small, fluted from base to apex and scarlet red when ripe. Its flesh is also red, juicy and sour. A round seed is embedded in the flesh (Fig. 12.61). Propagation is usually by seeds; grafting may be used.

Climate and soil

It is native to South America but is mainly grown as a crop in Sri Lanka and south China. It is quite cold-tolerant and can also grow in the sub-tropics. It will grow on most soils that are not waterlogged and adapts well to a hot and humid climate.

Cultivation and management

Two forms are grown; the more common one has scarlet fruits and the other has dark-red or black fruits. Both forms have red-tinged leaves. Fertilisers and water must be supplied in the drier seasons to obtain good yields. The plants are spaced 3–4 m apart or planted in hedges. Fruiting begins after 1–2 years. For maximum yield the shrubs should not be pruned.

Protection

As for rose apple

fig. 12.62 A tamarind tree bearing young fruit

Tamarind (*Tamarindus indica*)

Family Leguminosae (Caesalpinioideae)
Other names asam jawa (Malayan)

The brown pulp around the seeds is used for seasoning food and for making a delicious drink.

Botany and propagation

Tamarind is a tree reaching 20 m in height (Fig. 12.62). The leaves have 10–20 leaflets. Creamy red flowers grow in bunches. Pods are curved, brownish with 1–10 seeds (Fig. 12.63).

Propagation is usually from seed, sometimes from buddings or cuttings.

Climate and soil

Indigenous to the semi-arid Tropics of Africa, it may be grown in the humid Tropics provided the soil is well drained.

Cultivation and management

The tree is often planted as an ornamental or for shade. It is seldom planted on a plantation scale or given much attention. No distinct cultivars have been developed or found but the Indian tamarind is better than the African types.

Protection

Circle weeding is required. It is relatively free from pests and diseases.

fig. 12.63 Mature tamarind fruits

13 Grains

Maize (*Zea mays*)

Family Gramineae
Other names corn, Indian corn (general, English); mais (French); milho (Portuguese); maiz (Spanish)

Maize was first cultivated by Amerindians and is thought to have originated in Central America. It has been grown for thousands of years and many of the early civilisations of the Americas depended on learning how to grow maize; just as the early agriculture of western Asia and the Mediterranean depended on wheat and that of the east and Southeastern Asia on rice.

Maize is still one of the most important grain crops and is geographically the most widely planted species. It is grown from latitudes up to 50°N in central Europe, throughout the Tropics, to latitudes of about 45°S in New Zealand and the South American continent. The major producing regions are the USA, China, South and central Africa, Argentina and Brazil. It is very widespread in the humid Tropics.

Botany

Maize is an annual plant belonging to the grass family; it matures over a short period of a few months. Individual plants are fairly large and unlike many grass species, usually only one or two stems grow from each seed. Height varies from about 1·5 m in short varieties to over 3 m. Figs 13.1 and 13.2 show the essential features of a maize plant. Maturity time is usually 100–120 days, but some variation exists in this respect. Male flowers, called tassels, emerge after 50–60 days at the apex of the plant. Female flowers are formed in the axils of the lower leaves at about the same time. The stigmatic surfaces of the female flowers, called silks, emerge from the leaf axils shortly after tasselling. The seeds are formed on cobs which are the compacted stalks of female inflorescences.

Varieties and yields Maize is an out-pollinated species with many varieties most of which are adapted to or bred for particular geographical areas. Many tropical countries have breeding programmes producing maize seeds for planting by farmers, but yield still varies greatly from about 1 000 kg per hectare at the lower level up to levels approaching 8 000 kg per hectare. This variation depends not only on the amount of effort and skill of the maize breeder but also on climate and the standard of farming. Maize varieties have been classified on the basis of the grain. Three main groups are grown in tropical countries.

1 **Dent maize** These have large cobs and large grains which shrink on drying to produce a shallow hollow at the end of each grain. Most maize varieties of the USA central and southern African countries, are dent varieties. The 'Coastal Composite', from East Africa is a dent variety adapted to the humid Tropics.

2 **Flints** These generally have smaller cobs which bear hard round grains. They are commonly grown in Asia and tropical America and parts of tropical Africa. 'Metro' is a composite flint variety adapted to humid tropical regions.

3 **Sweet corn** The grains are soft and sweet with little starch and are grown for eating as a fresh vegetable. There are a number of superior sweet corn varieties from Hawaii and elsewhere.

Since maize is an outbreeding species, mass selection

fig. 13.1 A stand of a composite flint variety of maize in Malaysia

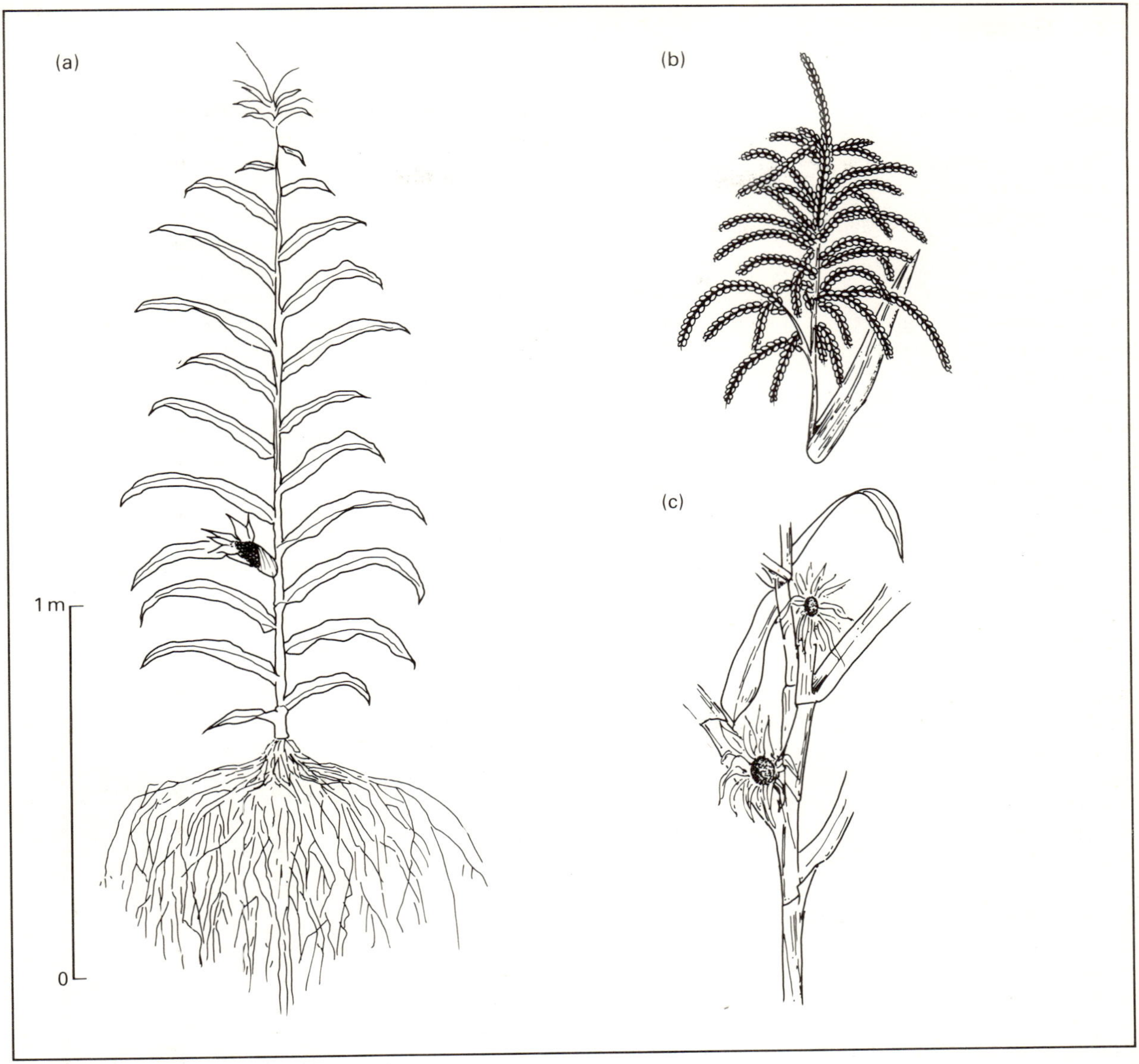

fig. 13.2 (a) Mature maize plant (b) Tassel (male flowers) (c) Young female inflorescences in the lower leaf axils. The threads protruding from the ends are called 'silks'. They lead to the female flowers and pollen falling on the silks fertilises the female flowers

has been used to develop particular open-pollinated composite varieties. Mass selection involves the selection of desirable ears from better plants which are then planted together in seed gardens to produce seed. Seed produced by such means often shows considerable variation, but close selection of specific types leads to inbreeding and

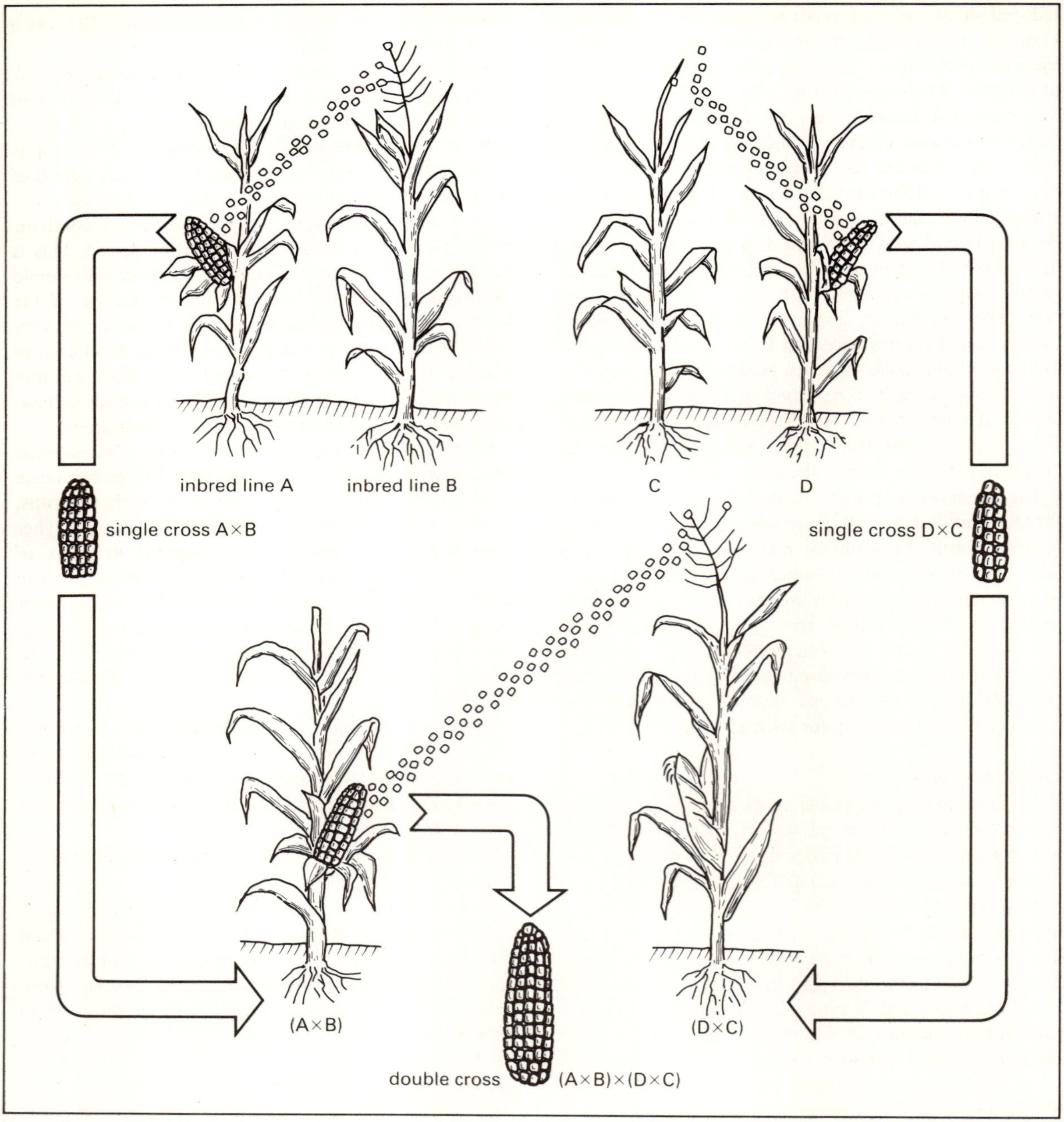

fig. 13.3 The production of single and double cross hybrids. (*Source:* **Berger, J.** (1962). *Maize production and the manuring of maize.* Centre d'etude de l'azote: Zurich.)

reduced yields, and such types are unsuitable for planting. However, the crossing of two closely inbred lines often leads to hybrid vigour. This characteristic is utilised in the production of hybrid seed (Fig. 13.3). Single cross seed is not practicable because the yield of the seed is low and therefore the cost of seed is high. Double cross seed is commonly produced for distribution to farmers. Maize hybrids often exhibit exacting requirements and should be grown only in the area for which they were developed. Because hybrid seed is produced in this special way the farmer must buy new seed for each planting. Composite varieties can be replanted with seed from the previous crop. For this reason, improvement of maize in many tropical countries has been and is being achieved by the use of new composite varieties. However, when selecting seed for replanting it is important that the farmer should choose the best cobs from the healthiest and highest-yielding plants and should also select them from the middle and not the edge of the field.

The relationship between yield ability, requirement for nitrogen fertiliser and plant spacing is important when growing maize. In addition, response to fertiliser and plant spacing depends on water supply and irrigation. High-yielding maize varieties are often smaller plants with vertical leaves rather than very large spreading leaves. They require more nitrogen fertiliser and, assuming moisture is adequate, they also require close spacing of up to 50 000 plants per hectare because maize does not generally tiller to make up for wide spacing.

Climate and soils

Maize is essentially a tropical species which is killed by frost but much of the breeding and selection work has been done in sub-tropical and temperate countries so that there is a shortage of good maize varieties for the humid tropical regions. Probably the best source of varieties for improving maize in tropical countries are the Central and South American countries. African maize varieties found suitable in Southeast Asia are coastal composite types from East Africa and 'Katumani' which matures quickly. Most maize varieties of sub-tropical origin have been selected for lower temperatures and would become too tall in a humid tropical environment. Temperature also affects moisture requirements and generally more moisture is required under tropical conditions than under temperate conditions. Maize is specially sensitive to moisture stress around the time of tasselling and fertilisation and under tropical conditions a 25 mm deficit of moisture at the two weeks around fertilisation will cause a significant yield reduction.

Maize also needs optimal moisture at planting when the soil should be near field capacity. Generally, in the humid Tropics, rainfall will be adequate or excessive for maize growing and the only benefit from irrigation, if the crop is planted when the soil is moist, might be if a dry period of more than 2 weeks occurs near the time of tasselling.

Very humid conditions are not considered good for maize, particularly during the maturation period. This is due to the low sunshine which is associated with humid conditions, and to the development of diseases of the flowers, cobs and leaves with high humidity. In very humid climates, the ripening period should be chosen to coincide with a seasonally dry spell if possible. Another problem of very humid and rainy regions is poor pollination which produces cobs with many missing grains.

Maize is grown on a wide variety of soils. One essential requirement however is good drainage. The performance of maize in relation to soil type is, as with all crops, closely related to climate, deeper soils with a higher moisture-holding capacity being required in areas of marginal rainfall. Generally speaking, however, maize cannot be regarded as a crop which will yield well on poor sandy soils. With heavy clays deep cultivation and ridging is necessary to improve drainage. Maize is suited for off-season cropping in paddy lands provided drainage is adequate.

Maize is not a good crop for conservation of the soil and crop rotation incorporating soil-ameliorating crops such as legumes in the cycle, is recommended. Conservation practices are important in maize growing on unstable soils on slopes.

In friable soils the roots will develop to a depth of about 60 cm (Fig. 13.2).

Maize will grow on moderately acid soils but does best on soils with pH values above 5·0. Liming is required for good yields on more acid soils, and many tropical soils require lime for good production. Maize can be grown well on alkaline soils provided nutrient deficiencies of iron and manganese are countered.

Cultivation and management

Great variation occurs in the yields of maize from place to place. These very large differences in yield cannot be accounted for by climate alone, since countries possessing similar climates often show markedly different average yields. Generally speaking, the less developed countries

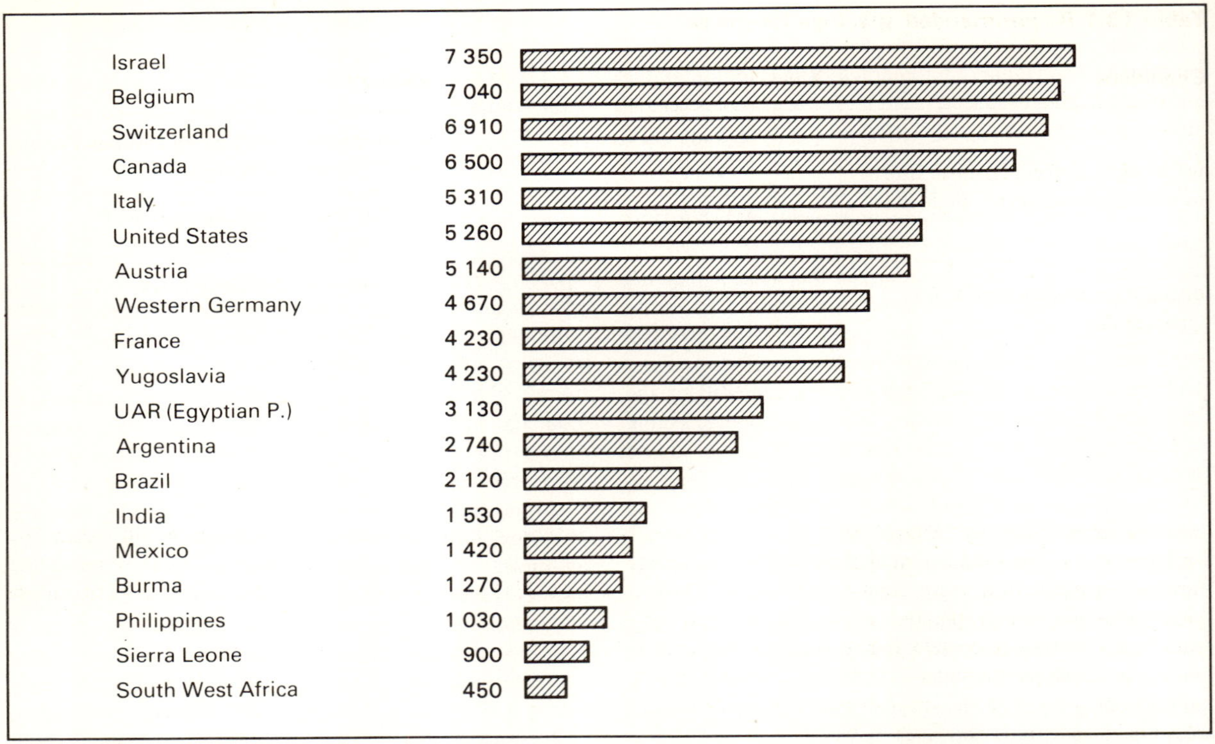

fig. 13.4 Approximate average yields of maize from different countries in kg per hectare. (*Source:* FAO Production Yearbook (1961). Vol. 15.)

produce the lowest maize yields; this may be attributed, to a large extent, to differences in cultivation techniques, including the use of fertiliser, and irrigation in sub-optimal rainfall areas (see Fig. 13.4).

It is generally found that tillage operations are important in maize production. In heavy soils, particularly in regions of high rainfall, deep cultivation to 50 cm and ridging is required to improve drainage. With light soils, shallow cultivation is satisfactory. Tilling the soil lightly and leaving the trash from the previous crop on the surface sometimes gives improved yields in light soils, for the previous crop residues form a mulch over the soil, conserving moisture.

Planting Seed should be drilled or planted 2–3 cm deep and may be planted on the flat, on the tops of ridges, or in the furrows of ridges depending on conditions. Planting in furrows is known as listing. Ridging is best in heavy soils and planting on top of the ridge improves drainage in such soils. This is to be recommended in low-lying soils such as off-season rice areas. Ridge planting must also be used when maize is irrigated by the furrow method. Listing can be used in well-draining soils and in dry areas. Listing is one of the most rapid methods of cultivation as the seed is planted at the time of ploughing. In tropical countries, where maize is grown as a subsistence crop, mound planting is often practised. The soil is built up into mounds about 50 cm high, and 1–1·5 m apart after incorporating burnt trash which contains much potassium, animal manure, top soil, etc. Several plants are grown per hill. This method is well suited to small scale cultivation and to hand tillage.

Maize is often grown on ridges 1 m apart in the USA. In Canada and in Europe, smaller varieties are often grown and these are best suited to closer spacing. Many

Table 13.1 Recommended spacings for maize

Conditions	Type	Spacing
no drought, correct fertiliser use, or new cleared and burnt jungle area	tall or spreading leafy type	60 cm rows × 45 cm along rows or 1 m rows × 30 cm along rows
	small erect leaves	60 cm rows × 25 cm along rows or 1 m rows × 15 cm along rows
drought possibility and/or no fertiliser use	tall or spreading leafy type	1 m rows × 60 cm along rows
	small erect leaves	1 m rows × 30 cm along rows 60 cm rows × 45 cm along rows

flint varieties such as 'Metro' (Fig. 13.1) are small varieties and are best planted at about 60 cm row spacing. Smaller varieties with more erect foliage have a higher yield potential, when planted more closely. Provided sufficient moisture is present maize is often planted too widely in developing countries. If drought is likely, maize will probably do better at lower densities. A guide to correct maize spacing is shown in Table 13.1.

The following minimal water requirements are needed to avoid drought on average soils for maize if planting is done in soil with full water capacity (Table 13.2).

Table 13.2 Annual rainfall for maize planting in mm

Month	Rainfall
First	85
Second	85
Third	100
Fourth	110

Where drought is likely it is better to use the wider spacings above if there is no possibility of irrigation, and in regions of a short rainy season planting must be done promptly so as to make full use of the rain. Where the rainy season is short, maize varieties maturing quickly should be grown such as the 'Katumani' composites from Kenya.

Fertilisers Maize usually responds well to fertilisers but is also exhausting to the soil. It is recommended that maize be grown after a legume crop such as groundnuts in humid tropical countries, this practice will reduce the amount of nitrogen fertiliser needed. Maize removes the following quantities of nutrients for a crop of 4 000 kg per hectare.

Nitrogen	100 kg
Phosphorus (P_2O_5)	40 kg
Potassium (K_2O)	80 kg

Some of the potassium will be returned to the soil from the stubble but most of the nitrogen and phosphorus is removed in the crop. A large crop will remove more and these nutrients and additional fertilisers must be applied to keep the soil fertile. In newly-burnt jungle soil however potassium may be omitted from the fertiliser for the first crop of maize. Because spacing and yield are affected by rainfall, fertiliser use should be adjusted for climate and expected optimal yield. Under tropical conditions and with the maize varieties available the fertiliser rates in Table 13.3 can be recommended.

Note that superphosphate and not rock phosphate should be used for maize; rock phosphate is more suitable for slow-growing plantation crops.

Fertilisers should be divided into three parts and given at the time of planting, when the maize is about 50 cm high and at the beginning of tasselling. Fertilisers may be sprinkled evenly along the planting lines.

When maize is grown immediately after a good legume crop the level of nitrogen fertiliser may be reduced to one-third the above rates on average soils.

In the Tropics, soil below pH 5·0 may benefit from a

Table 13.3 Recommended rates of fertiliser for maize (in kg per hectare)

Type of maize	Ammonium sulphate (VI)	Urea	Potassium chloride	Superphosphate
High plant density; high-yield types	200	200	150	150
Low plant density; unimproved vars.	100	100	75	75

dressing of about 1 tonne of magnesium limestone per hectare.

Maize sometimes needs zinc and molybdenum. Superphosphate with molybdenum added should be used. Zinc(II) sulphate(VI) may be applied at 5 kg per hectare.

Irrigation Water supply is not normally a problem in humid tropical countries. In sub-humid regions, however, such as many parts of eastern Africa, rainfall has great effects on the yields of maize from year to year. If irrigation water is available it should be supplied to keep the quantities of water up to the minimum levels indicated above. If only limited water is available it will best be utilised by bringing the soil up to field capacity at the time of tasselling of the maize.

Weed control Weeds greatly reduce all crop yields and maize is no exception. Fortunately good methods of weed control exist for maize. The seeds should be planted as soon as possible after tillage of the soil, so that they can grow before the weeds. When maize is planted in straight rows, machine cultivation can be used between the rows to combat weeds. This is usually best done at about three weeks, before the maize is too high for the tractor to pass over it. In tropical countries maize is often grown by hand and hand cultivation is so slow that weeds may catch up with the crop. Here herbicide can be used with a simple knapsack sprayer. 2:4:D amine at 2 litres per hectare can be used to kill non-grass weeds by spraying between the rows and when the maize is about 60 cm high, grammoxone may be used for spraying close to the ground, below the maize leaves. A very good herbicide for use in maize is simazine. This should be sprayed straight after planting at the rate of 2–4 kg per hectare, on the soil surface. If simazine is used promptly after planting probably no other weed control will be needed. However, simazine is long-acting and may affect many other crops (e.g. beans, vegetables) if they are planted on the same soil within about four months of application in the humid Tropics.

(a)

fig. 13.5 (a) Ripe maize cobs (b) Maize cobs hung up to dry

(b)

Harvesting and storage

Maize is harvested by hand over most of the tropical world. When required for eating as fresh vegetables, the cobs are picked before they are fully ripe, as with sweet corn. If required for the grain they are left till the grains harden. Maize cobs should be stored in a dry place. Seed cobs are usually chosen from the crop at the time of harvest as described before (Figs 13.5(a) and (b)).

Protection

Maize rust caused by the fungus *Puccinia polysora* is a serious maize disease which is controlled by seed dressing, by good field hygiene which includes removing diseased plants as soon as they appear and burning them, and by growing resistant varieties. The first signs are tiny water-soaked spots on the leaves that enlarge to form reddish-brown streaks about 4 mm long. Warm, wet weather promotes the disease. Various other leaf diseases occur on maize. They include *Helminthosporium maydis, Cochliobolus heterostrophus,* downy mildew caused by *Sclerospora sacchari* and others. All should be controlled by hygiene and growing resistant varieties when available. The stubble of the old crop should be burnt. Various diseases of the cob also occur especially in wet weather. One is called pink ear rot and is caused by a species of *Gibberella*. Burning stubble also reduces this disease. Harvesting ears promptly is also recommended or turning the ears down so that they shed the rain.

The chief insect pests are various stem and cob borers. They may be controlled by spraying with a suitable insecticide such as dimechron.

Rice (*Oryza sativa*)

Family Gramineae
Other names riz (French); arroz (Portuguese, Spanish)

Two groups of rice varieties are recognised: *indica* and *japonica*. There are certain morphological differences between *indicas* and *japonicas* which affect yield. *Indicas* are generally tall, with large light-green, drooping leaves, respond poorly to fertilisers particularly nitrogen and are sensitive to daylength in regard to flowering; they are also usually relatively low yielding. *Japonica* varieties are generally shorter with short dark-green erect leaves, respond well to nitrogen and are not sensitive to daylength. They are also generally higher-yielding. (The relationship between leaf type and yield has been considered in the first part of this book, see p. 34.) Throughout all rice varieties there is much variation in grain quality but in general, *indicas* tend to produce rice which is long-grained and *japonicas* to produce rice which is short-grained and of sticky consistency. Historically *indicas* have been developed in the Tropics under conditions of low fertility where they depend only on nutrients in the soil or flood waters and *japonicas* under conditions of high fertility and fertiliser use in the higher latitude regions of China and Japan. Most tall *indica* varieties lodge when fertilised but certain varieties possess stiff straw and have potential for fertiliser response and high yield, see p. 152. Other *indicas* are short-strawed and have been much used in breeding programmes to produce new high-yielding varieties.

Rice is of enormous importance as a food crop, being the main food source throughout Asia and many other parts of the world. Most varieties (both *japonicas* and *indicas*) are adapted to growth under flooded conditions and rice is virtually the only grain crop that can be grown under the monsoonal conditions of flooding that occur throughout much of tropical Asia and elsewhere. Other rice varieties can be grown under dry-land conditions and are known as hill varieties, but this section is concerned primarily with growing rice on swamps.

A second species of rice, *O. glabberima,* is of some importance in Africa and in recent breeding programmes.

Botany

The rice plant belongs to the grass family and is grown from seed either planted directly or transplanted as young plants. When the seed germinates it produces the first root, known as the radicle, and shoot, as shown in Fig. 13.6(a). The stem of the rice plant is at first very short and is located at the base of the shoot, and the main aerial parts of the plant are the leaves which are produced from the growing point located at the top of the short stem. Branch shoots or tillers are also produced from the short basal stem and this tillering proceeds throughout the early growth stages to give a dense population of leaves. The lower part of the stem also produces the root system which branches throughout the top 20 cm or so of soil. The production of leaves and roots is the vegetative phase of the life cycle.

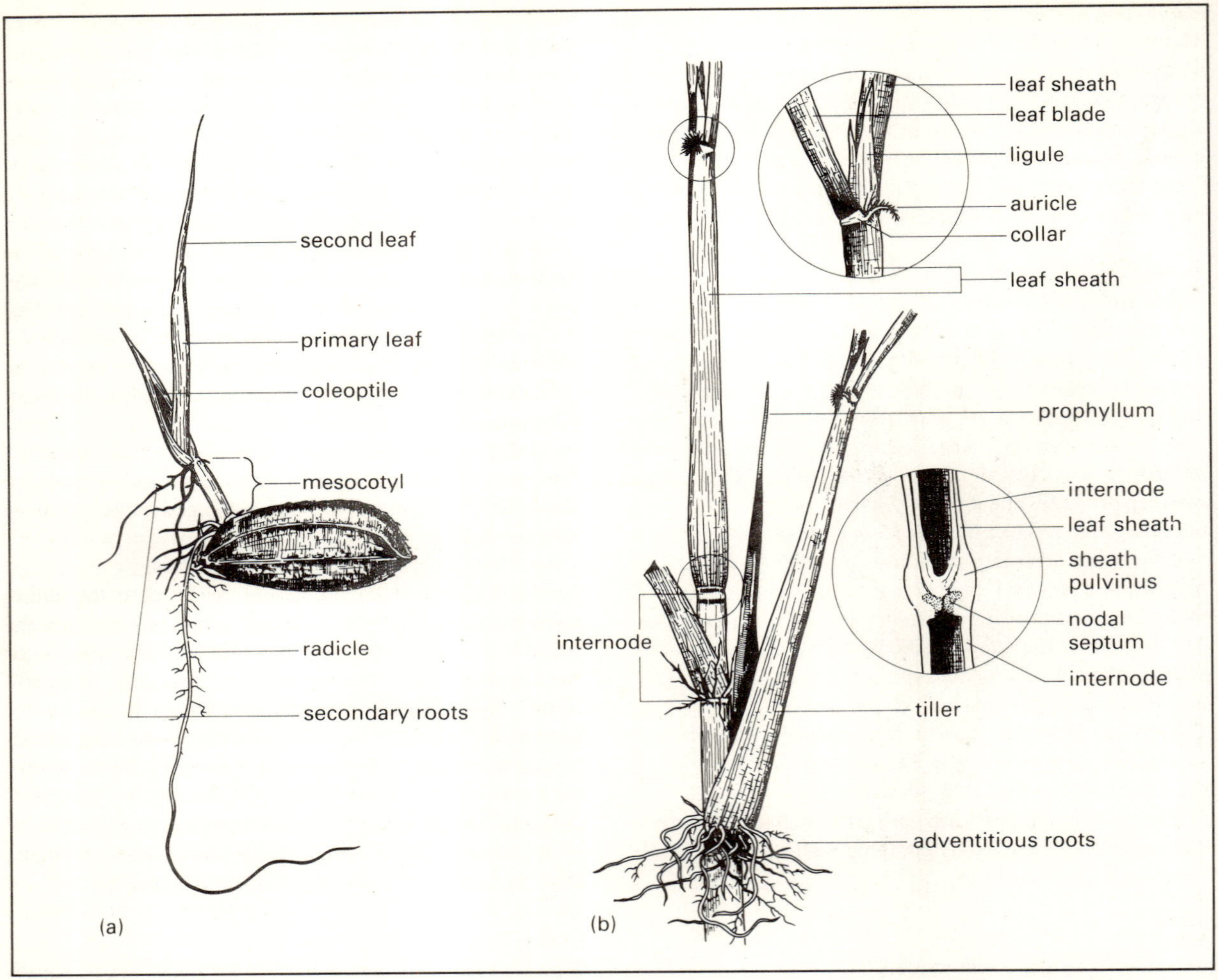

fig. 13.6 (a) Germinating rice seed showing the main parts of the young seedling. Magnification about × 10 (b) Parts of the main shoot (primary tiller) of a rice plant at the beginning of stem elongation

Later the stem begins to increase in length as shown in Fig. 13.6(b). The beginning of the elongation of the stem marks the end of the vegetative phase and this raises the growing points of the main tillers which cease producing leaves and form the young inflorescence or flowering shoot which will give rise to the rice grains of the crop. The flowering shoot is called the panicle and the last vegetative leaf which is produced is known as the flag leaf (see Fig. 13.6(c)).

Varieties, yields and photoperiodism From the point of view of yield, the differences in the botany of the typical low-yielding *indica* varieties and the *japonicas* and improved *indicas* mentioned above need to be further discussed.

Yields of 6 000 kg per hectare of paddy or unmilled rice are quite common in Japan, Australia, USA and elsewhere in more developed countries and at higher latitudes, whereas yields of unimproved native rice in

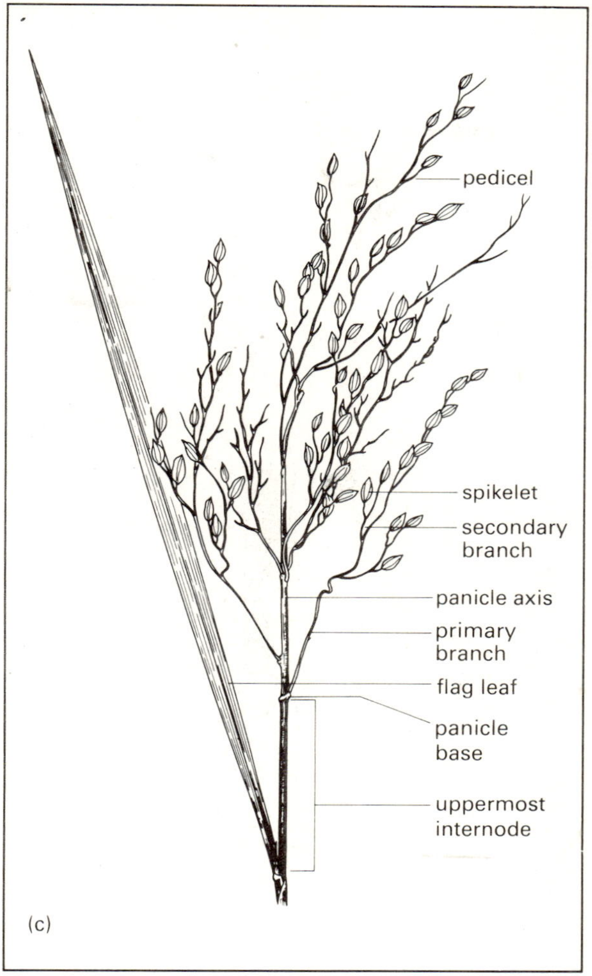

(c) The inflorescence (panicle) of a rice plant (in part). (*Source*: Rice Production Manual (1967) compiled by R.I.C.E. University of the Philippines.)

tropical Asia, Africa, Central and South America are generally only of the order of 1 000–2 000 kg per hectare and often less than this. The primary reason for the yield differences are related to the botanical differences between the typical *indicas* and *japonicas*. *Indicas* produce rapid early growth of roots and tillers but the fact that they are generally grown without fertilisers, limits their tillering so that they do not over-produce leaf to the point of inefficiency. Their response to nitrogen however is poor since this induces excessive tillering and leaf production and an even taller habit with a tendency to lodge. Some of these tall, lodging-prone *indicas* are well adapted to poor physical conditions and do quite well on acidic soils. Lodging occurs primarily as a result of the straw being too tall and weak, resulting in bending and breaking of the stems. However there are tall *indica* varieties that are lodging resistant. Hill varieties seldom lodge and some swamp varieties and varieties adapted to both drained and flooded conditions are also lodging resistant (see Fig. 13.7). These tall varieties have very large panicles and a potential for high yield and flood resistance. They respond to moderate quantities of fertilisers equivalent to about 30 kg of nitrogen per hectare.

In regard to maturity time and photoperiod some rice varieties, particularly *indicas,* show a marked response to daylength and, since the phenomenon of photoperiodism has not been described before, a brief explanation is given here:

Photoperiodism is the ability of the plant to respond to daylength and changing daylength, in changing from the vegetative to the flowering phase. Daylength or photoperiod varies throughout the year and photosensitive rice varieties respond to short or decreasing daylengths in flowering. When exposed to long or increasing photoperiods flowering is delayed. This is shown very clearly by the behaviour of a typical *indica* rice variety, 'Siam 29', grown in northern Malaysia at different times of the year. Fig. 13.8 shows the changing photoperiod (hours between sunrise and sunset) and the time taken to flower. Those plantings in which daylength was increasing during the major part of the life cycle took more than twice as long to flower than those planted during decreasing photoperiod or when photoperiod was minimal.

The main objectives in breeding higher-yielding rice varieties for the Tropics are:

1 To reduce maturity time and to reduce photosensitivity (which makes varieties more adaptable). Short maturity also enables multiple cropping and reduces water-use through a shorter time in the field.
2 To produce non-lodging varieties with erect leaves which are responsive to fertilisers and resistant to lodging and have a high ratio of grain to vegetative material.
3 Increasing pest and disease resistance.
4 Maintaining good grain quality and nutritive value.

A considerable degree of success has already been achieved in breeding high-yielding rice varieties for the

fig. 13.7 (a) A non-lodging tall *indica* ('PB 175') from Borneo. This bears heavy panicles and yields can reach 6 000 kg per hectare; it is one of the buloh varieties

(b) A tall lodging-prone *indica* from Borneo, 'GP 175'

Tropics, particularly through the programmes of the International Rice Research Institute (IRRI) based in the Philippines. The approach to producing high-yielding tropical rice varieties has been through breeding and selection of typical *indicas* to give improved yield and fertiliser response and by the use of certain *japonicas* known as the ponlai *japonicas* from Taiwan which are adapted to the higher temperatures of the Tropics (e.g. 'Tainan', 'Taichung 65', 'Chianung 242' and others) as well as certain dwarf *indicas* such as 'Taichung 1' and 'Dee-geo-woo-gen', also from Taiwan. The breeding programme of the IRRI produced the remarkable high-yielding tropical rice varieties 'IR 5' and 'IR 8' from crossing 'Dee-geo-woo-gen' with an improved tall *indica* 'Peta' from Indonesia. Comparative yields of 'IR 8' and other varieties are shown in Table 13.4.

By comparison local unimproved *indicas* may yield around 1 000–2 000 kg per hectare. Combination of *japonicas* and *indicas* has not been generally successful because of crossing problems but a number of good hybrids have been produced such as 'ADT-27' from India. Newer IRRI varieties which are more disease resistant and have

(c) 'Sri Malaysia' an outstanding, improved *indica*, of intermediate height and suitable for rain-fed conditions

Table 13.4 Comparative yields of some rice varieties. (*Source:* IRRI Reporter, 2(4) (1966).)

Variety	Yield in kg unmilled rice per hectare
'IR 8' (IRRI)	10 100
'Taichung 1' (short *indica*)	8 200
'Tainan 1' (ponlai *japonica*)	7 700
'Chianung' (ponlai *japonica*	7 400
'Peta' (improved tall *indica*)	5 200

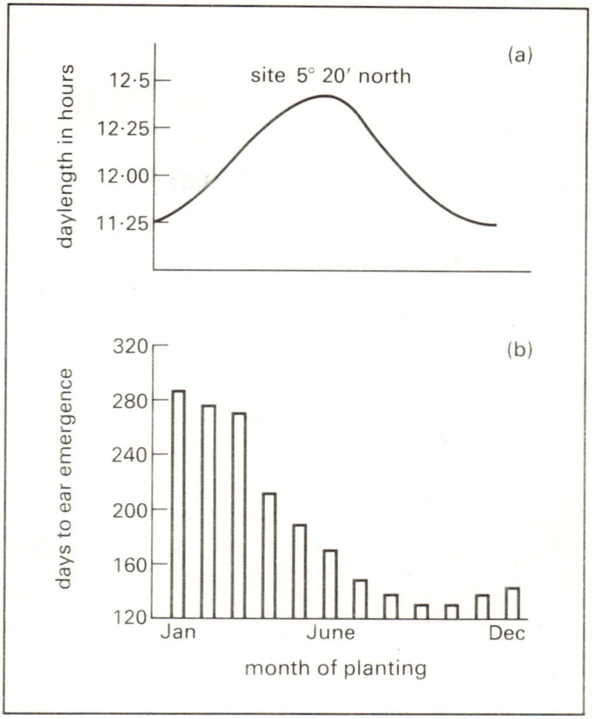

fig. 13.8 Graphs (a) and (b) compare change in daylength and the number of days to flowering related to the time of sowing in an *indica* rice variety 'Siam 29' grown in northern Malaysia. (*Source:* **Dore, J.** (1960). *Malayan J. Agr.*, **43**, 28.)

better grain quality are 'IR 20', 'IR 22' 'IR 32', 'IR 34' and others. Outstanding varieties produced under IRRI regional programmes are 'Bahagia', 'Sri Malaysia I' and 'Sri Malaysia 2' (see Fig. 13.7). Most of the improved *indica* rice varieties and *indica/japonica* hybrids have little or no sensitivity to daylength and are therefore relatively adaptable throughout the tropical regions. The variety 'IR 8' matures in about 125 days from seeding and requires 95–100 days in the field after transplanting. IRRI produces varieties suited for conditions of good water control such as 'IR 5' and 'IR 8', and taller varieties suited for rain-fed conditions such as 'IR 32' and 'IR 34', 'IR 442', etc.

Climate and soil

Different varieties of rice are adapted to a wide range of environmental conditions but varieties adapted to produce well in the Tropics require temperatures in the range of

fig. 13.9 Irrigated rice fields at Tanjong Karang, Malaysia. A system of drains leads water to the fields. Drainage gates control the level of water

25–35°C. In most tropically adapted varieties the optimal temperature for tillering and vegetative growth is about 30°C. Poor adaptation of some otherwise outstanding varieties such as 'IR 32' and 'IR 34' in Borneo has been attributed to the warm temperatures at night and it seems that these two varieties are best suited to somewhat higher latitudes. In the same region, varieties selected in the equatorial region of Malaysia such as 'Bahagia', 'Sri Malaysia 1' and '2' are extremely well adapted. Lower night temperatures, down to about 20°C, are thought to increase yield somewhat because of lower plant respiration.

Most of the rice areas of the Tropics depend on rain for irrigation, the crop being planted in the wet season. Under these conditions the onset of the rains is a critical factor. The uncertainty in the time of the rains, particularly in monsoon areas, means that the transplanting method of planting is mostly used because it offers more flexibility in planting out time. Plants can be kept in the nursery until there is sufficient water for planting to the fields. In rain-fed cultivation of rice the depth of flooding is also very uncertain. The fields are filled when flood waters rise and the flood waters are impounded by means of bunds. Because flood waters may rise too high and because the farmer tries to retain as much water as possible against the possibility of drought later in the season, this type of rice growing requires rather deep flooding. Unfortunately many of the higher-yielding tropical rice varieties require more controlled water use and do not do well with very deep flooding, and in fact, this problem of water control has been a limitation in using high-yielding varieties in many parts of the Tropics. The solution to this problem lies in irrigation which allows not only better water control during the wet season or seasons but also multiple cropping with up to three crops a year under tropical conditions. With newer high-yielding varieties such as 'IR 5' and 'IR 8' grown by the transplanting method a total of about 1 200 mm of water is needed to produce a crop. Many tropical countries are investing in irrigation schemes for all-year-round rice production. Fig. 13.9 illustrates such a scheme in Malaysia in which 25 000 ha is under triple cropping; short-maturity, improved, semi-tall *indica* varieties capable of yielding up to 8 000 kg of unmilled rice per hectare, are used, and average yields of around 6 000 kg per hectare may be obtained.

Moisture stress in the rice crop has devastating effects on yield and the most sensitive period is from the beginning of flowering to the beginning of grain-ripening. Since the aim of good agriculture is to produce the highest yields, the requirements of improved high-yielding rice varieties are considered on p. 157.

Soils Rice is grown in a wide range of soils in the humid Tropics and there is often little relationship between soil type, growth and yield. These factors are more dependent on variety, husbandry and fertiliser use than on soil type. In general, rice grows better on soil with some degree of drainage and lateral movement of water to prevent excessive reduction, than on very heavy soils. However, soils which are excessively well drained will be difficult to cultivate because of high water requirement and fertiliser leaching. Excessively drained soil will benefit from the addition of organic manures but very heavy clay soils should not be manured because they produce toxic reduction products from organic matter. Reduction is the process which takes place on flooding when oxygen is used up.

There are certain troublesome soils in humid tropical coastal regions which are sometimes used for rice culture. These include organic peat soils, saline soils and very acidic sulphur soils. Organic peat soils are usually friable and, if a good through-flow of water occurs, will not become excessively reducing. However, water control and efficient fertiliser use are difficult. Very acid sulphur soils cause lowered yields for two reasons. One is excessive acidity and if the pH is as low as 3·5 in the drained condition, prior to flooding, liming should be carried out to raise it to about 4·0–4·5. The other reason for low yields is the formation of hydrogen sulphide on flooding, a product which is toxic to plants. If these soils are very heavy, as they often are, then they may have to be temporarily drained once or twice during cultivation to prevent excessive reduction and hydrogen sulphide production. A dressing of manganese dioxide at 50 kg per hectare may also prevent excessive reduction. Certain *indica* rice varieties are more adapted to growth in sulphurous soils and peats, but new high-yielding varieties are also well adapted, mainly because they have more active roots towards the mid-period and end of the growth cycle which produce oxygen and delay reduction and formation of toxic products in the soil.

Saline soils contain sodium chloride and sometimes sodium hydrogen carbonate, so they are alkaline in reaction. Certain *indica* rice varieties are somewhat adapted to these conditions but such soils can never give very high yields.

Usually there is not much choice of soil in rice-producing areas and whatever soil is present must be cultivated. In transplanted rice culture, the puddling or cultivation of the soil to produce a thick even mud is important for development of the plant and subsequent yield.

Cultivation and management

Seed and nursery The nursery is used to raise seedlings for transplanting. For early-maturing varieties the nursery will last 2–3 weeks in the humid Tropics and should be prepared in advance of planting by this period. Since transplanting by hand may itself take some weeks, several successive nurseries may be required. About 500 m^2 of nursery is necessary to plant each hectare. The nursery beds are made about 1 m wide for easy weeding, etc., and as long as required, for the area to be planted. The nursery bed should be bunded and the soil puddled to a depth of 10 cm to an even consistency prior to planting. Use 50 g of urea and 50 g of a complete bag fertiliser (e.g. nitrophoska blue special) for each square metre, well mixed prior to sowing. The nursery should not be shaded. Only good seed of the desired variety from a reliable source such as a government agency or approved seed grower should be used. Seed should be treated with fungicide such as cerasan, soaked for 48 hours in clean water, drained and kept damp for a further 24 hours before sowing evenly on the prepared seed bed at a rate of about 100 g of soaked seed per square metre. The bed is kept surface wet during germination and the level gradually raised to a maximum of about 4 cm as the seedlings grow. Weeding should be carried out promptly and the seedlings sprayed with insecticide, etc. as necessary. Recently the insecticide carbofuran has been used for treatment of nursery soil.

When water control in the field is good, transplanting may be done when the seedlings are between 2 and 3 weeks old. Large seedlings may not establish so well with the newer high-yielding varieties. Large seedlings do better on acid soils. However, when water control is poor it may be necessary to keep the seedlings for an extended period in the nursery and in this case taller improved *indica* types should be grown.

A second type of nursery is prepared by sowing pre-germinated seeds on polythene sheets, old fertiliser bags, banana leaves or other material. For the first few days the germinating seeds are pressed firmly but gently on to the plastic or other surface using flat boards and are at all times kept wet. After the shoots grow the beds are kept covered with about 2 cm of water. This method makes raising of the seedlings for planting easier. In either type

fig. 13.10 Earth bunds 36–60 cm high are used to impound water for rice growing

fig. 13.11 One of the oldest and best ways of field preparation for the small farmer. The buffalo is drawing a wooden cultivating implement which puddles the soil into a thick mud down to about 20 cm

(a)

(b)

fig. 13.12 (a) Many small farmers in the Tropics are now using pedestrian tractors like this Kubota to prepare paddy fields. They are faster than the buffalo and more suitable for larger farms (b) These small tractors are very versatile; the one shown here has been fitted with a trailer to carry fertilisers and bags of rice

fig. 13.13 Transplanting rice seedlings. Here the rows have not been kept straight and inter-row weeding will be impossible using the weed-roller. Herbicides will have to be used to control weeds

of nursery the seedlings are lifted from the nursery a few at a time and stacked in bundles for transplanting. The roots should be kept intact for the best growth and not pruned.

Field preparation In preparing the field for planting, bunds of soil must be formed to keep the water on the field (Fig. 13.10). The soil should then be thoroughly puddled to a depth of at least 20 cm to mix in weeds and short stubble, level the area and break up soil clods (Figs 13.11 and 13.12). Straw from the previous crop should either have been burned or otherwise disposed of and should not be puddled in during field preparation as this will cause nitrogen starvation in the following crop. The only instance in which working in of rice straw is desirable is in excessively sandy soils when it is necessary to increase organic matter; if this is done extra nitrogen should be used at a rate of 500 kg of urea per hectare. Fertilisers should be incorporated just prior to planting.

Transplanting Transplanting should be done in straight rows to facilitate weeding between the rows. All too often transplanting is done very unevenly so that weeding is difficult (Fig. 13.13). Small groups, usually from 1 to 5 seedlings are planted at each point. If carefully done, two seedlings can be recommended and this is fairly economical in seed use. The correct spacing depends on variety, soil-fertility or fertiliser used and on the season. Improved high-yielding fertiliser-responsive varieties should be planted at 25–30 cm. In soils of poor fertility such as acid or saline soils a closer spacing of 20 cm should be used. This spacing can be recommended for 'IR 5', 'IR 8', 'Sri Malaysia 1' as well as the taller 'IR 32', 'IR 34', 'Bahagia', and 'Sri Malaysia 2'. Improved superior *indicas* can be planted at 30–35 cm on good soils and 25 cm on poorer soils.

Fertilisers Only urea or ammonium fertilisers should be used as sources of nitrogen on rice; never nitrate fertilisers. They should be incorporated just prior to transplanting and puddled well into the soil to the maximum depth. Fertiliser applied to the surface will be lost as nitrogen gas. Sometimes about 20 per cent of the fertiliser is kept for top dressing at the flowering stage, i.e. about 60 days after transplanting, but generally the best results are obtained by fertilising just before transplanting.

In general, nitrogen is the fertiliser which produces the most yield response for often there is plenty of phosphorus in flooded soils and the supply of potassium and other nutrients is sufficient in the irrigation water. However, sometimes dressings of complete fertilisers may also be necessary for optimal yield. The amount of nitrogen needed depends on the variety and also on the season; whether high sunshine or low. Table 13.5 indicates suitable nitrogen fertiliser levels per hectare.

Fertiliser recommendations based on the use of urea and a complete bag mixture are shown in Table 13.6. These can be recommended for situations in which little is known of the soil fertility conditions.

Irrigation When irrigation water is available, the depth of water in the field should be kept at 5–10 cm. When irrigation water is not available, deep flooding is sometimes necessary to allow sufficient water for development of the crop; tall varieties should be used. Under conditions of high rainfall the level of water in the irrigation system (as in Fig. 13.9) may be controlled by means of drainage gates and the water fed to the fields by gravity. When the water in the irrigation channels is lower than the field level then water must be pumped into the fields using diesel pumps (Fig.

Table 13.5 Levels of nitrogen fertilisers for rice (rates in kg per hectare)

Variety type	Dry season		Wet season (cloudy)	
	Urea	Ammonium sulphate (VI)	Urea	Ammonium sulphate (VI)
High-yielding types (e.g. 'IR 5', 'IR 8', 'IR 32', 'IR 34', 'SM 1', 'SM 2', etc.)	150	200	75–100	125–175
Improved tall *indicas* (non-lodging)	50	100	25	50
Traditional leafy *indicas* (poor soils)		40–80		0–40

Table 13.6 Fertiliser recommendations for rice varieties in different soil conditions for the main (wet) season. (Rates in kg nitrophoska blue special (nbs) and urea per hectare. *(Source:* Annual Report, Brunei Dept. of Agriculture.)

Soil	Fertiliser	Swamp rice varieties			Hill rice varieties
		Improved varieties	Non-lodging local indicas	Lodging local indicas	
Good soil conditions	nbs	60	60		35
	urea	120			
Intermediate	nbs	120	120	120	60
	urea	120	45		35
Poor soil conditions	nbs	120	120	120	120
	urea	180	67	20	35

13.14). When drainage is necessary to improve soil aeration, this should be done during the active tillering phase, not before about 30 days or after 60 days. Fields should be drained for about 3 days to ensure good aeration, and then reflooded.

Weed control Flooding will control most dry-land weeds but sedges and water plants will grow. Planting in straight rows will enable early weed control using the roller shown in Fig. 13.15. Broad-leafed weeds may be controlled by spraying with MCPA or 2:4-D at 1 litre of commercial preparation in about 400 litres of water per hectare. The spray should be concentrated between the rows. An effective pre-emergent herbicide for flooded rice fields is TOK or CNU which is used in granular form at a rate of 25–30 kg per hectare with an active ingredient content of 7 per cent. It should be broadcast a few days after transplanting. Another very effective pre-emergent herbicide is amiben, sprayed at the rate of 1–1·5 kg per hectare.

Harvesting and processing

The most time-consuming operation in the cultivation of rice is harvesting and threshing and this must be done fairly quickly to obtain a good grade of rice. Generally harvesting is best done about 25–35 days after heading and the field should be drained gradually over some 10 days prior to harvesting, beginning when the grains begin to yellow.

Throughout the Tropics harvesting is usually done by hand labour. In the most common method the rice is cut in clumps, and carried in bundles for threshing. Fig. 13.16 shows one of the best methods of threshing by hand. Sometimes powered or hand-operated threshers are used (Figs 13.17 and 13.18). The grain is removed by the threshing operation and after this it is winnowed to remove the chaff. This may be done by pouring the chaff and grain mixture on to a mat; in the presence of a light wind the chaff is blown away. Winnowing machines are also used such as the one shown in Fig. 13.19.

fig. 13.14 A bank of diesel pumps for pumping water from the irrigation channels to the field of rice

fig. 13.15 Weed-roller to control weeds in inter-row areas in rice fields

fig. 13.16 Threshing rice by hand in Malaysia. The rice plants are threshed against a grid or bars running across the inside of the tub. The net prevents loss of grains

fig. 13.17 A pedal-operated thresher. The sheaves of rice are held against the wire tynes on the drum as it rotates

fig. 13.19 A hand-operated winnowing machine. The circular drum contains rotating fan blades. Paddy is fed through the hopper and chaff is blown out at the front (left)

fig. 13.18 A small, powered thresher. Sheaves of rice are fed into the toothed groove (centre right) and are ejected at the opposite side. A fan blows away chaff

fig. 13.20 A small-scale rice mill, owned by a farmer's cooperative in Borneo. The rice is milled by the unit on the left, next to the operator. The unit on the right polishes the milled rice if required

fig. 13.21 A small rice mill

Drying must be done promptly to prevent spoilage of the grain. It is done by spreading the grain in the sun, usually on mats, or in grain dryers.

Generally hand harvesting will occupy a family for several weeks to complete 2 ha of good rice.

Milling of rice which is grown for sale is often done in large mills which will serve several thousand hectares of land; small village mills are also widespread throughout rice-growing regions (Figs 13.20 and 13.21). Rice which is milled without polishing has the highest nutritional value and is recommended in preference to polished rice.

Protection

'Blast' which is caused by the fungus *Piricularia oryzae* is probably the most widespread rice disease in the humid Tropics and high humidity and nitrogen fertilisation favour the disease. Resistant varieties occur but there are many strains of the fungus so that resistant varieties are usually effective only in the region where they were bred or selected. Local varieties usually have better broad-spectrum resistance than improved, high-yielding varieties which need good hygiene and protection. Use of clean seed and dressing with a mercurial fungicide can effectively control blast. The use of high levels of potassium fertiliser reduces the disease. Treatment of the soil with silicon as 100 kg calcium(II) magnesium(II) silicate(IV) per hectare may reduce the disease. Diseased plants should be removed as soon as they are observed. The leaves of diseased plants in the early vegetative stages become covered with spindle-shaped spots.

Brown spot caused by *Helminthosporium oryzae* is another widespread rice disease. It is also more severe under conditions of low potassium nutrition. The symptoms may be seen in the young stages in diseased plants. Leaves develop round or oval spots which are dark brown or purplish brown in colour. Resistant varieties should also be used where the disease is severe and mercurial seed treatment is carried out.

Cercospora oryzae leaf spot also occurs in rice, producing elongated spots which reduce yields. This disease may be controlled by planting resistant varieties and removing diseased plants.

Bacterial leaf diseases caused by *Xanthomonas* also occur in humid locations both in the nursery and field. The diseases are promoted by deep flooding. Copper and mercury compounds may be used in the nursery.

The chief insect pests of rice are stem borers which may be controlled by use of insecticides. Granular carbofuran is now widely used to control borers and should be applied at 30 and 60 days. Other pests of rice are birds and rats. Birds appear at the time of grain ripening and are frightened away using various types of scares including cloth on strings, rattles made with stones in tins and so on. They can cause great losses in areas near forest and are particularly damaging at the 'milt' stage prior to grain ripening. Recently it has been found that ammonium hydroxide (1 per cent solution) dispersed with a motoblower can control birds. Rats are best controlled by baiting using anticoagulant poisons. In rat baiting a whole area must be treated at once and cooperation between farmers is essential. It is not satisfactory if only a few farmers do baiting because rats will simply move in from surrounding farms.

14 Industrial and miscellaneous crops

Major crops

Rubber *(Hevea brasiliensis)*

Family Euphorbiaceae

Other names para rubber, india rubber (general, English); caoutchouc (French); seringuereira (Portuguese); caucho (Spanish)

Almost all commercial rubber is produced from the para rubber trees and, although the tree originated in tropical South America, the most important region producing natural rubber is Southeast Asia, as shown in Fig. 14.1. Cultivation of rubber in tropical America is very seriously limited by the occurrence of South American leaf blight, caused by the fungus *Dothidella*. However, the potential for the expansion of rubber production in the humid zones of West Africa is also great. At the present time, natural rubber is sold in competition with synthetic rubber, but the synthetic rubber industry depends on the crude oil

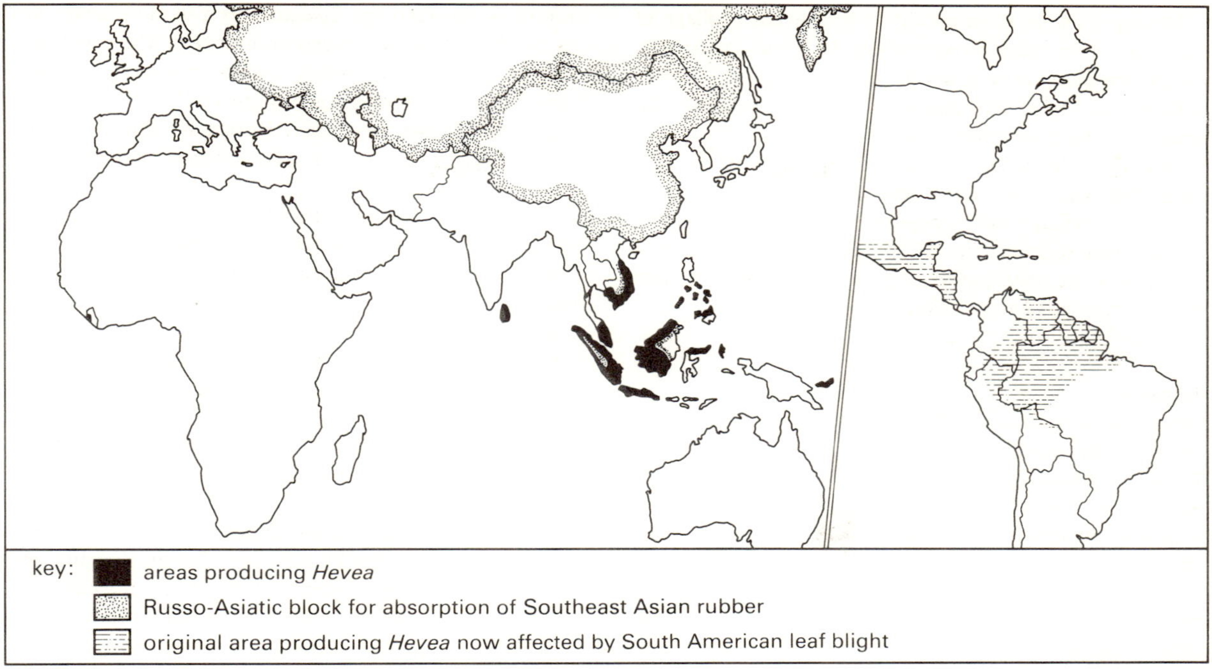

fig. 14.1 Areas producing natural rubber at present and areas where production is affected by disease. (*Source:* **Djikmar, M. J.** (1951). *Hevea.* University of Miami Press: Florida.)

from which it is made, and with declining world oil reserves the future of natural rubber seems promising.

At the present time most of the knowledge concerning the planting of rubber comes from Southeast Asia. The rubber tree when mature is fairly large (Fig. 14.5) and it is grown either from seedlings or from buddings. There are many different varieties and clones of rubber which have been bred for particular regions. Two of the most widely planted clones in Asia, from Sri Lanka to Vietnam, are the Indonesian clones 'GT 1' and 'PR 107'. Both these clones are highly wind resistant, tolerant of rather drier conditions and resistant to most of the rubber diseases that occur in Southeast Asia. Loss of leaves at the time of the dry season (called wintering) is not so severe with these clones and yields of more than 2 000 kg per hectare can be obtained with correct planting, fertilisation, weeding and so on. Other clones which have performed well are 'PB 5/51', 'PB 28/59' and 'PB 86' (all from Malaysia) and 'LCB 1320', 'PR 255' and 'PR 261', although 'PR 255' is fairly susceptible to wind damage and 'PB 5/51' is susceptible to *Oidium* leaf disease. Important Malaysian clones are 'RRIM 527', 'RRIM 600', 'RRIM 623' and 'RRIM 628'. All of these are capable of yields of 2 000 kg per hectare or above under good conditions. However, most of the Malaysian 600 series are susceptible to wind damage. A Malaysian clone 'RRIM 501' was high yielding and at one time very widely planted, it is no longer planted because of severe susceptibility to wind damage when tall. A new, promising but not yet widely-tested Malaysian clone is 'RRIM 703'. A Sumatran clone, 'AVROS 2037', has also been found to be high yielding. A clone which is much planted in Sri Lanka is 'PB 86' which has the advantage of being tolerant to *Oidium* leaf disease as has the Sri Lankan clone 'RRIC 102'. All the above are recommended to be planted as clonal material for budding on to suitable stock seedlings. Sometimes clonal seedlings, that is seeds produced in isolated clonal plantings, which are presumed to be genetically pure, are planted instead of budded seedlings. Two clonal seedlings still recommended for

fig. 14.2 Young rubber seedlings a few months after planting. Strip weeding has been carried out and the inter-row areas planted with cover legumes *Stylosanthes, Pueraria* and *Calopogonium*

fig. 14.3 Severe wind-damage in a susceptible rubber clone, 'RRIM 501'. This clone has heavy top branches which make it unstable

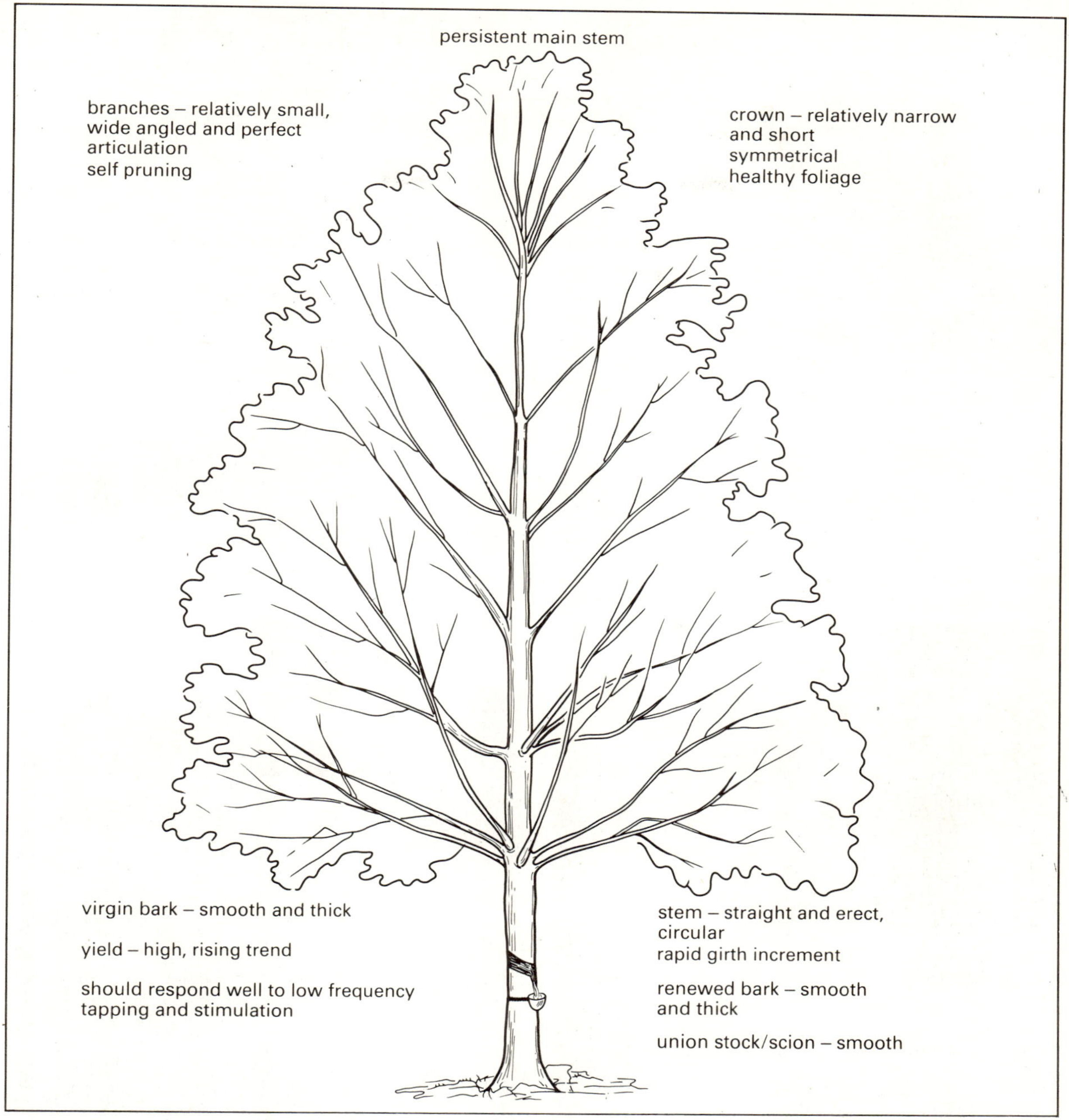

fig. 14.4 Characteristics of the 'ideal' rubber tree. (*Source:* **Sekhar, B. C. and S. Subramaniam.** (1972). In: *Third international symposium for sub-tropical horticulture,* Bangalore.)

planting are 'PBIG/GG 1' and 'PBIG/GG 2', both from Malaysia. Planting recommendations change from time to time and the prospective rubber planter should always seek the latest information from appropriate sources such as research stations, experienced planting organisations and so on.

In tropical America, the fungus leaf disease *Dothidella* greatly limits rubber production and a standard technique for growing the crop there is by crown budding. This involves the budding of a diseases-resistant crown on a normally budded tree to produce a three-part tree consisting of a seedling rootstock, a trunk from a high-yielding clone and a crown of the resistant variety. Resistant crowns which have been used in tropical America are 'FX 25', 'FX 4037', and others.

All rubber must either be planted from seedlings or from buddings on seedling rootstocks because these have strong wind-resistant root systems. Cuttings are unsuitable for rubber propagation because the superficial root system cannot support the tree when fully grown and because the trunk does not develop well.

Botany

Rubber seedlings (Fig. 14.2) produce successive flushes of leaves, the lower ones fall off as they age and the main stem increases in height. Later the main branch system is formed which varies considerably in different rubber clones. Fig. 14.3 shows a clone in which the upper branches are very heavy. This is undesirable because such trees are easily broken by strong winds. Fig. 14.4 shows a hypothetical rubber tree with a good branching habit and other desirable characteristics.

(a)

(b)

fig. 14.5 (a) Old rubber trees grown from seed. Note the tapered trunk characteristic of rubber trees grown from seed (b) Budded rubber trees. Budded trees have a uniform cylindrical trunk and tapping panel

The nature of the trunk and bark is very important in rubber because it is this which is tapped to obtain the rubber latex. Trees grown from seed differ from budded trees in the shape of the trunk as shown in Fig. 14.5. Budded trees have a more even trunk which does not taper as sharply as that of a tree grown from seed.

On tapping, the latex comes from the bark of the tree; in high-yielding rubber trees the bark is thick and has a large latex system. Figs 14.6 and 14.7 illustrate the nature of the bark and the trunk area from which the latex is obtained. The latex vessels are actually formed by the cambium and consist of lacy cylinders of cells fused into an intricate system of tubes which contain the latex. The process of tapping cuts into this lactiferous tissue and the rubber-containing latex exudes from the cut.

An important feature of a high-yielding rubber tree, as indicated in Fig. 14.4 is that the rate of growth of the trunk (or the girth increment) is not greatly reduced by tapping. If girth increment is not high the trunk will be too thin and susceptible to wind breakage, and also the area of bark to be tapped will be small.

Rubber trees grown from seedlings or buddings have substantial tap roots and spreading roots.

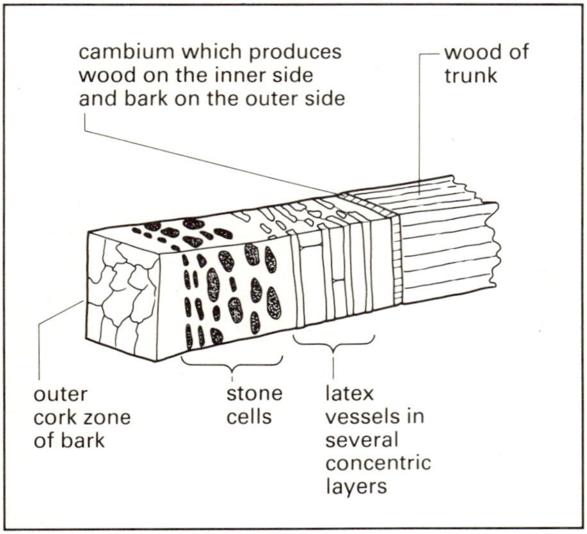

fig. 14.6 Section of the bark and outer wood of the rubber tree. The latex vessels occur near the cambium which is the layer of cells which gives rise to all the tissues of the trunk. In tapping, the cambium must not be disturbed by cutting too closely since this will prevent regeneration of new bark

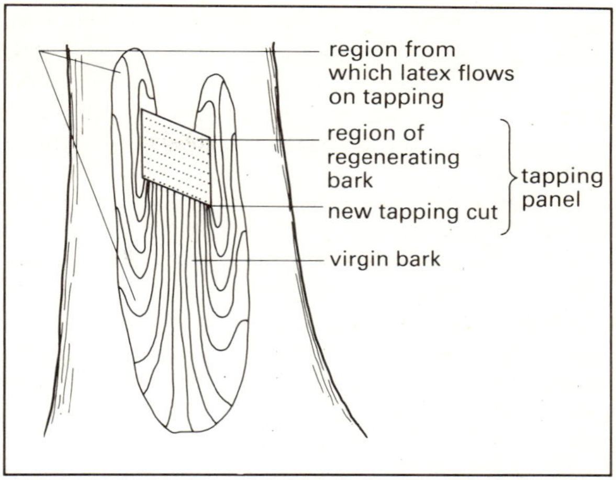

fig. 14.7 Tapping panel showing the region of the trunk from which latex flows on tapping. (*Source:* **Djikman, M. J.** (1951). *Hevea.* University of Miami Press: Florida.)

Flowers of the rubber trees are produced irregularly or in certain seasons. They occur in small panicles with the female flowers towards the end. Flowering is unimportant to the yield of rubber but seed production is necessary to produce rootstocks for budding. Seeds may also be a by-product since they yield an oil and the cake can be used as a livestock feed.

Climate and soils

Rubber cultivation is largely confined to the humid Tropics as can be seen from Fig. 14.1, and largely also, because of disease, to the Asian and African Tropics. For the highest yields there should be no dry season with a minimum rainfall of 1 500 mm, well distributed. However, excessive rain and humidity also affects rubber production because of interference with tapping schedules and also because the occurrence of leaf diseases is worse under humid conditions. Dry seasons produce a condition known as 'wintering' (Fig. 14.8) in which leaf fall occurs and yields drop to very low levels for several weeks or months depending on the degree of wintering. The yield decline is associated with the period of new leaf development. Clones differ in the extent of wintering and clones in which wintering is only slight such as 'GT 1' and 'PR 107', are recommended for regions with a marked dry season.

fig. 14.8 Many rubber clones show an annual loss of leaves called 'wintering'. This usually coincides with a dry season. During wintering latex yields decline and the trees are not tapped

Rubber cultivation is attempted over a fairly wide range of latitudes, up to about 23°N and S in Indo-China and Brazil respectively but low temperatures greatly delay maturity, or the time between planting and the opening of tapping. A drop in mean temperature of 1°C results in a delay in maturity time of up to six months and an increase in elevation of 300 m may result in a delay in maturity of about eighteen months. Consequently rubber does best in the equatorial lowlands.

Strong winds cause great damage to rubber (Fig. 14.3) but again there are major clonal differences in susceptibility, for instance clones 'GT 1' and 'PR 107' are relatively wind resistant.

A great deal of rubber is grown on well-drained inland tropical soils. Under conditions of high humidity and rainfall such soils are to be preferred. Soils of higher water-holding capacity are to be preferred under sub-humid conditions. In alluvial low-lying areas field drains of 1 m depth should be made. Much rubber is grown on peat soils in Malaysia but the trees tend to lean and shrinkage of the peat exposes the roots.

Rubber tolerates acidic soils and liming is not normally carried out unless the pH is below 4·0.

Cultivation and management

The nursery seedlings Rubber may be planted from seed or as budded material on a rootstock. Budded material gives a more uniform mature stand of trees. In either case the most convenient type of nursery to adopt is the polythene bag nursery. Generally rootstock seedlings and seedlings that are to be planted as unbudded trees are grown from clonal seed of a recommended clone. Polythene bags of size about 25 cm by 40 cm high, when laid flat, filled with a suitable soil are used for rubber. A rather heavier soil, adequately drained is more suitable than an excessively sandy soil. Organic manure may also be mixed in for rubber seedlings if available, at a rate of about 150 g per bag. Seeds should be planted lying flat at a depth of about 2 cm. If the seeds are not planted flat the main tap root will be twisted.

For a ground nursery, a site with a fairly heavy type of soil should be chosen and ground rock phosphate worked into the top 20 cm of soil at a rate of 400 kg per hectare and ground magnesium limestone at a rate of 100 kg per hectare. Mulching with grass or other material is advantageous in a ground nursery. After about 4 to 6 weeks, when germination has occurred, manuring with an

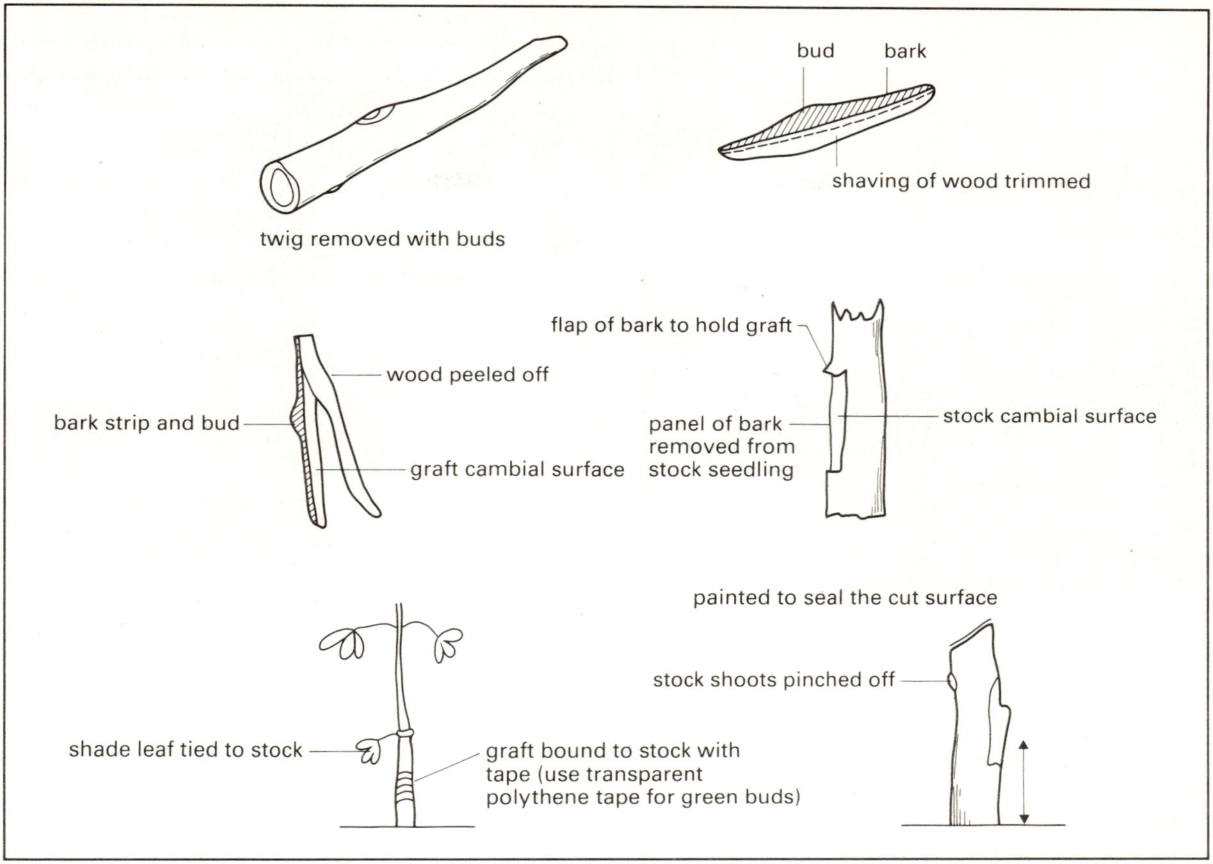

fig. 14.9 Details of patch-budding of rubber. The sliver of wood of the patch is very carefully peeled off to prevent rupture of latex vessels and exudation of latex on to the cambial surfaces

N:P:K:Mg compound fertiliser at a ratio of about 18:10:12:2 should be done, at a rate of 5 g per plant every 3 weeks, up to 3 weeks before planting out.

Budded stumps Budding of rubber seedlings is now mostly carried out at the green-wood stage a few months after planting when the stem is about 15 mm thick. Patch budding is used for rubber; the buds are obtained from multiplication nurseries. A small panel of bark is removed leaving a flap at the top to hold the bud-patch, the bud is inserted and bound in place with transparent polythene tape as shown in Fig. 14.9. After a few weeks, if successful, the bud union should have occurred and with care a high percentage of success should be obtained. After a further week or two the stock is pruned back a few inches above the bud-patch and the bud should begin to grow. At this stage, and until the budded scion has formed a main shoot, any other buds developing on the stock stump should be pinched off. Budding should only be done when the plants are in a vigorous state of growth and not during adverse weather conditions. Buds from the maintenance nursery should only be taken during active growth and are best when the top whorl of leaves has just matured.

Planting out can be done after a few months when several whorls of leaves have formed and should be carried out during favourable weather conditions when the

fig. 14.10 Bedded rubber seedlings in a multiplication nursery. The seedlings have been pruned to produce green bud wood

youngest whorl of leaves has just hardened. With ground nursery seedlings, lifting must be done using a core extractor. In due course the developing bud becomes the main aerial portion of the plant.

Stumped buddings Stumped buddings are simply overgrown buddings which may have reached a girth of about 15 cm near the base. They are normally used for replacement planting of buddings which have failed in the field, and are prepared usually from ground-nursery seedlings. They are extracted from the ground, trimmed of side roots to form a club-like base, topped and the base is white-washed for protection. Because of the large size and high content of food reserves in such buddings, they normally establish very quickly in the field. Stumped buddings may also be recommended for planting in remote areas where field husbandry is poor.

Multiplication nursery The multiplication nursery is maintained to supply bud wood for the production of budded stumps and is itself planted from buddings of the desired clones. Spacing is wider than in a conventional nursery: 1–2 m in rows, and the plants are pruned as shown in Fig. 14.10 and maintained for as long as required. One hectare of multiplication nursery should provide enough buds for planting about 1 000 ha of rubber per year. When bud wood is cut, a few buds at the base of the shoot should be left to allow branching. For green bud wood the main stem is allowed to produce 4 to 6 secondary branches at any one time and each of these produces about the same number of green-wood shoots which are used as the source of bud wood. Multiplication-nursery plants are manured with a similar mixture as for buddings but at a rate of 25 g per plant per month. However, fertilisation should be stopped about three months prior to taking bud wood.

Crown budding Crown budding is mainly used to produce a three-part tree for the purpose of disease or wind resistance, and in experimental plantings to investigate higher-yielding combinations. Crown budding is carried out in the nursery when the budded stump has reached a height of about 3 m. Budding of the crown bud is done in the same way but to prevent the crown bud from growing too rapidly, producing a top-heavy tree, one or two of the trunk shoots are also allowed to grow for a period and later pruned off.

Covers Cover crops or legume intercrops are normally planted with rubber. Legumes should be fertilised with ground rock phosphate on acidic soils at a rate of 100 kg per hectare or with superphosphate on soils of pH above 6·0. After germination a light dressing of an N:P:K:Mg mixture at 50 kg per hectare should be given along the planting lines. At one month 100 kg each of a P:K:Mg mixture of ratio about 2:1:0·5 should be applied on poor soils. On newly-cleared and burned land a top dressing of 50 kg of phosphate only is given. Cover crops may require further similar dressings at six-monthly intervals as long as they continue to survive. Covers and crop legumes should be inoculated with a suitable *Rhizobium*. The most suitable crop legume for rubber is the cowpea (*Vigna unguiculata*), particularly the cultivar 'TVu-335' from Nigeria.

Planting and fertilising Rubber is planted at densities between about 320 and 480 trees per hectare. Generally a higher density is more suitable for smallholders using family labour and a lower density for estates in which hired labour is used. This is because the yield per tree, is higher at the lower density, and a hired tapper can tap a larger volume of latex. A higher density will give higher yields in the first few years of tapping. If interplanting of the rubber with legume crops such as cowpeas is intended, then an avenue planting arrangement of trees in 10 m rows planted at 2–3 m within the rows can be recommended; if not, then a more rectangular arrangement of about 4 m by 6 m should be used as this will give more rapid ground shading and reduce weeding costs during establishment.

Planting is carried out by digging appropriately-sized holes and firming the seedlings to prevent wind lodging.

Recommendations for fertilising young rubber up to the time of commencement of tapping are:

1 For acid latosols, assuming that a legume cover or crop legumes are grown, about 100 g of ground rock phosphate in the planting hole. Subsequent fertilisation is best done at about two-monthly intervals along the weeding strips, using N:P:K:Mg mixtures of the following ratios: years 1–3, 15:15:10:2; years 4–6, 18:10:12:2, at the rates shown in Table 14.1.

2 In windy locations or on shallow soils or peat soils the fertiliser for 4–6 years should have a lower nitrogen level and a mixture of ratio 15 : 10 : 15 : 2 is suitable.

For rich coastal clay soils nitrogen is the main fertiliser required, this may be supplied in the form of urea at rates increasing up to 150 kg per hectare per year, given at two-monthly intervals. Sometimes a dressing of potassium chloride may also be required on such soils. For wind-susceptible clones and windy locations the quantity of urea should be halved. For mature rubber, manuring should depend on foliage analysis data. If such advice is not available mixtures of N:P:K in the range 15:10:15 to 15:7:18 to 12:6:22 may be used (the higher potassium mixtures being more suitable for windy locations and wind-susceptible clones) at rates up to 150 g per three months per tree, depending on soil status, latex yield, rubber price and so on. In mature rubber, magnesium fertiliser and lime are not recommended unless deficiency occurs, as these encourage latex coagulation.

Table 14.1 Total yearly applications of fertiliser in rates of g per 2 months per tree

	Poor inland soils previously cropped	Good inland soils newly cleared and burned
Year 1	15	8
Year 2	30	15
Year 3	100	50
Year 4	150	75
Year 5	200	100
Year 6	200	100

Weed control Rubber is normally strip weeded rather than circle weeded (Fig. 14.2) and the inter-row spaces are planted with cover legumes or cropped. In newly-planted rubber weeding is carried out by hand but when the canopy has elevated and the stem has reached the brown-wood stage, herbicides can be used for strip weed control. Traditionally sodium arsenate(V), MSMA and DSMA have been used as contact herbicides in strip weeding of rubber using a directed spray jet such as the 'polijet'. Paraquat and sodium chlorate(V) now often replace arsenical herbicides. For long-term pre-emergent weed control in the weeding strip, diuron or simazine at a rate of 1 kg per hectare may be used but higher concentrations of diuron are toxic to rubber seedlings. If grass weeds predominate, dalapon at 10 kg per hectare may be used, as a directed spray; glyphosate (roundup) may also be used and gives a long period of control.

In mature rubber, the dense canopy greatly reduces weed growth and weed control normally consists of periodic spot weeding of invading weeds using the above herbicides or using rubbing oil for grasses.

Harvesting and processing

Tapping Tapping is certainly the most exacting

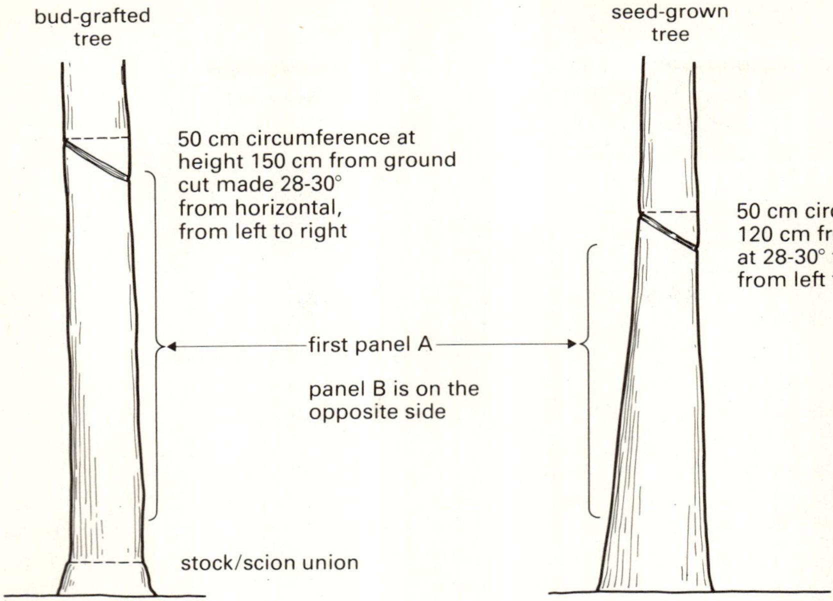

fig. 14.11 Opening of tapping in bud-grafted and seed-grown rubber trees

operation in rubber production. Trees are opened for tapping when they reach a girth (circumference) of 50 cm at a height of 150 cm from ground level in the case of budded trees. Trees grown from seed have more tapered trunks and these are usually opened at a height of 120 cm when the girth has reached 50 cm (Fig. 14.11). Depending on the standard of planting and husbandry, the climate and the clone used, the time of opening will generally be between 4 and 7 years from planting. The opening tapping cut should be carefully made by a skilled tapper at an angle of 28–30° from left to right as shown in Fig. 14.11. The most common method of tapping is the half spiral which taps half the circumference of the trunk.

Tapping must be carefully done to conserve the bark and also to allow good bark regeneration. Tapping is carried out in the very early morning when the trees are turgid because latex flow is best under these conditions. The usual time of tapping is between about 0500 and 1000 hours and a good tapper may manage about 400 trees in this time. However, tapping is not carried out on rainy days or when rain has occurred on the previous evening because this causes much spilling of latex from the cut as shown in Fig. 14.12. In tapping, a thin layer of bark is removed keeping the cut at right angles to the bark surface, see Fig. 14.13, and the latex guided down a vertical groove to a thin metal spout inserted in the bark just above the cup, as shown in Fig. 14.14. The cup may be held below the spout by means of a spring wire surrounding the trunk. Correct tapping should use up about 18 mm of bark per month when tapping is carried out on alternate days. The rate of bark consumption can be shown by painting a small spot on the bark each month (Figs 14.14 and 14.15). The depth of tapping is also important, if too shallow, latex yield will be low but if too deep, the cambium will be disturbed (Fig. 14.6) and regeneration of the bark will be poor.

When a cut is newly made latex oozes from it and flows into the cup. At first the flow is fairly rapid but after an hour or two flow slows down and plugging of the exposed latex vessels at the cut surface occurs. Collection of the

fig. 14.12 Tapping is not carried out in wet weather because latex spills from the cut when the surface of the tree is wet. Note the metal spout used to channel latex to the cup

fig. 14.13 Removing a thin layer of bark during tapping

fig. 14.14 The tapping cut and latex flow channel to metal spout and cup. The second cup is used on retapping

fresh latex into pails is made at the end of the tapping task and the cups replaced to collect any after-flow that occurs. This after-flow is usually collected on the next tapping occasion by which time the latex has usually solidified into a lump.

With good tapping the first tapping panel is consumed over about 5 years. The half spiral may be tapped every 2 days, minus any rainy days. If daily tapping is attempted a one-third spiral may be recommended, that is, one-third of the circumference of the tree so that the tree

fig. 14.15 Tapping higher on the panel with the re-curved type of knife. Spots are painted on the tapped panel at monthly intervals and indicate the rate of consumption of bark

is divided into three vertical panels. When the rubber grower is not entirely dependent on rubber for a living, as in mixed farming, tapping for profit is best undertaken by confining tapping to the periods when the rubber price is relatively high and resting the trees when the price is low. Then the full half spiral may be used on a daily system during the tapping period.

After the first panel has been tapped, at about 5 years, the second is opened which is on the opposite side of the tree in the half-spiral method, and when this is tapped after a further 5 years, bark regrowth should be sufficient on the first panel to return to this. If each of the lower panels is tapped twice for 5 years a tapping period of twenty years should be obtained after which new bark may be opened at a height of about 3 m using a short ladder. This upper tapping or ladder tapping is usually the last undertaken before replanting and is often tapped on a full spiral system and often daily because the trees are approaching the period when they should be replanted. This intensive tapping may, in fact, exhaust the tree and for this reason it has been called 'slaughter' tapping. In the first 20 years of the tapping life of a tree, however, when the lower panels are being tapped, care should be taken not to over-tap the tree since this will lead to dryness and a shorter productive life. Dryness is also caused by the incorrect use of stimulants.

Stimulation is the use of chemicals which promote the flow of latex and gives initially a higher latex and rubber yield. Continuous stimulation, however, if the same tapping schedule is used, soon overtaxes the tree and induces dryness and a condition known as 'brown bast' in which the renewed bark is corky and has none or very few latex vessels. For this reason stimulation is only recommended at present on old trees which are soon due for replanting. Stimulation may also be possible on younger trees when its use is accompanied by a lower intensity of tapping either by reducing the tapping frequency or reducing the size of the cut.

Several chemical substances induce stimulation of latex flow. Two of the most widely used are the herbicidal substances 2:4-D and 2:4:5-T which are produced in various commercial preparations usually containing about 1 per cent of the active ingredient in a neutral carrier such as palm oil. To induce stimulation an area of a few centimetres of the bark below the cut surface is scraped to remove the outer cork layers and the compound is painted on the scraped surface. Application will induce greater latex flow over a period of a few months. A newer stimulant is 'Ethrel' which is applied in the same way. The long-term effects of 'Ethrel' however are not so clearly established.

All stimulants should be used with care and should not be regarded as a means of obtaining a higher total yield of rubber but rather as a means of reducing the tapping task and the labour required in tapping. Properly used they should have a great effect on the future of the rubber industry as they provide the means of reducing labour intensity and paying higher wages to the tapper. Recent experiments have shown that extremely small cuts are possible with the use of 'Ethrel', which completely eliminates the problem of bark consumption. Stimulation is most likely to be useful at present on clones which have a high plugging index and cease flow early. **Plugging index** is defined as the percentage flow in the first five minutes of the total flow over the whole tapping period.

Preparation of sheet rubber In highly organised rubber industries, latex is handled by large factories which prepare it in a number of ways, of which the main are crumb rubber and latex concentrate. The former is a new type of prepared rubber that requires special machinery; the latter is latex which has had part of the water removed by centrifugation and is preserved with ammonia. For the small producer, however, located some distance from organised rubber factories, latex is best made into ribbed smoked sheet, which if well prepared, fetches a good price and is also easily stored and transported.

To prepare ribbed sheet the latex is placed in tin trays as shown in Fig. 14.16, of about 5 litre capacity, and methanoic (formic) acid solution (about 10 ml of a 10 per cent solution) is stirred in as a coagulating agent. After the rubber has coagulated it is removed from the tray, pressed to remove some of the water and then rolled in rollers of the type illustrated in Fig. 14.17. First it is rolled several times in the smooth roller and finally in a ribbed roller. The result is ribbed sheet such as that shown drying in Fig. 14.18. This ribbed sheet may be sold after sun-drying or further prepared by smoking in a smoke-house.

fig. 14.17 Smooth and ribbed rollers used to prepare sheet rubber

fig. 14.16 Newly-collected latex coagulating in trays after addition of methanoic (formic) acid. Bubbles are removed to produce high-grade sheets

fig. 14.18 Newly-made ribbed sheet drying in the sun

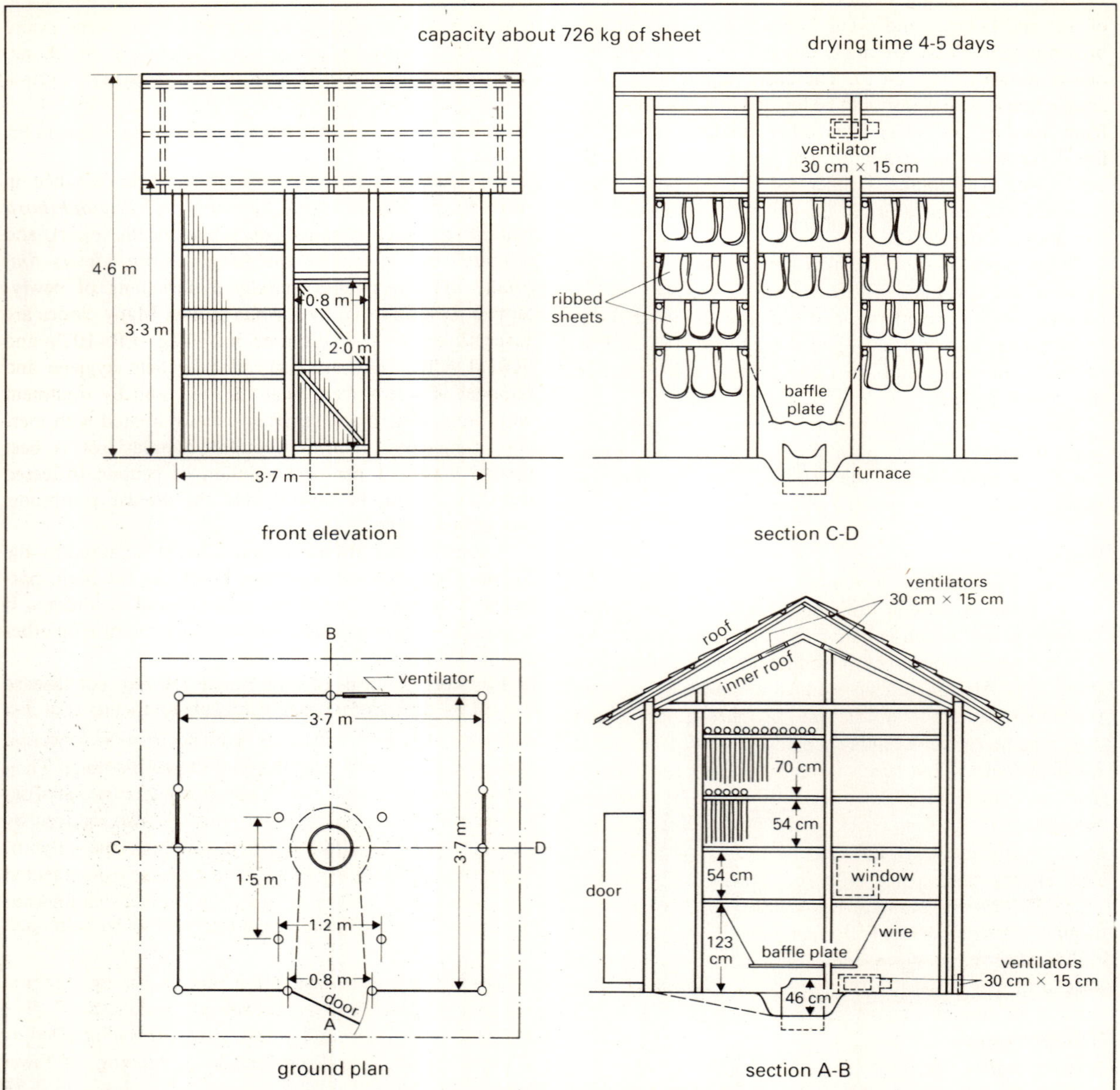

fig. 14.19 Design of a simple smoke house. (*Source:* **Edgar** (1937) (full reference not available)

For the best grade and highest price the sheet must be free of dirt and bubbles and evenly smoked to a clear amber brown colour. There are many designs for smoke houses; one is shown in Fig. 14.19. The basic requirements of a smoke house is that it should be located in a dry area, free from flooding and away from other buildings in case of fire. Except for the ventilators the drying area should be well sealed. The smoke is generated in a fire box or furnace and the temperature in the drying racks should not rise above about 40°C.

Other types of rubber which are produced in rubber growing are cup lump which hardens in the cups and is removed prior to retapping, and tree lace and other scrap from spillage and overflowing of the cut (Fig. 14.12). This rubber is normally sun-dried and sold at a lower price than smoked sheet.

The water from latex preparation contains numerous plant nutrients. This is normally discharged into drains, but if it is piped back to the trees the amount of fertilisation required may be greatly reduced.

Protection

Rubber is affected by a number of serious diseases. Of these, South American leaf blight caused by the fungus *Dothidella ulei* which is native to tropical America, is by far the most serious. The introduction of this disease to Asian and African rubber-producing countries has been prevented by strict quarantine procedures. In South America, protection of nursery plants is possible by use of fungicidal sprays but this is impractical in mature rubber. The only satisfactory way of establishing rubber in South America is by grafting resistant crowns, but even this has been largely unsuccessful because the fungus produces new strains that attack so-called resistant clones. The clone 'FX-25', for example, has lost its resistance to some strains of *Dothidella*. Field planting of mixed clones rather than large areas of a single clone reduces the spread of new strains of the fungus. The disease is most serious in humid regions since the fungus requires a humidity of over 95 per cent for a period of at least 10 hours to cause infection.

Two important fungal leaf diseases occur in Asia; one caused by *Oidium heveae* and another by *Phytophthora*. The latter is more confined to regions with a monsoonal climate, including Sri Lanka, south India and Thailand and the former to very humid regions and at higher elevations. *Oidium* occurs most frequently after wintering when the new leaves become infected. *Phytophthora* infects the young seed pods and spreads to the leaves and elsewhere. Both diseases can be controlled to some extent by sulphur spraying and dusting. Sulphur is commonly used for *Phytophthora*, with aerial spraying. Certain clones are resistant to *Oidium*; 'PB 86' and 'RRIC 102', for example, are recommended for elevated regions in Sri Lanka.

Two tapping-panel diseases of some importance in rubber are black stripe, caused by *Phytophthora palmivora*, which produces dark lines on the panel, and mouldy rot caused by the fungus *Ceratocystis fimbriata* which produces initially a darkening of newly-tapped-bark followed by a white mould. Many clones are susceptible to these diseases including 'PR 107', and 'RRIM 600'. They are controlled by field hygiene and removal of weeds from near the tree, and by treatment with fungicides. Black stripe is usually treated with mercury fungicides or with captafol. Mouldy rot is best treated with a 1 per cent solution of benlate. Affected trees should not be tapped until the disease symptoms have disappeared.

A stem disease known as pink disease is caused by the fungus *Corticium salmonicolor*. It attacks the bark, particularly at branch junctions, under humid conditions. It is satisfactorily controlled by Bordeaux mixture or other copper sprays.

Two root diseases of importance are red-root disease caused by *Ganoderma pseudoferreum* and white-root disease caused by *Fomes lignosus*. Both occur in old roots and stumps in the soil and from these infect new plantings. They are best controlled in replanting by thorough removal of old rubber stumps during clearing. *Fomes* is also reduced by dusting the planting holes of new seedlings with sulphur. Isolation trenches are used to prevent spread from infected trees and the main roots and collar of *Ganoderma*-infected trees may be treated with a preparation of drazoxolon dissolved in grease.

A physiological disease of the bark caused by over tapping and stimulation is known as 'brown bast'. The renewed bark becomes dry and lumpy. It is controlled by resting the trees and prevented by tapping at lower intensity.

Rubber is not generally affected by insect pests at epidemic levels. Termites sometimes occur and are treated by spraying around the trunk with dieldrin or a similar insecticide. Cockchafer grubs sometimes occur in the soil. They may be controlled with endrin, poured into holes in the ground.

Minor crops

Castor (*Ricinus communis*)

Family Euphorbiaceae

The oil from its seed is a purgative but is mostly used in industry, e.g. for making plastics, soap, aeroplane lubricants, paints, etc. An insecticide is extracted from the leaves.

fig. 14.20 Leaves and flowers of castor

fig. 14.21 Leaves and fruit bunch of castor

Botany and propagation

Castor is an erect annual plant 1–3 m tall. It is the only species of the genus *Ricinus*. The stems may be green or reddish and have many leaf scars; they become hollow when mature. Leaves also have long hollow stalks and large laminae 10–75 cm across with 5–11 lobes and saw-teeth margins (Fig. 14.20). Flower bunches appear at the top of the plants and consist of unisexual flowers; yellowish male flowers at the base and green, spiny female flowers on top. Fruits (Fig. 14.21) are dark brown when ripe and consist of three chambers containing three seeds which look like smaller rubber seeds. The seeds contain 35–55 per cent oil.

Castor is usually grown from seeds directly sown in the field.

Climate and soil

About 60 per cent of the world's castor is produced in Brazil and India. It grows well in the humid Tropics, particularly in places of lesser rainfall, as too heavy rainfall and waterlogging reduce growth and yield. It needs a rich and moist but well-drained soil for successful production.

Cultivation and management

There are many varieties of both wild and cultivated castor, varying in plant height, colour of stem and seeds, bean size, etc. Preferred varieties do not shatter when ripe and mature evenly. The tropical tall types are planted at spacings ranging from 90 cm by 60 cm to 150 cm by 120 cm. Seeding rate is 10–20 kg per hectare. Tall cultivars are often topped to produce more branches. Complete fertilisers should be applied as the crop has high nutrient requirements. The bunches of fruits are cut manually when ripe. Yields of up to 900 kg of seeds per hectare are obtained. Seed damage should be avoided as this causes the oil to go rancid easily. The meal after oil extraction contains poisonous substances and should not be fed to livestock; it is a useful fertiliser.

Protection

Leaf spots which also attack the capsules sometimes occur and may be controlled by spraying with fungicides. Attacks of leaf and capsule caterpillars should be controlled by spraying with insecticides, e.g. malathion. Weeds should be kept under check until the crop is 60–90 cm tall but cultivation should be shallow because the roots are near the soil surface.

Gambier (*Uncaria gambir*)

Family Rubiaceae
Other names white cutch (English)

Gambier is a tanning material obtained from the leaves and branches of the crop. It is also used as a dye for silk and as a masticatory together with betel leaf.

Botany and propagation

It is a large, woody climbing shrub about 3 m tall, with spear-shaped leaves in opposite pairs (Fig. 14.22).

Propagation is usually from freshly gathered seeds sown first in a shaded nursery.

Climate and soil

The crop grows quite well in the humid Tropics. Sandy to medium loams rich in organic matter or virgin jungle soil appear to be the most suitable for gambier.

Cultivation and management

Seedlings 3–6 months old and about 8 cm tall are transplanted at spacings varying from 2–4 m. Fertilisers are needed on poorer soils. It is first harvested about 15 months after planting and later at intervals of 4–6 months. Yields are highest after about 8 years of growth and yields may be expected to continue for about 15 years. About 12 tonnes per hectare of leaves or 1–1·3 tonnes of gambier per hectare are produced yearly.

Harvesting and processing

Leaves and young branches 30–60 cm long are harvested by pruning the side branches, leaving the main shoot. The gambier is extracted by boiling in water and evaporating by further boiling. The cooled liquid sets into solid gambier which is cut into cubes or blocks for drying.

Protection

Weeds should be controlled. Leaf-eating caterpillars can be controlled with suitable insecticides. No serious disease of gambier has been recorded.

Gutta-percha (*Palaquium gutta*)

Family Sapotaceae
Other names gutta (English); taban merah (Malayan)

This tree is grown for its greyish-red rubber, especially useful in the manufacture of golf balls, deep-sea cables, dentistry products, etc.

Botany and propagation

Gutta-percha grows to a height of about 30 m. The leaves are covered with brown hairs on their undersides. Flowers are pale green in colour and fruits are pointed.

It is usually grown from seed planted in the field. Until recently the latex was gathered only from wild trees.

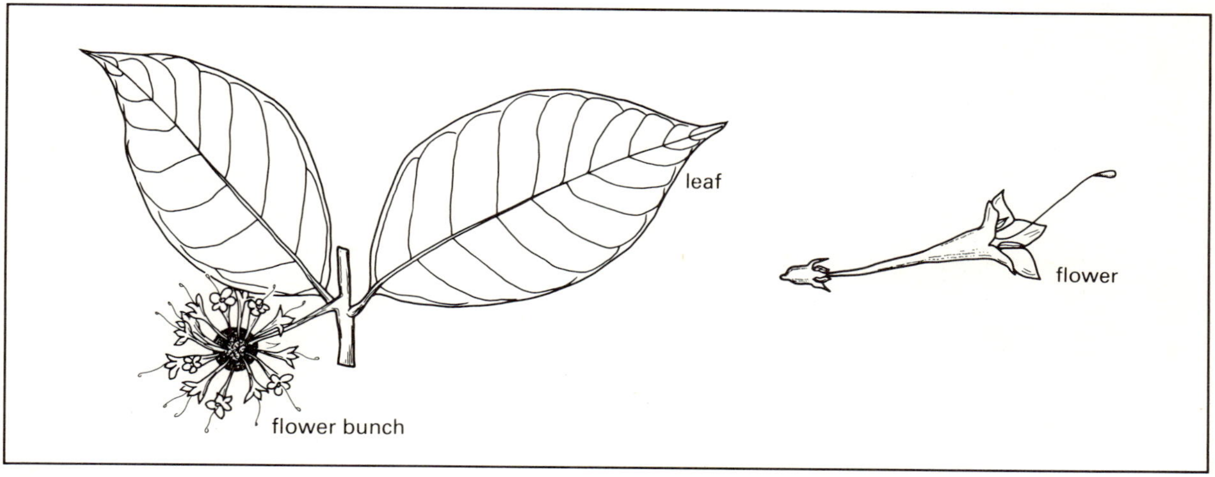

fig. 14.22 *Uncaria gambir*

Climate and soils
The tree is indigenous to Malaysia and Indonesia where it thrives in the wild in hot and humid tropical forests. Little is known about its soil requirements but they are probably similar to those of *Sapodilla*.

Cultivation and management
The tree grows slowly at first, its girth increasing about 4 cm yearly, until the 6th year when growth becomes much faster.

Harvesting and processing
The latex may be tapped from the bark of the trunk and branches after felling or extracted from the leaves. Leaves of about 10 months old are collected for extraction as they are richest in gutta-percha. Tapped trees yield up to 4 kg of the dry rubber per tree. It is now more common to tap living trees as for *Hevea*. The coagulated latex is boiled; on cooling, the finished product is a hard rubber that can soften on heating.

Nipa palm (*Nypa fruticans*)

Family Palmae

Nipa leaves are used for roofing. The flower stalk is tapped and the sap used for making sugar, palm wine or alcohol.

Botany and propagation
The nipa palm tree has an underground creeping stem up to 60 cm thick with fronds about 7 m high (Fig. 14.23). Male and female flowers are produced; the female flower stalk is about 1 m long; the inflorescence is orange brown.

The plant is usually wild and rarely planted; it can be grown from seeds.

Climate and soil
Nipa palm grows in brackish tidal swamps on clayey soils high in organic matter, in the humid Tropics where rainfall exceeds 1 500 mm a year.

fig. 14.23 Nipa palms growing in tidal swamp

Harvesting and processing
Mature leaves are cut, dried and the leaflets folded over a pole for thatching; five-year old palms are tapped daily for two to three months. The tree can yield 3 000 kg of sugar per hectare. Nipa fruits are also harvested for food.

Protection
Nipa is generally quite free from pests and diseases.

fig. 14.24 Rattan palm

Rattan (*Calamus* spp.)

Family Palmae

There are more than one hundred species comprising a number of genera of which *Calamus* is the most important. Rattan is collected for its stems, or 'canes' which are used in the manufacture of furniture and many other items in which a flexible cane is required.

Botany and propagation
Rattan is a climbing palm reaching 100 m in length (Fig. 14.24). The stems grow up to 7 cm thick; they climb by means of recurved thorns.

Climate and soil
Native to Southeast Asia and New Guinea; they grow wild near streams, swamps or wells in humid tropical forests at low elevations. They are found particularly in regrowth secondary rain-forest and on river banks.

Cultivation and management
Rattan is not usually grown except under shifting cultivation. It is planted at a spacing of about 9 m by 15 m and allowed to grow up with secondary jungle. Planting in cut lines in jungle at 6 m by 2 m is often successful. Harvesting begins in 6 to 15 years when each clump bears fifty or more stems 25–30 m long. Yields of about 135 kg good canes per clump per cutting are obtained.

Protection
Rattan is relatively free from pests and diseases.

Sago (*Metroxylon sagus*)

Family Palmae

Sago is a palm tree grown for the sago flour obtained from its trunk. Sago flour is also produced from other palms, e.g. rough sago palm (*Metroxylon rumphii*), the dwarf date palm (*Phoenix acaulis*), the fish tail palm (*Caryota urens*), sugar palm (*Arenga saccharifera*) and the cabbage palm (*Roystonea oleracea*).

fig. 14.25 Sago palm

Botany and propagation

The erect trunk grows up to 1 m in diameter and 10 m in height (Fig. 14.25). Young suckers arise from main roots and the palms generally form clumps. The crown consists of leaves up to 2 m long with many leaflets. Flowers are male or female and grow in a large tree-like inflorescence at the tree-top (Fig. 14.26), at a height of about 7 m. After flowering the trunk dies.

Propagation is from young suckers.

Climate and soil

Native to Southeast Asia, sago grows in the humid Tropics in freshwater swamps and bordering rice fields, in regions where rainfall exceeds 2 000 mm annually. Sago does well on peat soils.

Cultivation and management

Sago palms grow wild but may be planted to a limited extent. They are cut at the beginning of flowering (about 15 years of age). The trunk is cut into 1–2 m lengths, split, soaked and macerated in water to extract the starch which is washed, sieved, dried and sold as pellets. A tree yields 100–500 kg of starch or about 20 per cent of the trunk. The sago waste can be fed to livestock.

Protection

Sago palms are generally free from pests and diseases.

fig. 14.26 Fruits of sago palm

15 Leguminous vegetables

Bambara groundnut (*Voandzeia subterranea*)

Family Leguminosae (Papilionoideae)
Other names earth nut, Madagascar groundnut

A pulse crop important in some African countries.

Botany and propagation

The bambara groundnut has an annual, bunchy or trailing habit. The leaf has three leaflets. Yellowish-white flowers grow in pairs. Pods are round, usually with one seed; the seeds are hard and of many colours (Fig. 15.1).

It is planted from seed.

Climate and soil

Native to Africa, it is now grown all over Africa. It can grow in a hot climate. It does better than most other bean crops in poor soils and grows best with moderate rainfall and sunshine.

Cultivation and management

A number of cultivars are found in Africa, especially in Togo and in Zambia. Planting is at a spacing of about 50 cm by 15 cm (seed rate of 35 kg per hectare). The rows are ridged once the crop is established. Pods mature in 4 months and are harvested as for groundnut. Yields of 500–1 000 kg of dried nuts per hectare may be obtained.

Protection

Weeds must be removed regularly by hand. The crop appears remarkably free from pests and diseases.

Cowpea (*Vigna unguiculata* or *V. sinensis*)

Family Leguminosae (Papilionoideae)
Other names black-eye pea or bean, marble pea, China pea, southern pea, tonkin bean; paayap, kibil (Tagalog, in the Philippines)

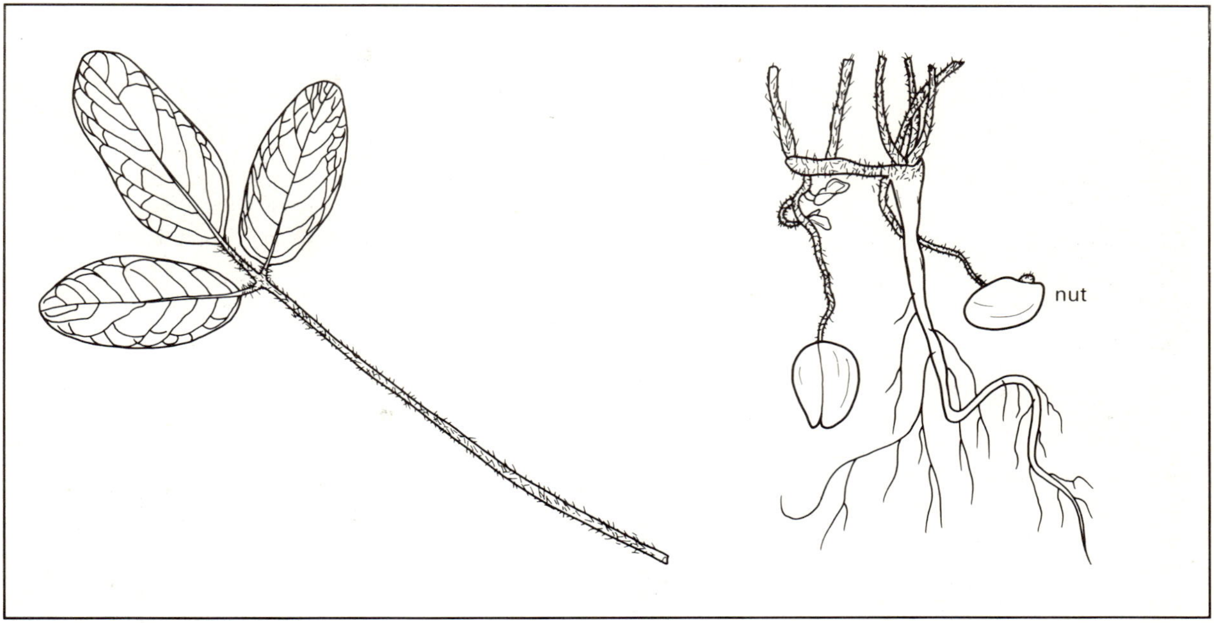

fig. 15.1 *Voandzeia subterranea*

fig. 15.2 Cowpea plants

Used as a pulse or pot herb and as forage.

Botany and propagation

The plants are annuals with spreading and trailing semi-erect stems. The leaves have three leaflets. Flowers are white, yellow or purple, in bunches, with long stalks. Pods are long with seeds of different sizes, shapes and colours (Fig. 15.2).

Climate and soil

The cowpea originated in tropical Africa and is found now all over the Tropics and sub-tropics. It requires less rain than the long bean and can grow on most soils provided they are well drained.

Cultivation and management

There are two classes of cultivars: the catjang group, e.g. 'Cream Lady' and the common cowpea group, e.g. 'Brown Crowder', 'Chipewa's Black Eye', 'Cream', 'Purple Hull' and 'New Era'. Seeds are planted at a spacing of 75–90 cm by 10–15 cm (seed rate of 15–25 kg per hectare) for beans. For forage, seeds are broadcast at 90 kg per hectare. A low-nitrogen complete fertiliser is applied. The crop matures in 3 to 5 months. Yields of up to 2 000 kg of dry seeds per hectare can be obtained in the humid Tropics.

Protection

Fungus diseases attacking cowpea are cowpea wilt (*Fusarium oxysporum)* and charcoal rot (*Sclerotium bataticola*). Other diseases are cowpea mosaic due to a virus spread by bean leaf-beetle, and root-knot nematodes. The most serious pest is cowpea curculio *Chalcodermus aenus*. Bean flies damage seedlings; they may be controlled with insecticides.

Regular weeding is needed in early stages.

Jack bean (*Canavalia ensiformis*)

Family Leguminosae (Papilionoideae)

Other names sword bean, horse bean; awara-kai (Tamil)

It is used as a pulse, and grown for forage or as a green manure or cover crop.

Botany and propagation

The jack bean is a much-branched semi-erect or trailing plant with a woody stem. The leaves are alternate with three leaflets. Flowers are reddish-purple arising from leaf axils. Pods are 20–50 cm long and straw-coloured; they contain 8–20 white seeds each 2 cm long (Fig. 15.3).

The crop is planted from seed.

Climate and soil

Native to tropical America, jack bean is now found all over the humid Tropics. The hardy and drought-resistant plant can grow in shade, and on poor soils.

Cultivation and management

The seeds are planted when rain is sufficient. For use as a pulse crop, spacing is at 1 m by 0·3 m; for green manure, seeds are drilled at a rate of 40–60 kg per hectare. A trellis should be provided for climbing types. Complete

fig. 15.3 Jack bean plants

(a)

fig. 15.4 (a) Long bean fruits
(b) Long beans planted with stakes

(b)

fertilisers are given on poor soils only. Pods are produced four months after planting. Yields of 1 500 kg of dried beans per hectare and 40–50 tonnes of green matter per hectare are obtained.

Protection
Weeding and spraying with insecticide may be necessary.

Long bean (*Vigna sesquipedalis*)

Family Leguminosae (Papilionoideae)
Other names yard-long bean, asparagus bean, snake bean; sitao (Tagalog, in the Philippines)

Usually grown as a green vegetable, it is closely related to cowpea.

Botany and propagation

Long bean is an annual with a climbing stem. The leaf consists of three leaflets. White to violet flowers arise from leaf axils. Pods are 30–100 cm long (Fig. 15.4(a)) with kidney-shaped seeds whose colour varies with the cultivar. It is grown from seed.

Climate and soil

Native to tropical Africa and Asia, it is an important market-gardening crop in Southeast Asia. Long bean needs a warm and moist climate but will grow on a variety of well-drained soils.

Cultivation and management

Many cultivars are available such as 'Yard-long' and 'Long White' from Hong Kong. Two seeds per point are sown at a spacing of 60–90 cm by 60 cm. Wooden sticks for climbing should be provided (Fig. 15.4(b)). Compound fertilisers with low nitrogen (e.g. 12:12:17:2) should be given at a rate of 400 kg per hectare. Young pods may be picked from the 8th to the 14th week, 2 or 3 times per week.

Protection

Regular weeding by hand is necessary. The most serious pests are bean fly and sucking bugs which may be controlled with diptrex or a similar insecticide. After the beginning of harvest use lannate or other non-persistent insecticide. Other pests are root-knot nematodes. Wilt (*Fusarium oxysporum*), charcoal rot (*Sclerotium bataticola*) and anthracnose (*Colletrotrichum* sp.) are the chief diseases.

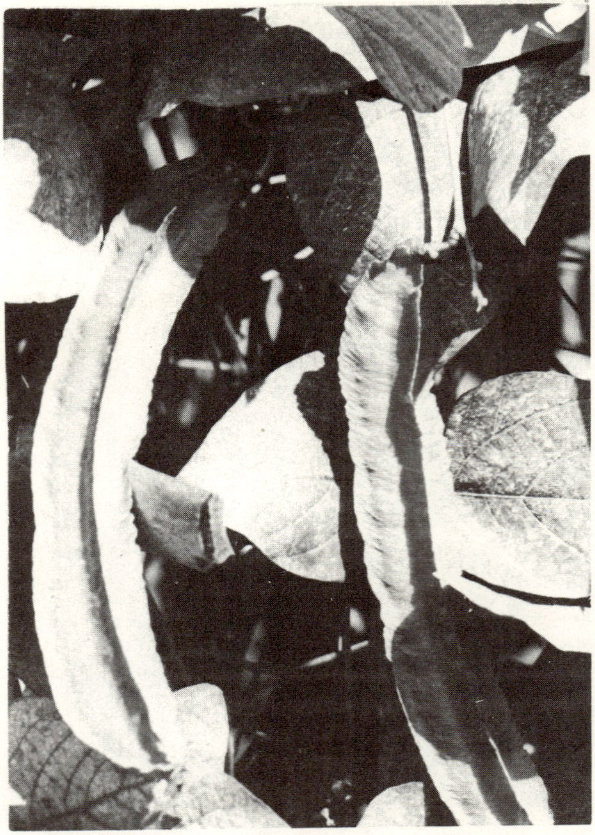

fig. 15.5 Fruits of wing bean

Wing bean (*Psophocarpus tetragonolobus*)

Family Leguminosae (Papilionoideae)
Other names four-angle bean, Goa bean, Manila bean, princess pea; seguidilla (Tagalog, in the Philippines)

It is grown as a green vegetable and sometimes for the tubers.

Botany and propagation

Wing bean is a perennial legume with a twining stem. The roots are tuberous with large root nodules; the leaves are trifoliate. Flowers are blue or white borne in bunches at the end of branches. Pods are 15–25 cm long, angular and roughly square in section with 4 wavy wings along the length (Fig. 15.5). They bear brown and round seeds containing 30 per cent crude protein.

They may be planted from seed directly in the field.

Climate and soil

Native to India it is now grown as a vegetable all over the Tropics. Requirements are as for long bean but it can withstand more extreme conditions, e.g. dryness.

Cultivation and management

The seeds are planted at a spacing of 1·3 m by 0·6 m; supports are provided for climbing (Fig. 15.6). For tuber production, a spacing of 60 cm by 100 cm is used and the plants are left to trail. Green pods are harvested at 3 months for about a year. Tubers may be harvested at 7–8 months. A complete fertiliser is needed at planting.

Protection

Bean fly is sometimes a problem and may be controlled by use of a suitable insecticide.

Yam Bean (*Pachyrrhizus* spp.)

Family Leguminosae (Papilionoideae)
Other names potato bean; sengkuang (Malayan); sinkama (Tagalog, in the Philippines)

The young pods of *P. erosus* are eaten as beans; those of *P. tuberosus* possess irritant hairs and cannot be eaten. The roots of both species are eaten raw or cooked as a vegetable. The ripe seed and mature roots contain rotenone and may be used as an insecticide or fish poison.

fig. 15.6 Vines of wing bean in bearing on a trellis

Botany and propagation

Yam beans are herbaceous climbing plants with hairy stems, leaves and pods. The leaves have three leaflets. Flowers are white to violet, carried in bunches. Tubers are turnip-shaped with white flesh and brownish skin (Fig. 15.7). The plant is grown from seed.

Climate and soil

Native to tropical South America, the yam bean requires a well-tilled, loose and sandy soil. Climatic requirements as for long bean.

Cultivation and management

Seeds are planted at spacings of 1 m by 0·3 m on ridges or on hills 1–1·3 m apart with 3 to 5 seeds per hill. The plants are best staked up but often left to trail. Flowers should be removed to get better tuber yields and harvested 4 to 5 months after planting. Tubers are fibrous when harvested late.

Protection

Leaf caterpillars are often a problem; they can be controlled by spraying with insecticides. Regular weeding is needed.

fig. 15.7 *Pachyrrhizus tuberosus*

16 Oil crops

Major crops

Coconut (*Cocos nucifera*)

Family Palmae

Other names niu (Fijian); cocotier (French); côco (Portuguese); cocotero (Spanish)

Coconut palms are among the most important plants in the tropical world. Table 16.1 indicates approximately the area of coconuts in the producing regions of the world.

Most of the coconuts grown in Africa are on the eastern side of the continent, while most oil palms occur on the western side.

Coconuts have been grown throughout tropical countries of the world for so long that their origin is obscure. Some botanists consider that they originated in tropical America because other closely-related species occur in that continent; others consider the origin to be in the Pacific, because fossils have been found in New Zealand.

Wherever the coconut originated, there is no disputing its importance in tropical countries, for not only is it widely used as a food and in trade, but various parts of the coconut are also used for building, roofing, making coir and rope, as fuel, and for making charcoal, alcohol and other products useful to man.

Coconuts are of special importance in trade in most of the small islands of the world where transport by ship is

fig. 16.1 A tall coconut variety about 12 years from planting in Malaysia

Table 16.1 Approximate areas planted with coconuts in different countries and regions

Country	Area (million ha)
Philippines	1·00
India	0·60
Indonesia	0·60
Indo Pacific	0·55
Caribbean and tropical America	0·50
Sri Lanka	0·40
Malaysia	0·22
East Africa	0·22

irregular. This is because the 'meat' of the coconut can be dried and stored until such time as a ship arrives. Most other crops cannot be stored for long periods. The dried 'meat' is called copra and oil is extracted from it. In other parts of the world where other more profitable crops can be grown coconut is more important as a food. Coconut is not very profitable as a plantation crop, mainly because of its low yield, which is generally less than 2·5 tonnes of copra per hectare, and the high cost of husking the nut and extracting and preparing the copra. The main problem with coconuts is that in the past there has been very little yield improvement by breeding. Some varieties of coconuts give higher yields than others.

fig. 16.2 A Malayan dwarf coconut at 5 years; it is already yielding heavily

fig. 16.3 A young coconut inflorescence growing from the axil of a leaf

Botany

Like the oil palm and betel nut palm coconuts have only one growing point which produces a single crown of fronds on a single upright trunk, as shown in Figs 16.1 and 16.2. The growing point is located deep within the heart of the crown. It produces a series of primordia which grow into fronds. The fronds emerge from the top of the crown so that the youngest leaves are always at the top with oldest leaves displaced downwards. When a leaf first emerges from the top of the crown it is folded into a 'spear' but after this it unfolds to produce the typical palm frond with a central thick midrib and a row of green leaflets on either side. Old leaves fall off when they die, leaving scars which are visible on the trunk.

The roots of coconuts all grow out of the swollen base of the trunk. The main roots are about 1 cm thick throughout and reach a length of 10 m or more. They are mostly in the top 50 cm in clayey soils but in sandy soils may grow much deeper until they reach the water table beneath the soil surface.

The nuts are produced from inflorescences which grow from the axils of the leaves, as shown in Fig. 16.3. The inflorescences consist of a central stalk bearing numerous

Table 16.2 Comparison of stages in fruit growth for tall and dwarf coconuts (age in months)

Tall coconut		Dwarf coconut	
Age	**Stage**	**Age**	**Stage**
0·1	male flowers ripen	0–2	male and female flowers ripen
1–2	female flowers ripen		
3–4	fruits set	2–3	fruits set
4–10	kernel growth	3–8	kernel growth
10–12	shell forms	8–9	shell forms
11–12	water sloshes	10	water sloshes
14–15	nuts ripe	12	nuts ripe
16	nuts fall		

male flowers at the tips and female flowers at the base (Fig. 16.4). The male flowers ripen first and in tall varieties female flowers do not become ready for pollination until all the males have shed their pollen, so that cross-pollination with other trees always occurs. In the Malayan dwarf coconut the female flowers ripen while the male flowers are still producing pollen, so that self-pollination often occurs.

After pollination the female flowers develop into coconuts as shown in Fig. 16.5. Tall varieties take about 15 months for the nuts to ripen and dwarfs take about 1 year. The stages in fruit growth for a typical tall coconut and for the dwarf type coconut are shown in Table 16.2. At the stage of fruit set a high proportion of the female flowers will drop off leaving only a few to ripen (Fig. 16.5). This abortion of nuts is what causes the main peaks and troughs in yield. Under the best conditions about ten nuts will develop on an inflorescence in a tall variety and fifteen in a dwarf.

The main commercial varieties of coconuts are now present in most tropical countries. Tall varieties, however, are very locally adapted and show a great range of characteristics because of cross-pollination. Three colour varieties can be recognised, they yield yellow, green and red nuts. Tall varieties produce nuts of an average weight of 1·5–1·8 kg containing about 0·4–0·5 kg of 'meat', but some yield very large nuts of about 3 kg in weight with up to 0·8 kg of 'meat'. These large-nut types often occur in small islands such as Rotuma in the Fiji group, and are thought to occur in such places because the original

fig. 16.4 An inflorescence removed to show its structure; female flowers are at the base of each branch

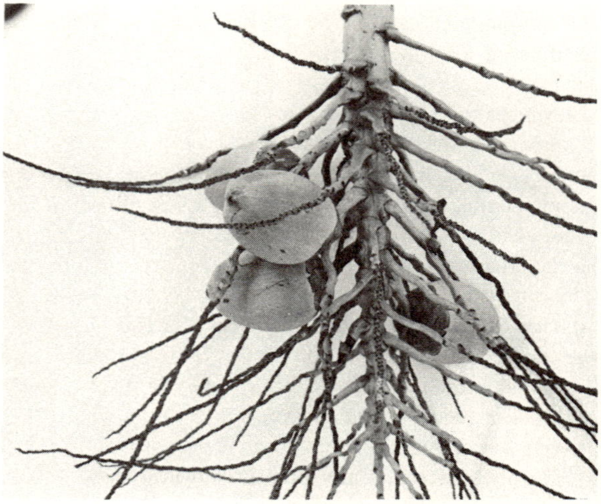

fig. 16.5 An old inflorescence long after the male flowers have fallen. The female flowers have grown into ripe nuts

settlers carried coconuts in their canoes for drinking as well as planting and naturally chose large-nut types to take with them.

There are two important types of dwarf coconuts; the Fiji dwarfs and the Malayan dwarfs. The Fiji dwarfs are rather like tall coconuts in structure being large plants with large crowns, but with shorter trunks. They are notably leafy and produce nuts similar in size to the normal tall coconut. The Malayan dwarf is a smaller tree and should be planted at a closer spacing. It produces smaller and more numerous nuts. Tall coconuts require 4 000–5 000 nuts to make a tonne of copra. The dwarfs mature sooner and produce after 4 years, while talls require 6 to 7 years, but dwarfs usually last only 25–45 years, while talls will live up to 90 or more years. However, because dwarfs are planted closer they yield more copra and are therefore a better type to grow for plantation copra production. The Malayan dwarf is also resistant to a serious coconut virus or mycoplasma disease known as 'lethal yellows' which affects tall coconuts in many countries in the Americas. Yellow, green and red-fruited types occur and the red type is said to be higher yielding. In the Caribbean and in Malaysia, hybrids of Malayan dwarf and local talls are found to be vigorous and high yielding. The Malaysian hybrid is referred to as the 'Mawa' coconut.

In addition to the above types, various others occur. Some have long thin fruits and are grown to get coir for rope making. These are common in parts of New Guinea. Others have an edible type of husk which is eaten young and tastes a bit like palm heart.

Climate and soil

Coconuts only really do well in the humid Tropics and most coconuts are grown near the sea. For good yields annual rainfall should exceed 1 200 mm and dry seasons should be minimal. However, long hours of sunshine are also required for good yields. The most suitable climates are within the equatorial latitudes, about 10°N and S and have 2 000–2 400 mm of rainfall evenly distributed, and sunshine hours in excess of 2 000 per year.

Sometimes, on small islands with steep hills inland, rainfall is less than the above, but coconuts growing on sandy sea shores thrive on seepage water from inland.

Because coconuts are often grown in coastal regions, sandy soils predominate; they may be either coral or quartz sands. Clayey alluvial soils are also suitable. Sandy soils, in general, support poorer growth and yields are generally low. The palms often depend for their nutrients on underground water. On quartz sands potassium and nitrogen deficiencies are common. On coral sands iron and manganese, as well as potassium and nitrogen, may be deficient. Clayey alluvial soils which are not too acid, above pH 4·5, and rich in nutrients are the best coconut soils provided they are drained to about 60 cm. Coastal clays formed under marine or estuarine conditions are generally the most nutrient-rich. Coastal clays, however, with a high sulphur content will become too acid for coconuts when drained. Occasional inundation with brackish sea water with a salt content of about 1 000 ppm generally increases coconut yields on very acidic soils.

Cultivation and management

Choice of variety The main problem with coconut growing is that very few countries operate an improved seed programme. If improved or hybrid seed nuts, obtained from controlled crossing of suitable mother palms are available, they should of course be used. If not, then some attempt at selection should be made. For both tall and dwarf coconuts, take seed nuts only from the best trees yielding more than 150 nuts per year under good climatic conditions. Select the correct nut type as shown in Fig. 16.6 and mark the tree clearly.

If coconuts are being grown for the production of copra, rather than for domestic use, it should be remembered that the Malayan dwarf or the dwarf/tall hybrid will give higher yields; 5 tonnes of copra or more per hectare can be obtained. However, the palms should be locally adapted and where the dwarf has not been introduced and become known to grow and yield well, it may be wiser to grow locally-adapted tall varieties.

Crossing of two high-yielding palms, each producing over 150 nuts per year, will produce better seed nuts than the simple selection method illustrated in Fig. 16.6. To do this the male flowers are cut off the inflorescence as soon as they appear and the female flowers are pruned leaving about 30 on the inflorescence. The inflorescence is then sprayed with 1 per cent methanal (formaldehyde) solution and covered with a polythene sleeve or a thick paper bag such as an empty but clean cement bag. When the flowers are ripe, pollen from another tree is introduced by puffing it into the bag with a rubber puffer bulb.

Nursery and planting Coconuts should be planted in loose sandy soil beds about 20 cm deep and about 1 m wide (Fig. 16.7(b)). It is unnecessary to plant them in

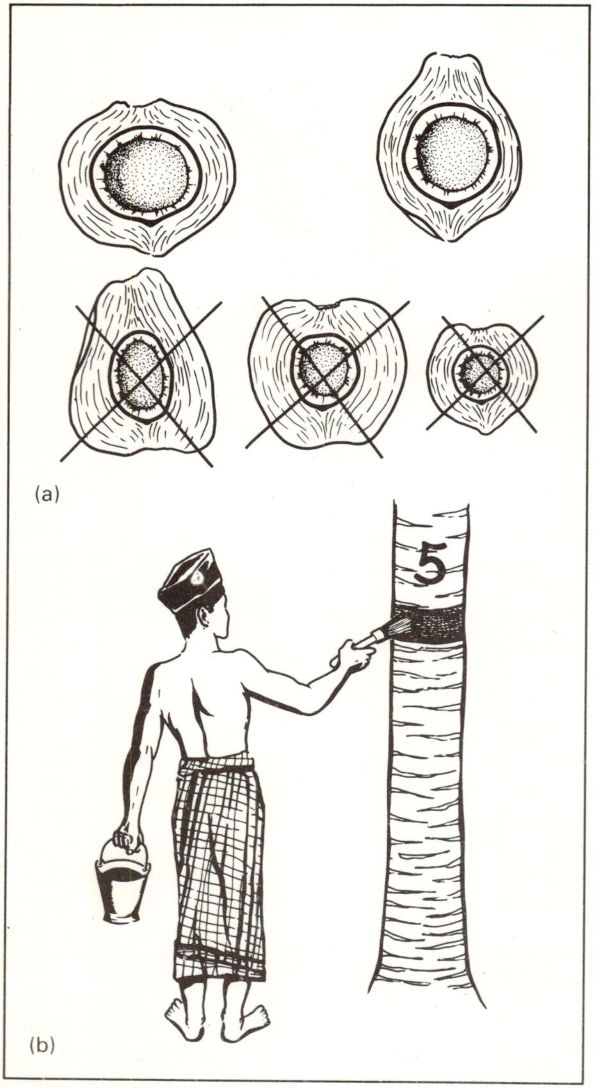

fig. 16.6 (a) The best seed nuts are as round as possible (b) The selected seed palms should be marked and numbered like this (*Source*: Ministry of Agriculture leaflet, No, 41, Malaysia.)

polythene bags as is done with many other tree crop seedlings. The nuts are laid horizontally Fig. 16.7(a) in shallow troughs about 20 cm apart and lightly surrounded with soil. The husk may be left on or removed, whichever is convenient. Usually if nuts have to be transported for any

fig. 16.7 (a) Method of laying the nut horizontally in the soil

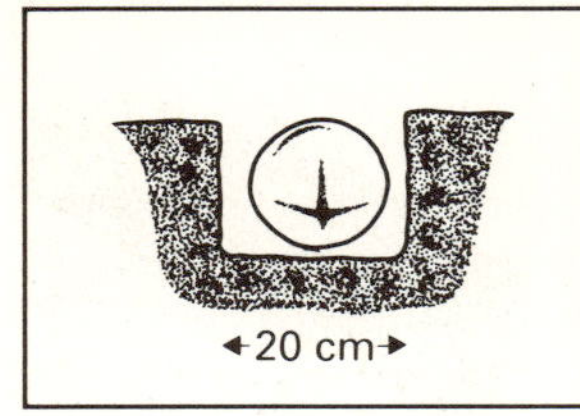

(b) Coconuts planted just below the surface in loose sandy soil in a nursery

(c) A germinated coconut seed

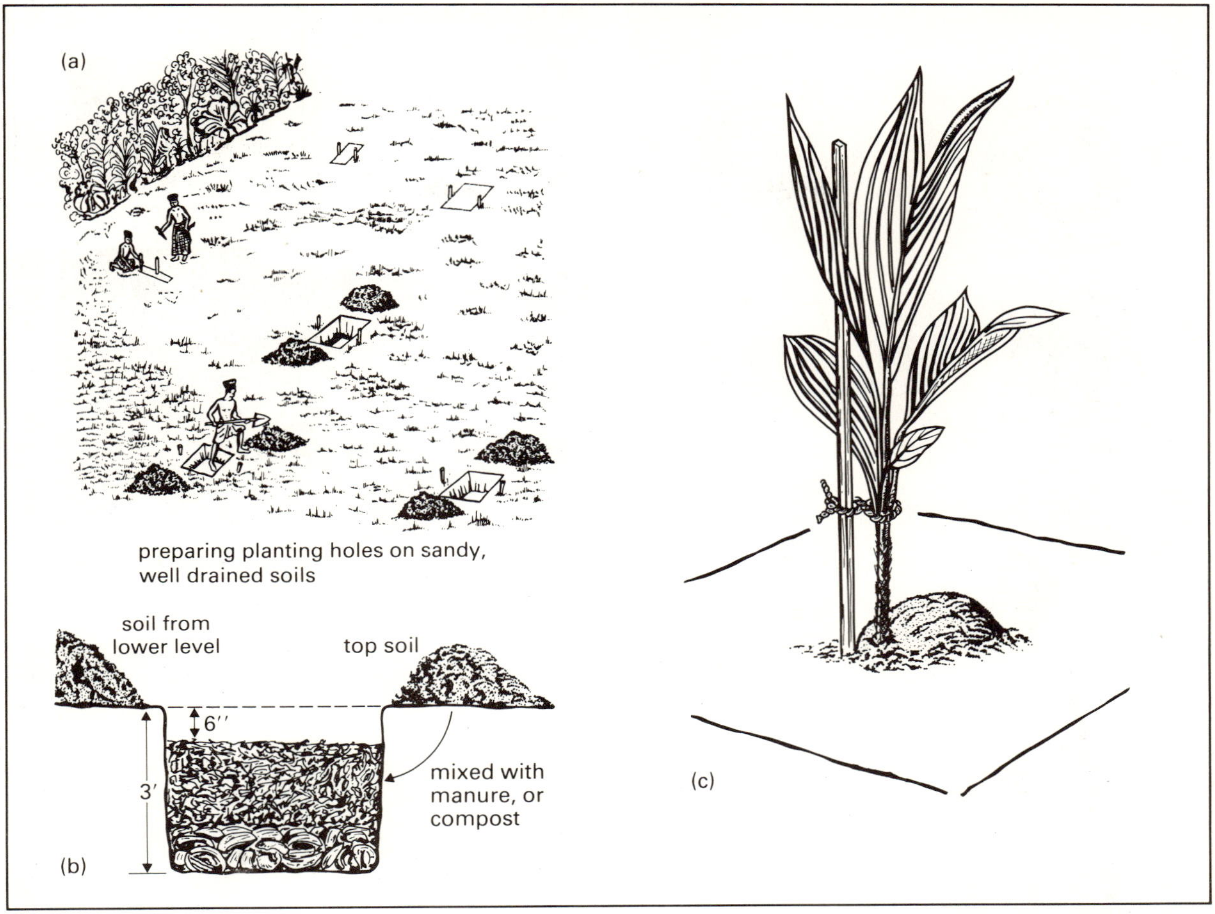

fig. 16.8 The method recommended for planting coconuts in sand in Malaysia (a) preparing planting holes on sandy, well-drained soils (b) composition of planting hole (c) coconut planted in soil. (*Source*: Ministry of Agriculture leaflet No. 41 Malaysia.)

distance the husk is removed. Most coconuts have about 10–15 weeks dormancy before the shoot appears. There is no need to shade the nursery but after the shoot appears the soil should be watered in dry weather. The young seedlings should be planted to the field after about 5 to 6 months when the shoot has appeared. At this stage there will be 6 to 10 large, thick, pencil-like roots which are raised by loosening the soil. If the soil is poor, then a large planting hole of 60 cm diameter and depth can be recommended, this being filled with top soil mixed with some manure. In good soils a 30 cm planting hole is satisfactory. On sands even larger planting holes, with a layer of husk as shown in Fig. 16.8, can be recommended.

Coconuts are usually planted on a square pattern but recent evidence suggests that they are better planted on a triangular pattern when they will give higher yields per hectare. The following planting distances are recommended. Talls (also Fiji dwarf): on good soils, 10 m; on poor sandy soils, 8 m. Malayan dwarf: on good soils, 6 m; on poor soils, 5 m.

Fertilisation Potassium is the most important nutrient in terms of quantity and as a lot of this occurs in the husk, these should be spread around the trees after removal. The nut water also contains much potassium so that it is a good practice to husk and crack nuts in the field as shown in Figs 16.9 and 16.10. Ash from burning shell or husk in

fig. 16.9 Husking coconuts which have fallen from the trees. The best way to remove the husk is on an iron spike in the ground

fig. 16.10 Cracking the husked huts. The meat will shrink as it dries and pull away from the shell making it easy to remove the copra

copra drying should also be spread around the trees. In addition to this, coconuts will show a yield response to potassium chloride on all but the richest soils. For most soils, with the exception of rich coastal clays, potassium chloride at rates up to 250 g per tree every two months will produce a response in yield. On acid soils below pH 5·0, magnesium limestone is often beneficial. Phosphorus responses are rare in coconuts. Nitrogen is best supplied by undersowing the trees with legumes.

On coralline soils, iron and manganese should be supplied at rates of 10 g iron(II) sulphate(VI) and 5 g manganese(VI) sulphate(VI) injected into a hole in the trunk once a year.

Weed control and covers Cover legumes such as *Stylosanthes* spp., *Pueraria* spp, and *Calopogonium* spp. should be grown under coconuts (p. 4). These legumes usually fix sufficient nitrogen for the crop. In seasonally dry areas, covers may reduce coconut yields if allowed to grow uninhibited and therefore should be cut back (usually by light disc harrowing) at the beginning of the dry season. They will regenerate naturally. In humid climates covers should be left uncut.

Covers help to control weeds but circle weeding around the palm trunks should be done periodically. The circles should be sprayed first with paraquat or sodium chlorate(V) and then with diuron at 2–3 kg per hectare on clayey soils but not sands. This will control weeds for up to six months. Generally weeds will not be such a problem on sands. Cattle grazing also reduces weed growth and generally eliminates the need for weeding, though fertiliser rates should be increased by 50 per cent.

Intercropping and grazing On account of low profits from copra production intercropping or intergrazing is often practised. Only tall coconuts are suited to long-term intercropping or grazing, not dwarfs. There are many possibilities for intercropping grasses and legumes that can be used for grazing. Maximum livestock productivity could be of the order of two to three head per hectare. Further details may be obtained from a companion volume, in the ITAS series, entitled 'Tropical Pasture and Fodder Crops'.

Young coconuts can be interplanted with bananas for one to two years but not subsequently.

Extra fertiliser should be supplied at a rate of up to 1 kg

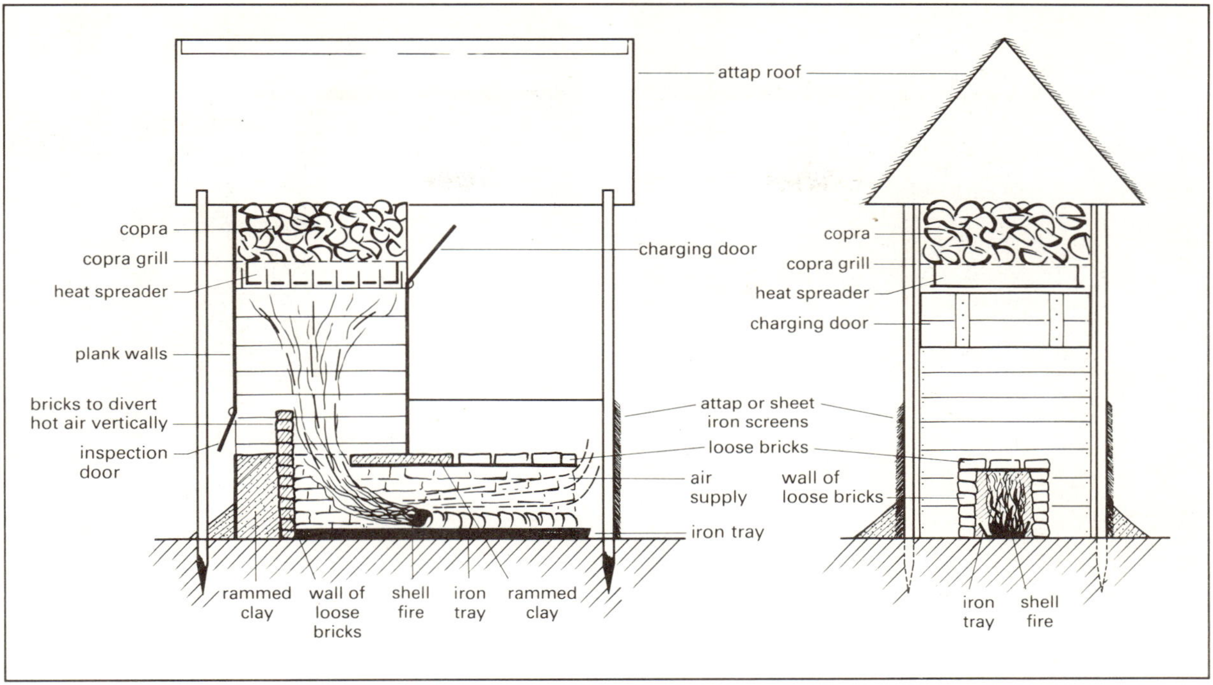

fig. 16.11 Cooke-type kiln (a) cross-section (b) front view. Attap can be replaced by any kind of thatch

fig. 16.12 (a) Coconut crowns damaged by rhinoceros beetles. They appear as if whole pieces of frond have been cut out

(b) Grub of rhinoceros beetle in bole of coconut

fig. 16.13 (a) Young oil palms growing in a plantation with legume cover crops

(b) Mature oil palms about 20 years old. Bunches are cut by the use of a curved knife on a long pole

of N:P:K mixture per palm every three months when bananas are grown. Cocoa is sometimes grown under tall coconuts as described in Chapter 9.

Preparation of copra

Copra is prepared by extracting the 'meat' (Figs 16.9 and 16.10) and drying it to a moisture content of about 7 per cent. The simplest method is sun-drying. The pieces of copra are spread over a cement or other surface and require about five sunny days to dry. Kiln-drying is more reliable in humid climates. Fig. 16.11 shows the construction of the simplest type of kiln. It should be located away from other buildings to minimise fire risk. Coconut shell may be used as a fuel. Drying should be done promptly to preserve quality.

Protection

The rhinoceros beetle sometimes causes severe damage to coconuts as shown in Fig. 16.12. It chews up the young leaves which are deformed when they expand. These beetles are controlled by clearing up all old stumps and other rubbish in which the grubs breed. Rhinoceros beetles also spread a nematode disease called 'red-ring disease' which is fatal to coconuts.

Seedlings are sometimes infected with leaf diseases such as *Helminthosporium*. They can be controlled by spraying with fungicides. In the American Tropics a serious mycoplasma disease called 'lethal yellows' affects tall coconuts. The Malayan dwarf coconut is resistant. Another serious disease which occurs in Asia is called 'cadang cadang' which is a general decline in vigour with the production of smaller and smaller fronds; its origin is unknown. There is no known cure, but some coconuts seem to be resistant to this condition.

Oil palm (*Elaeis guineensis*)

Family Palmae

Other names palmier à huile (French); denden (Portuguese); dihola (Portuguese, in Angola); palmera de aceite Africana (Spanish)

The oil palm is an important producer of vegetable oil and is found growing wild in the humid lowlands of West Africa from Cape Verde to Angola. It is grown commercially in plantations in West Africa, Zaïre, Southeast Asia and Central America, and is generally regarded as the highest yielder of oil of all the oil-producing plants.

The fruit has a red-orange pericarp and mesocarp, a shell and a kernel. The palm oil of commerce is contained in the mesocarp of the fruit while a different kind of oil called kernel oil is present in the kernel.

There is a second species of oil palm, *Corozo oleifera*, sometimes referred to as *E. melanococca*. It is grown mainly in Central and South America extending into the Amazon basin, preferably in swampy areas and river banks subjected to flooding. *C. oleifera* palms are not grown to any extent on estates but the oil of wild trees is used domestically in certain areas. However, in Central and South America hybrid *E. guineensis* × *C. oleifera* palms are now being tested on a commercial scale as they appear to flourish better than *E. guineensis*. The fruits of *C. oleifera* are small and there is a tendency for parthenocarpy or formation of fruit without fertilisation.

Botany

The oil palm has a regular growth pattern, producing 25–35 fronds annually from a single terminal apex. These fronds remain green for more than a year. A mature palm usually has about 40 green fronds at any one time which are arranged spirally. Each frond bears 100–200 leaflets on each side of the rachis arranged in upper and lower ranks (Fig. 16.13). In *C. oleifera* the leaflets are in one plane.

The roots of the oil palm grow from the bole of the stem base. Primary roots extend radially from the base of the palm up to a distance of about 20 m from the stem. Secondaries arise upwards and downwards from the primaries and these bear the tertiary and quaternary roots that form the main absorption centres for water and nutrients. In spite of the length of the horizontal primaries, the majority of roots are to be found near the palm but sometimes with a second concentration about 2 m from the base. The greatest concentration of feeding roots is in the surface soil up to a depth of about 50 cm in normal soils. In friable, deep soils the oil palm also produces a number of descending roots reaching a depth of 2 m or more.

The oil palm has a straight rough trunk, formed from the leaf bases and the central stem tissue. It grows 30–40 cm in height every year depending on the climatic conditions, more under evenly-distributed wet conditions. Old oil palms reach 15 m or more in height and may live for at least 60 years. In commercial estates, however, they

fig. 16.14 (a) Male inflorescence producing pollen
(b) Young female inflorescence with flowers ready for pollination. The stigmas are pale yellow, after pollination they turn pink

are usually replanted after 25–30 years because of declining yield and because the increasing height makes harvesting more difficult.

The apical meristem is buried in the crown of the palm which is surrounded by soft tissue of young leaves and leaf bases. This soft tissue is edible.

The palm produces separate male and female inflorescences in the leaf axils in cycles of a few months' duration (Fig. 16.14). The female inflorescences develop into compact bunches of fruit. A variable number of bunches is produced annually ranging between 5 and 15, and mature palms in good climates usually yield in the range of 20–25 tonnes of bunches per hectare per annum.

The commercial oil palm has three genetic forms depending on the shell thickness of the kernel (Fig. 16.15). The *dura* has a thick shell with a fibre ring; the *pisifera* is shell-less and the *tenera* is a hybrid between the above two and has a medium shell thickness with a fibrous ring in the mesocarp around the shell. The *pisifera* flowers are often sterile and therefore do not produce fruits. However fertile *pisiferas* with fully-developed fruits also occur. In commercial estates the *tenera* is usually planted now.

fig. 16.15 Fruits of oil palm. From left to right: *pisifera* (without shell); *tenera* (thin shell); *dura* (thick shell)

Seed Removal of the oily mesocarp will reveal the nut or endocarp which is hard and generally dark brown to black in colour. The seeds vary in size from 1·5 to 3 cm in length.

fig. 16.16 (a) Newly-germinated oil palm seedlings planted in 20 cm bags at 6 weeks (pre-nursery)

(c) Oil palm seedling at 1 year, producing divided leaves

(b) Oil palm seedlings transplanted into 50 cm polythene bags, at 6 months; they have undivided leaves

Inside the endocarp is the greyish-white, hard oily kernel or endosperm, containing the embryo.

For the oil palm industry high-yielding, selected seed material should be used when planting. Seeds are sold germinated or ungerminated. Seeds from selected parents for germination are stored in shelves at an ambient temperature of 25–30°C until required for sale or germination. Seeds do not germinate well without heat treatment. Prior to 1960 the germination of seeds was by the 'wet heat' technique in which seeds were heated continuously at 38°C and 20–22 per cent moisture content, and germinated seeds removed as germination occurred. Since about 1965, the dry or pre-heating technique has been used. The moisture content of the seed is maintained at about 17 per cent so that the seed will not germinate during the heat treatment of about 40 days but will do so when the seed is subsequently soaked after heating is complete.

Seedlings When the seeds first germinate, the shoot produces undivided leaves but after about 6 months divided leaves are produced (Fig. 16.16). After a few months typical fronds with rows of leaflets are produced.

Climate and soil

The oil palm is best suited to a humid tropical climate in which rain occurs mostly at night and days are bright and sunny. There is a direct relationship between the amount of sunshine received and yield when there is sufficient moisture for growth. Minimum monthly rainfall for optimal yield is around 150 mm and the best climates are those in which there are no dry seasons and sunshine is evenly distributed and exceeds 2 000 hours per annum. Dry periods and cloudy periods produce a decline in yield through inducing male inflorescences anything up to 2 or more years after the event.

The oil palm also requires warm temperatures. Elevated sites in the Tropics are unsuitable. The highest latitude in which oil palms are commercially grown is around 15°N in Honduras.

The best soils are rich alluvials in which, with proper use of fertiliser, yields of up to 28 tonnes of fruit or 6 tonnes of palm (mesocarp) oil, are possible. The oil palm is tolerant of temporary flooding when mature, that is, more than 3 years after planting. In good climatic locations high yields of 20 tonnes (4·5 tonnes of oil) per hectare may also be obtained from acid latosols with moderate water-holding capacity and a rooting depth of about 1 m. Oil palms do not yield well on shallow soils. The palm is tolerant of acid conditions down to about pH 4·0.

Cultivation and management

Planting material and nursery In commercial planting, oil palm seed or seedlings are generally obtained from government or commercial organisations who are constantly upgrading the quality of the palm, to produce maximum yields of oil. The selection of both *dura* and *pisifera* parent palms to produce the *tenera* hybrid is a long and involved process. Palms with highest yield together with high oil to bunch ratio are selected to be potential mother palms. The selection of the *pisifera* palm for the supply of pollen is more time-consuming since the *pisifera* palm is generally sterile and its yield potential cannot be assessed; its quality can only be determined by the performance of its *tenera* progeny.

Germinated seed or bare-root seedlings can be planted into perforated black polythene bags filled with free-draining loamy soil (Fig. 16.16). It is essential to plant the seed on its side with the shoot side upwards otherwise a twisted seedling will result. Daily watering is necessary in the drier months. Overhead shading although sometimes used is not usually necessary. One month after planting small quantities of compound fertiliser mixture are added at a rate of 5 g per bag in the 1st month rising to 50 g per bag in the 6th to 9th months. A weekly spray of 0·5 per cent urea is also beneficial. Palms remain in the nursery for 6 to 9 months but can be held longer if spaced at 1 m.

Sometimes seedlings are grown in a pre-nursery of smaller polythene bags of size about 20 × 25 cm for three months prior to planting in larger bags. The polythene bag nursery should be free from weeds. Hand weeding when labour is plentiful is best. However, a combination of hand weeding together with the use of the pre-emergence herbicide diuron has been shown to be highly successful. Translocated herbicides such as 2:4-D are lethal to oil palm seedlings and should not be used anywhere near them.

When planting out, all stunted and deformed seedlings should be rejected.

Field planting In planting newly-cleared jungle a cut and burn technique is used as discussed on p. 2. In replanting from old oil palm, rubber or coconuts, complete stump removal is necessary and the ground should be ploughed, harrowed and rotavated. Lines for positioning of the oil palms are then marked at 8 to 10 m

Table 16.3 Schedules for fertilising mature palms (rates in kg per palm per annum)

Age of palm (years)	Ammonium sulphate (VI)	Potassium chloride	Rock phosphate	Kieserite
3–5	1–2·5	1–2·5	0·3–0·7	0·3–0·7
5–7	1–2·3	1–2·3	0·3–1	0·3–1
7	2–5	2–5	0·5–1·5	0·5–1·5

triangular spacing, giving a density of 116–180 palms per hectare. The density of planting will depend on the performance of the palm in that environment; in good soil with adequate water the optimal density will be less than in an environment where the soil is poor.

Inoculated leguminous cover seeds (consisting of a mixture of *Pueraria phaseoloides, Calopogonium mucunoides* and *Centrosema pubescens* are then sown at the rate of about 10 kg per hectare.

Holes about 45 cm diameter and 60 cm depth are dug at planting points. 200 g of rock phosphate should be added to each planting hole prior to planting the palms. Planting is best done at the beginning of a rainy season but flood periods should be avoided.

Applications of mixed fertiliser high in nitrogen and potassium should be made 3–4 times per year, starting at 100 g at initial stages and rising to 1 kg or more when the palm is about 2 years old. Oil palms growing on replanted land do not require so much phosphorus or magnesium as these nutrients tend to accumulate in the soil from previous applications. Of the trace elements, boron appears to be the most important for oil palm, especially in the first 5 years of growth. Annual application of disodium tetraborate(III)-10-water (borax) at about 100 g per palm is necessary, especially under good climatic and soil conditions. Table 16.3 indicates a range of fertilisers that can be applied to oil palm after maturity. When oil palm bunches are being formed and harvested from about 3 years old, potassium manuring becomes more important. The higher rates may be recommended for good physical environments such as occur in Southeast Asia and the lower rates for poorer environments with a pronounced dry and/or cloudy season.

Yield The yield of oil palm is highly variable depending on the environment, management, planting material and the fertilisers used. Palms growing in environments with an even annual rainfall and suitable soil will yield twice or three times as much as palms on similar soil but where there are dry periods and inadequate applications of fertiliser. Under good physical conditions palms may come into commercial bearing from as early as 2·5 years from planting. The yield increases from about 1 tonne per hectare to more than 20 tonnes per hectare of fruit bunches containing about 22 per cent oil at about 10 years, after which there is a gradual decline with age.

Pollination Oil palm is predominantly wind pollinated. Because of the superior planting material which has been bred to produce more female inflorescences than male, assisted pollination has been shown to be of great advantage in producing higher yields especially on palms below five years of age. Pollen from male flowers (Fig. 16.14) is collected and stored under dry conditions. The pollen is then mixed with talc and puffed on to female flowers at the anthesis stage, with a rubber puffer or a dusting machine. Once-weekly rounds of assisted pollination may be recommended. Assisted pollination will prevent bunch failure (fruits that do not develop) and will give more oil per bunch.

Protection

In Southeast Asia and Africa the fungus *Ganoderma* is of economic importance especially on replanted areas. The main source of infection is old stumps and it occurs by root contact, therefore removal of old infected stumps by burning though costly, is recommended. In Africa *Fusarium* wilt often affects old plantations. The disease is endemic and can only be controlled by the use of resistant types. In dry seasons a serious seedling disease known as seedling blast occurs if watering is not carried out adequately.

Pests of oil palm are localised and often short-lived; rats, however, damage young developing fruits and remove ripe fruits from the bunch. Regular baiting should be practised to keep down the rat population. Other pests of oil palm are bagworms and other caterpillars, rhinoceros beetles (*Oryctes* sp.) and the bunch moth (*Tirathaba mundella*). Lead(II) arsenate(V) can be used to control bagworms and caterpillars. Removing breeding grounds of old stumps will control the rhinoceros beetle and removing old and rotting branches will control the bunch moth. Pollination to reduce the incidence of rotting bunches also reduces the incidence of the bunch moth.

Minor crops

Groundnut (*Arachis hypogaea*)

Family Leguminosae (Papilionoideae)

Other names peanut, monkeynut, goober pea; arachide (French); arnendoim (Portuguese); mani, cacahuete (Spanish)

The seeds are eaten raw or cooked. Crushed they produce an edible oil and livestock concentrate feed.

Botany and propagation

The groundnut is an annual reaching 20–40 cm in height. It has a tap root with round nodules. The stem has many low branches, and the leaf has four leaflets (Fig. 16.17). Bright yellow flowers are produced singly from the leaf axils near the base of the stem. The flowers peg into the soil after pollination. The pods are yellowish-brown with 1 to 4 seeds (Fig. 16.18); the seeds contain 40 per cent oil and 30 per cent crude protein.

The plant is grown from seed.

fig. 16.17 Foliage of groundnut

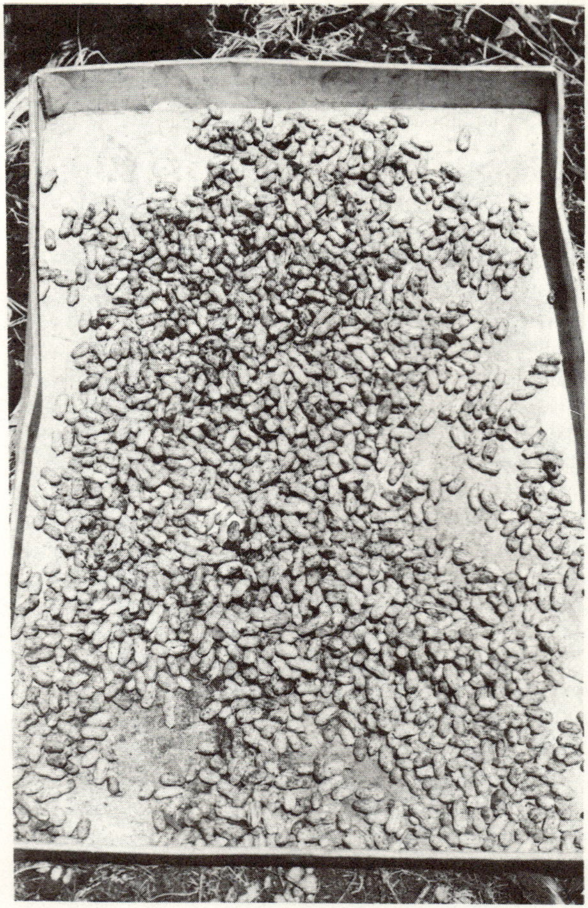

fig 16.18 Freshly threshed groundnut pods drying in a tray

Climate and soil

The groundnut originated in Brazil. It is grown extensively in the sub-tropics but is of minor importance in the humid Tropics. A minimum rainfall of 300–500 mm, depending on the variety, is required during the growing season, and dryness is needed during harvest.

Well-drained, loose sandy soils rich in calcium are most suitable. On heavy soils, fruit formation and harvesting are hindered.

Cultivation and management

There are two groups of varieties: Virginia runner type with large seeds and the Spanish-Valencia erect type with smaller seeds. A fine seedbed, sometimes with ridges, is made. Planting is at a spacing of 30 cm by 15 cm or 60 cm by 10 cm for erect types with a seed rate of 60–80 kg per hectare; 60 cm by 30 cm for runner types at a seed rate of 40 kg per hectare. Seeds should be treated with fungicide and *Rhizobium* bacteria. Groundnuts may need phosphorus, calcium and sulphur; foliar analysis can show the requirements. Gypsum or lime to supply calcium is also needed to prevent empty pods. Erect types mature in about 3 months but runner types take 5 months. Delay in harvest results in losses of yield due to shedding or sprouting. Plants are dug up, dried, threshed and shelled (Fig. 16.18). Yields up to 2 500 kg per hectare of dried nuts can be obtained in the humid Tropics.

Protection

Leaf spot due to the fungus *Cercospora arachidicola* is most common; it can cause defoliation which usually occurs near to maturity. Wilt due to the fungus *Sclerotium rolfsii* is also common. *Aspergillus flavus* attacks seeds giving rise to aflatoxin. This is of importance when peanut meal made from infested nuts is fed to young livestock as it is toxic. Groundnut rust can be devastating to the crop. This has recently spread throughout Malawi and other important groundnut areas. Clean seed and sanitation are essential to prevent spread of rust. Rosette virus is also a serious disease, spread by aphids. Leaf-eating beetles and caterpillars, aphids and thrips are the main pests; they may be controlled with suitable sprays. Weed control is needed in early stages. The pre-emergence herbicide 'Lasso' is very effective.

Soya bean (*Glycine max*)

Family Leguminosae (Papilionoideae)
Other names soy, soybean; feijão soja (Portuguese); soja (Spanish)

The soya bean is of major importance as a source of oil and relatively high quality protein for human and livestock feeding; it contains all the essential aminoacids required by man. The seeds contain 16–24 per cent oil and 30 per cent crude protein.

Botany and propagation

It is an erect legume 60–100 cm tall (see Fig. 16.19). The leaves have three leaflets which they shed at maturity (Fig. 16.20). Small white or purple flowers grow in bunches from the leaf axils. Pods are short, hairy and brown. Seeds are of different colours and sizes.

It is grown from seed.

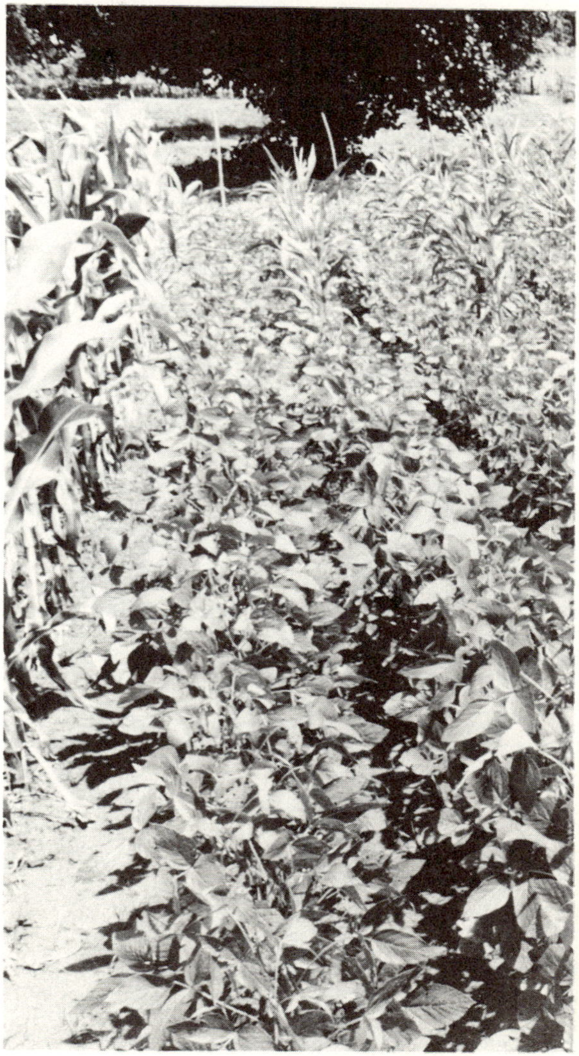

fig. 16.19 Young soya-bean plants

fig. 16.20 Soya bean ripening, note the natural defoliation

Climate and soil

Soya bean is indigenous to the warm temperate or subtropical regions but is now grown extensively from the Tropics to about 52° latitude north or south. Heat resistant and less photo-sensitive varieties are grown in humid Tropics. In equatorial regions maturation occurs in about 90 days and at all times of the year daylength seems to be below the critical level for flowering of the photo-sensitive types. Soya bean grows best on rich sandy loams.

Cultivation and management

Plant breeders are breeding new varieties for the humid Tropics. The variety 'Chipewa' from Nigeria has been found to be well adapted to humid lowland tropical conditions. In the humid Tropics, seeds are sown at a spacing of 45 by 15 cm at rates of 40–60 kg per hectare at a depth of about 2 cm. Lime should be applied on acid soils. 15–30 kg per hectare of phosphorus and 25–40 kg of potassium per hectare are also given. Yields of up to 2 000 kg of dry seeds per hectare and 20 tonnes of green forage per hectare can be expected. After harvest, soya bean is threshed and dried; it is stored at about 12 per cent moisture.

Protection

The most serious pest is bean fly (*Ophiomyia phaseoli*). Spraying with an insecticide such as diptrex will effect control. Other pests are pod borers, blister beetles and leaf-eating caterpillars. Diseases caused by bacteria are bacterial blight, leaf pustule and stem blight. Fungal diseases are wild fire, downy mildew and frog-eye spots; spray with fungicides to control. Virus diseases are soybean mosaic and yellow bean mosaic.

17 Roots

Major crops

Cassava (*Manihot utilissima* syn. *M. esculenta*)

Family Euphorbiaceae
Other names tapioca, manioc; yuca (Spanish).

The crop originated in the American Tropics but it is now widely grown throughout the tropical world. It is an important food crop in many South and Central American countries and in parts of West Africa, where it is made into a product known as 'garri'. The tubers are used as animal feed, for making glucose and sodium glutamate and in the preparation of high grade starch.

fig. 17.1 A cassava stem cutting

fig. 17.2 Cassava stems about 1·5 m long have been planted close together to form a fence which will keep out goats and other intruders. This is a good type of fence to put round a garden. It is strong and, if times are bad, it provides food

Botany and propagation

Cassava is planted from stem cuttings which sprout shoots and roots and produce the new plant. Fig. 17.1 shows a short stem cutting and Fig. 17.2 shows large stem cuttings which have been planted close together to form a fence. When growing the crop, cuttings 20–30 cm long are usually planted. As the shoots grow the cutting also grows, it becomes large and woody and gives off numerous roots. Some of the roots start to swell after about 2 months and grow into tubers as shown in Fig. 17.3. The stem of cassava grows to a height of 2·5 m or more before harvest. It may be unbranched as in Fig. 17.4 or may branch several times as in Fig. 17.5. Some varieties branch regularly and some remain unbranched. Branching is also affected by planting distance, thus widely-spaced cassava will often branch while closely spaced plants will remain unbranched. The leaves of cassava usually have between 5 and 8 lobes and, as shown in Fig. 17.6, they vary a lot in shape

fig. 17.3 Starchy tubers on the cassava plant. The tubers grow in the soil to a depth of 50 cm, they are not easy to harvest unless grown in light peat soil

fig. 17.4 A field of closely-planted cassava at 3 months. The stems do not branch

with different varieties. Quite often varieties with long thin leaf lobes are higher yielding. Although leaf and stem growth continues throughout the life of the plant most of the leaf growth occurs in the first 6 months and after this the leafy canopy may become thinner. Higher yielding varieties often retain their leaves for a longer period.

The tubers of cassava begin to grow after about 2 months and continue to grow for a year or more. The starch is formed in the cells of the tubers. The tubers shown in Fig. 17.3 are about a year old. When left unharvested for more than about a year in the humid Tropics, the tubers start to become woody and the starch content decreases. Yield of starch is greatest at the usual planting distance of about 1 m, after about one year in equatorial regions and after about 15 months in subtropical regions. The flowers and fruits of cassava are unimportant except in breeding new varieties.

Varieties There are hundreds, perhaps thousands, of varieties of cassava which have been given different local names in every region where they are grown. Different varietal characters can be recognised; they include branching habit, shape of leaf and characteristics of tubers. The tubers are covered with a rind which varies in thickness and colour, from pale yellow to pink to dark brown, from variety to variety.

Some varieties have high concentrations of toxic substances (hydrocyanic glucosides) which form cyanide when the tubers are harvested or the tissues are damaged.

These are suitable as food only when especially prepared. Other varieties have 'sweet' tubers with only a little or no toxic content at all. Care must be taken therefore in preparing and eating cassava. Generally varieties with large-sized tubers will be highest yielding; the yield varies from as low as 5 tonnes to more than 60 tonnes per hectare.

Climate and soil

Cassava is grown throughout the Tropics but high yields are only obtained in fairly humid climates. Although the crop is quite resistant to dry conditions, the plants take much longer to produce a moderate yield of tubers and yields are lower. It is widely grown in drier regions, as a reserve or famine crop in case of failure of maize or other crops. It forms the main food crop or is grown as a cash crop in more humid regions. Within the humid Tropics cassava can be successfully grown up to about 1 000 m elevation.

Many soils are used for growing cassava but high tuber yields can only be obtained in friable, light soils. Heavy clays give low yields. Heavy soils should be well cultivated and built into high ridges to improve yield. Organic matter improves most soils for tuber growth. Within the humid Tropics peat soils are especially suited for cassava growing if limed to raise the pH above 4·5. Peat soils are very light and make for easy harvesting.

Cultivation and management

Most soils need to be well cultivated for cassava. If grown on a large scale, deep cultivation by tractor should be undertaken to loosen the soil to about 50 cm. Any previous

fig. 17.5 This variety of cassava branches regularly. Branching is encouraged by widely-spaced planting

fig. 17.6 Two cassava varieties with different leaf types. The type with long, thin leaflets is higher-yielding

crop stubble or vegetation should be ploughed in, to improve the level of organic matter and the soil should be ridged as high as possible. Ridges of 1 m are suitable. If cultivated on a small scale by hand, the mounds of soil should be built about 1·2 m apart to incorporate as much organic matter and top soil as possible. However, if peat soils are used, ridging or mounding is unnecessary but drainage to 60 cm should be undertaken.

Planting material and planting Stems of the desired variety should be obtained in advance of field preparation. They may be stored for several weeks before use, in a dry place. Cuttings of 20 to 30 cm are normally used. If spacing is 1 m, 20 cm cuttings may be used, if wider, then 30 cm cuttings are better. However, if plenty of planting material is available, cuttings of up to 50 cm may be used and this may increase yield at either spacing. Generally any part of the old stem except the green section at the top can be used.

Cuttings should be planted horizontally 3–5 cm below the soil surface at 50 cm spacing on ridges, or in the mounds of prepared soil. Sometimes cuttings are planted at an angle with one end sticking out. This has been recommended for heavy soils and very wet conditions but if the soil is correctly prepared horizontal planting is best.

Fertilisers The quantities of fertilisers required depend on whether a high-yielding or low-yielding variety is grown. Some sweet-tuber types are low yielding but are especially grown as a market vegetable. Usually a high-yielding variety is grown and for a variety yielding 40 tonnes per hectare the fertilisers shown in Table 17.1 may be recommended for humid tropical regions.

Soils with a pH below 4·5 should be limed with magnesium limestone.

Fertiliser should be spread along the ridges but not applied with the cuttings.

Weed control Since soil cultivation is required for growing the crop, weeds would normally be killed prior to planting. Planting should be done promptly to give the cassava a good start before weeds grow. When the crop is fully developed it will shade out most weeds. Mechanical cultivation is usually done after about 4 weeks before the cassava plants become too high for a tractor to pass.

Chemical weed control may be undertaken by using a contact herbicide followed by diuron at 1·5–2 kg per hectare. In peat soils a contact herbicide alone (e.g. paraquat) is recommended because diuron can be harmful on peat.

Harvesting and processing

Cassava is mostly harvested by hand which means a lot of digging. In peat soils however tubers can be pulled with very little effort. Tractor harvesting using a heavy lifting blade is sometimes done but a high proportion of the crop is lost. The stems which are not required for planting elsewhere and the leaves, should be left on the field. The leaves may be used as animal feed or cooked as a vegetable (Fig. 17.7).

Many varieties of cassava contain a toxin and must be processed in special ways. In West Africa the tubers are placed in deep ponds where a type of fermentation neutralises the toxin. After this they become soft. They are then broken open and grated, and the white starchy material, garri, is dried in the sun. For industrial purposes cassava may be processed into starch. Fig. 17.8 shows a starch factory. The crushed tubers are washed and filtered, the starch passing through the cloth filter with water. The starch suspension is settled in tanks. The starch settles as a

Table 17.1 Recommended rates of fertiliser for cassava (in kg per hectare)

	Urea	Ammonium sulphate (VI)	Superphosphate	Potassium chloride	Kieserite
First 5 months	200	50	100	100	100
From 6 months to maturity at 1 year (divided into monthly applications until 3 months before harvest)	100		50	200	

fig. 17.7 Harvesting cassava

(a)

fig. 17.8 Processing cassava into starch
(a) Crushed roots are passed through a rotating filter, the starch then passes through cloth filters
(b) The starch settles in tanks; after a few hours the water is drained off and the starch is dried. This process removes the toxin present in many cassava varieties

(b)

cake and the liquid is drained off to remove any toxin present.

Cassava tubers may be shredded and dried and then pelleted. Much of the cassava used in industry or for animal feed is processed and sold in this way.

Protection

Cassava does not have many insect pests, mainly because of the toxin which is present in all parts of the plant. The major disease is caused by the mosaic virus, which is spread by aphids. The disease is widespread in Africa and elsewhere and causes much loss. Resistant varieties should be grown and any diseased plants removed and burned to prevent spread.

In humid regions bacterial wilt caused by *Xanthomonas manihotii* is sometimes serious. Resistant varieties should be grown where the disease occurs; care should be taken to ensure good soil drainage especially in heavy clay soils.

Cocoyams (*Colocasia* spp. and *Xanthosoma sagittifolium*)

Family Araceae

Other names dasheen, eddoe, tannia; dalo (Fijian); arri, kachu (Indian languages); talla (Malayan); taro (Polynesian); gabi, yautia (Tagalog, in the Philippines)

One species, *Colocasia esculenta,* has two edible varieties *esculenta* and *antiquorum*; the former is harvested as one cylindrical main corm whereas the *antiquorum* variety has a number of smaller corms around the main central corm at the time of harvest. The *esculenta* variety is also called dasheen and taro, while the *antiquorum* variety is often called eddoe (Fig. 17.9). Dasheen and eddoe are thought to originate in Southeast Asia from where they have spread throughout the Tropics. The other species of cocoyam, sometimes also called tannia or yautia, is *Xanthosoma sagittifolium*; this plant also produces corms. Tannia originated from tropical America (Fig. 17.10).

The common names for the three types of cocoyam are often used interchangeably, depending on where they are grown. Cocoyams form the main subsistence crop in many parts of the Pacific and are also popular in the West Indies, West Africa and Asia.

Botany

The plants are 1 to 2 m tall, tannias being the largest, with leaf stalks produced at the top of the central corm (underground stem). Leaves are large, somewhat heart-shaped and the roots are fibrous and shallow.

Propagation is by vegetative means using side corms. In the case of dasheen and tannia, the leaf-bearing shoots of the main corms are planted and the yields have been higher than with lateral tubers. The side shoots of the *esculenta* variety must be pruned to encourage growth of the corm.

Many cultivars of all three species exist; differing in colour, shape, acridity (due to their ethanedioic (oxalic) acid content), fibre content and palatibility. They seldom flower.

Climate and soil

Cocoyams are crops of the humid Tropics; they are adapted to flooding. In drier areas, they can be grown in swampy places or with irrigation. They usually grow best as a first crop after jungle clearing. Soils should be somewhat heavy and moist, with good fertility and sufficient organic matter. The eddoes, are hardier than the dasheens and can be grown with less rainfall on poorer soils but they require good drainage.

fig. 17.9 Dasheen or taro plants. The eddoe plant is similar

fig. 17.10 Mature tannia plants with clusters of corms

Cultivation and management

Cocoyams may be planted in pure stands or interplanted with other field and tree crops. They are planted in appropriately-sized holes to which organic manure may be added. Dasheens and eddoes are planted 60–90 cm apart while tannias are spaced wider at 1·2 m by 90 cm. Mulching may be practised. On other than newly-cleared soil a compound fertiliser with a high potassium content should be used at rates of 200–500 kg per hectare in three split applications approximately 1, 2 and 3 months from planting. Nitrophoska blue special (12:12:17:2) is a suitable compound fertiliser. Cocoyams also respond well to animal manure which may be used at rates up to 25 tonnes per hectare.

Cocoyams mature 5–10 months from planting. Yields of up to 40 tonnes per hectare have been obtained.

The corms of eddoes are obtained by careful digging in order not to leave some of them in the ground, although during dry periods they may be stored in the ground until required.

Tannias produce clusters of tubers to a depth of about 25 cm with a spread about 50 cm. They are first harvested at about 9 months by digging around the base, cutting mature tubers, replacing the soil and leaving the plant to continue growing. Further harvests may be made later, so that it is possible to obtain yield throughout the year.

Dasheens and eddoes produce a very fine-grained starch flour; larger grain flour is made from tannia. The young leaves, both leaf blades and stalks, are used as vegetables and as animal feed. The calcium(II) ethanedioate (oxalate) which causes acridity in both tubers and leaves may be removed by cooking.

Protection

Dasheen and eddoe are seriously affected by a leaf blight, due to the fungus *Phytophthora colocasiae* which causes wet rotted spots on the leaves followed by death of the plant, it is especially rife in wet weather. Fungicides such as dithane may be of use. Serious defoliation is often caused by the larva of the moth *Hippotion celerio* and root-knots by nematodes (*Meloidogyne* spp.)

Tannias are often seriously affected by a root rot due to a virus and another due to the fungus *Pythium irregulare*. The growing tubers are sometimes attacked by the beetle *Lygyrus ebenus* while the stored tubers are damaged by another beetle.

Sweet potato (*Ipomaea batatas*)

Family Convolvulaceae
Other names yam (American English); kumala (Fijian); batata douce (French); batatas (Spanish); camote (Tagalog, in the Philippines)

The crop is grown for its edible tubers. It is also used as a filler in canning and sauces and for the manufacture of starch flour, glucose, syrup and alcohol. The young leaves are used as a vegetable and the fresh vines as well as the tubers are often fed to livestock.

Botany

The sweet potato plant is a creeping annual. The stems are trailing or twining vines, thin, dark green to brown in colour and contain latex. The leaves are held on long stalks and have different sizes and shapes, depending on the variety. They may be simple and entire or divided. In the humid Tropics sweet potatoes flower freely producing large purplish bell-shaped flowers at the base of the leaf stalks. Insects visiting the flowers bring about cross-pollination. The fruits break up when dry to reveal four small black seeds. Tubers are usually produced in the top 25 cm of the soil by thickening of certain roots. The tubers vary greatly in size, shape, colour and quality or taste, depending on variety. Mature tubers contain latex. They are rich in starch and also contain some sugar, protein and fat. Pigmented types of sweet potato are rich in vitamin A and contain some vitamins B and C.

The sweet potato is not known as a wild plant. It originated in tropical America where it has been grown for many centuries. It is today grown all over the Tropics, particularly in Africa and New Guinea. About 100 million tonnes of tubers are produced each year throughout the world. It is also grown to some extent outside the Tropics: in Japan, China, USA and New Zealand.

There are hundreds of different cultivars or varieties of sweet potato. Most cultivars have trailing or twining stems which may be very long indeed but there are bunch varieties which have shorter, more erect stems. The latter are often higher yielding. There are two main groups of varieties, those which produce hard and usually yellow-fleshed tubers and those which produce softer tubers with a sweeter and more watery flesh which becomes soft and gelatinous when cooked, the latter are whitish in colour. Sweet potato varieties also differ in depth of rooting and time of maturity. Well-known American varieties include

fig. 17.11 A variety of sweet potato from Malaysia called 'Biru', with yellow to purple flesh and skin and flesh which is firm and sweet when cooked

'HES 107', 'Red Nancy', 'Orlis', 'Kanda', 'Nemagold' and 'Allgold'. In the West Indies, 'Black Rock' and 'Red Nut' are the most well-known varieties. In Malaysia, the best yielding variety is 'Large-white' giving up to 20 tonnes of fresh tubers per hectare in 4 months. Other Malaysian varieties include 'Serdang', 'Biru' (Fig. 17.11), 'Kangkong' and 'Centennial'. The latter is American in origin with orange flesh and is very good to eat but yields are generally low, of the order of 4–10 tonnes per hectare.

Climate and soil

The sweet potato grows best where the mean temperature is around 25°C or higher with high sunshine and a rainfall of at least 300 mm over the growing period of about 4 months. In places where rainfall is high it is often planted at the end of the rainy season. It cannot stand dry conditions for long periods and irrigation may become necessary. In some places sweet potato is grown in the drier parts of the year only to provide planting materials and out-of-season tubers. In equatorial regions, the crop is grown at altitudes of up to 2 000 m, as in New Guinea.

The crop can be grown on most soils but drainage to prevent waterlogging is important and a sandy loam is ideal. It is a good crop to grow on tropical peat soils as formation of roots is not hindered by the soft light soil; it is a crop that does not demand high nutrients from the soil.

Cultivation and management

The planting material normally used consists of stem cuttings about 30–40 cm long from the tips of the growing mature plants. The bottom leaves of each cutting are removed and the lower half of the cutting is pushed into the soil at an angle. Often partly wilted cuttings are used, for instance, in Uganda. In India it is customary to bury only the central part of the cutting, leaving some part of the stem exposed at either end. In some countries shoots or slips obtained by planting small or medium-sized tubers close together in nursery beds are used for planting. The sprouts reach a length of about 30 cm in 4–6 weeks and are then pulled out for planting.

Sweet potato should be planted on ridges or mounds, see Fig. 17.12. The ridges are usually about 1 m apart and up to 50 cm high. Cuttings are planted on each mound.

fig. 17.12 A small field planted with sweet potato on ridges

The material planted should be carefully selected to ensure that it is free from pests and diseases and that the desired varietal characteristics are present.

The amount and type of fertilisers to apply will depend on the soil type. The crop gives higher yields of tubers when organic manures are applied. Artificial compound fertilisers are applied at rates up to 500 kg per hectare. Too much nitrogen in the fertiliser increases growth of vines and decreases tuber development so a high potassium fertiliser such as nitrophoska blue special (12:12:17:2) should be used.

The crop is weeded when needed and the vines are turned back on to the ridges from time to time to prevent rooting at the nodes, this ensures fewer small-sized tubers and a more uniform crop. The crop matures in 3–6 months depending on the variety. The leaves turn yellow and begin to drop when the crop is ready for harvest. The mature tuber when cut, produces a milky sap which quickly dries up and does not change colour. The tubers are usually harvested all at once and may be stored for several weeks, although they lose much of their moisture in storage. Yields of up to 40 tonnes per hectare may be obtained but the average yield is 15–20 tonnes per hectare.

Protection

The most serious pest of sweet potato is the sweet potato weevil. Adult weevils injure the plants by feeding on the leaves, stems and roots; they drill small holes in the tubers. The grubs do even greater damage by tunnelling through stems and tubers. Two sprays of gamma-BHC or heptachlor at 2 month intervals give good control of this weevil. One other important pest of sweet potato is the caterpillar of the moth *Omphisa anastomosalis*. It bores into the main stem leading to the tubers. Regular sprays of gamma-BHC and carbaryl 'Sevin' give effective control. The leaves of sweet potato are often eaten by caterpillars. These are easily controlled by sprays of gamma-BHC or other insecticides.

There are a few fungal diseases of sweet potato but they are not very serious in the Tropics. Black rot, for example, is caused by the fungus *Ceratocystis fimbriata*. Tubers can be affected by a soft rot on storage caused by *Rhizopus* spp. The fungal diseases can be easily controlled by treatment with suitable fungicides such as dithane.

Yams (*Dioscorea* spp.)

Family Dioscoreaceae

Other names The word 'yam' refers to many species of plants bearing rhizomes or tubers belonging mainly to the botanical genus *Dioscorea*. The genus contains about 600 species of which only about 8 produce edible yams. The botanical and common names and the regions of the origin of these are listed in Table 17.2.

Table 17.2 Edible yams of the humid Tropics and their regions of origin

Botanical name	Common names	Area of origin
D. rotundata	white yam, white Guinea yam	West Africa
D. alata	greater yam, water yam, winged yam, ten-month yam	Southeast Asia
D. cayenensis	yellow yam, yellow Guinea yam, twelve-month-yam, cut-and-come-out-again	West Africa
D. esculenta	lesser yam, Asiatic yam, potato yam, Chinese yam	Indo-China
D. bulbifera	potato yam, aerial yam, bulbil-bearing yam	tropical Asia and West Africa
D. dumetorum	bitter yam, cluster yam	tropical Africa
D. hispida		tropical Asia
D. trifida	cush-cush yam, aja, yampi, mapuey, India yam,	South America

In the humid Tropics, yams together with cassava (*Manihot utilissima*) provide the staple starchy food of millions of people. Yams are also an important secondary food in many other tropical countries. Wild yams are also often used in times of famine or scarcity. Some yams contain organic compounds used to produce steroid chemicals used in birth-control pills and medicines.

The eastern part of West Africa is the most important yam-producing region of the world. About half the world's production comes from Nigeria. Here yams are a staple food and more than 12 million tonnes are produced annually.

D. rotundata (the white yam), *D. alata* (the greater yam) and *D. cayenensis* (the yellow yam) are the most widely grown. In West Africa, *D. rotundata* is the most common species. *D. alata* takes second place in total production but it is the most important yam of the Southeast Asian region, where it first originated, and together with *D. rotundata* accounts for the greater part of the world's yam production. Although popular in most parts of the world, it is less highly regarded in West Africa as, when cooked, it does not give a satisfactory 'fufu' (pounded and steamed yam), an important local dish. *D. cayenensis* is somewhat like the white yam but is also considered inferior in West Africa. Of the remaining 5 species of yam, *D. esculenta* is the most widely cultivated. It is, however, not popular in West Africa because of its poor cooking properties but it is often higher yielding than the first 3 species of yams. The cultivation of *D. bulbifera* is quite limited but it is grown to some extent in most yam-growing countries, especially in tropical Asia and Africa. *D. dumetorum* is said to yield more than most other yams and is grown mainly in tropical Africa. Many wild species are poisonous. Boiling or soaking removes the toxic substance known as dioscorine.

Botany

The cultivated yams have weak stems which climb by twining up supports (Fig. 17.13). Some of them, for example, *D. rotundata*, have stems that remain erect up to about 1 m high before twining. The stems usually produce spines, wings or ridges which help the plants to climb. The vines vary greatly in length. *D. esculenta*, for example, seldom grows to more than 3 m but *D. rotundata* may reach up to 10 m under good conditions. Most of the yams have simple leaves held on long leaf-stalks. Leaves may also be lobed as in *D. trifida* or made up of three leaflets as in *D. dumetorum*. The cultivated yams seldom flower. When they appear, the flowers are very small and white to brown in colour. Male and female flowers are found on separate plants. Fruits look like three-pointed stars in cross-section and break open when dry. Seeds are small, hard and winged to help in their dispersal. Roots normally arise from the top of the tubers and develop for some distance before branching.

The tubers are really underground stems; new ones are formed by the food yams every year, although some wild types produce tubers that last for the whole lifetime of the plant. The cultivated yams produce tubers that differ greatly in number, size and shape. The important species of yams usually produce 1 to 3 tubers per plant; yields of up to 50 kg per plant have been obtained. The usual tuber weight is 5–15 kg. The yams *D. dumetorum* and *D. hispida* form medium-sized tubers in clusters (Fig. 17.14). The tubers of *D. esculenta* and *D. trifida* are small, thin, long and regular shaped; a large number is produced on each plant. *D. bulbifera* produce bulbils at their leaf

fig. 17.13 Leaves of a yam plant

fig. 17.14 The bitter yam

bases; these are also storage organs and the underground tubers are reduced in size. Each species of the cultivated yams has many cultivars which vary in size, number, shape and taste of tubers and also in leaf type and other plant characteristics.

Climate and soil

The food yams need a hot and wet tropical climate to grow well. Temperatures below about 30°C reduce growth rate. During growth the yams need a lot of water, good yam-producing areas have at least 1 500 mm of rain per year. Most important yam species need at least 6 months of rain to produce well.

Yams require a light well-drained but fertile soil in order to allow the tubers to grow freely. When the soil is too wet the yam roots do not develop freely and tubers that form are likely to rot.

Cultivation and management

Yams are usually the first crop planted after land is cleared for cultivation and a crop rotation is often followed, e.g. in West Africa the rotation may consist of yam-maize-cassava-bush fallow. Yams are frequently intercropped; in West Africa the usual intercrops are pumpkins, melons, chillies, maize, okra and other vegetables. Yams are grown as the sole crop only occasionally.

The crop is almost always grown from vegetative planting materials which are small tubers, tuber cuttings or bulbils. Small tubers are used for *D. esculenta* and *D. trifida* as they produce quite a large number of the small tubers. Bulbils are used for planting *D. bulbifera* but the most important yams are grown from tuber cuttings of about 0·25 kg weight.

Where the soil is soft and deep yams can be grown on flat land. The soil is usually raised to form beds, mounds or ridges and the yam setts planted in holes dug in them. Plant spacing is usually about 1 m square or 2 m by 50 cm when planted on ridges (i.e. about 10 000 setts per hectare). The crop is planted together with rich earth and compost in the planting hole. As yams are seldom irrigated, the setts are normally planted at or near the onset of the rains and harvested one year later. The soil is often mulched with a cover of grass, straw or leaves.

Most farmers do not use manures or fertilisers for their yam crop. However, as yams have fairly high requirements for plant nutrients fertilisers are usually recommended; e.g. an N:P:K mixture (12:12:18) is applied from 1 to 5 months after planting at a total rate of around 600 kg per hectare. In the West Indies 5–20 tonnes of dung or compost is recommended to fertilise yams, which may be planted after a legume green manure crop.

As yields are low when the yam vines are left to creep on the ground, they require artificial support. This may be provided by bushes and trees but the usual cultural practice is the use of stakes which should be about 2 m high and preferably branched. A few stakes may be tied together to give them greater strength. Sometimes a wood and bamboo trellis is used or the stems of the previous crop e.g. sorghum or maize, may be used to support the yam vines.

In edible yams, the seed tuber dies and rots away as the vines develop. New tubers appear about 4 months after germination becoming quite large only at 8 to 9 months. Yams are actually perennials but are treated as annuals. They are sometimes harvested more than once a year and replanted, usually in the early part of a dry season when the vine growth is reduced. However, so long as the vines are still growing, the tubers will continue to increase in size and harvesting should be delayed for as long as possible to obtain maximum yield. Sometimes yams are left in the ground and harvested only when required. When yams are harvested more than once, the earth is carefully dug from the tuber in order not to damage the roots too much, the tubers are cut and the soil is returned to the root area of the yam plant. Such an early harvest is often called 'topping', 'beheading' or 'milking' the yam. When the main crop is harvested, care should be taken not to bruise or cut the tubers which could cause them to rot during storage.

Widely differing yields of yams have been obtained. In West Africa yields of 6–15 tonnes per hectare are normal. In Southeast Asia yields are 10–20 tonnes per hectare and in the West Indies even higher yields of 16–25 tonnes per hectare are obtainable. About 2 tonnes per hectare of tubers are needed to replant the crop.

Protection

The food yams are generally fairly free from pests and diseases. The only serious pests are yam beetles of which *Heteroligus meles* is the most common species. The adult beetles make round lesions on the tubers, 1–2 cm in size. Control is by dusting the setts with gamma-BHC. Mealy bugs and scale insects also attack the plant but sprays of malathion or BHC give good control. Nematodes can be controlled to some extent by nemagon etc. but the best control is by rotation and use of pest-free setts for planting. Leaf spots are common, usually due to the fungus *Cercospora carbonacea.* Stored tubers often rot due to infection by a fungus; tubers become watery and soft. Treatment with Bordeaux mixture gives quite good control. In West Africa and the Caribbean, yams are sometimes affected by mosaic virus disease; plants are stunted, leaves mottled and many side shoots develop. Control is by removing and destroying diseased plants.

Minor crops

Arrowroot (*Maranta arundinacea*)

Family Marantaceae
Other names Bermudan arrowroot; West Indian arrowroot.

Arrowroot is grown for its underground rhizome that contains about 20 per cent high quality starch.

Botany and propagation

A herbaceous perennial 1 m tall, arrowroot has numerous aerial stems arising from fibrous roots. The leaves are

fig. 17.15 Arrowroot plants growing on peat soils

fig. 17.16 Some uprooted rhizomes of arrowroot

spear-shaped 20–30 cm long and the leaf stalks clasp the stem. The white flowers grow in loose bunches and are usually removed. The white rhizomes, grow 20–30 cm long and 3–5 cm across, they are covered with brownish scales.

Arrowroot is propagated from bits or the tips of rhizomes.

Climate and soil

Native to tropical America arrowroot is now chiefly produced in St Vincent in the West Indies. It needs a warm climate and rainfall of 1 500–2 000 mm yearly. It may be grown up to 1 000 m elevation and grows best on sandy volcanic soils.

Cultivation and management

There are two cultivars from St Vincent: 'Creole' and 'Banana'. Soil should be ploughed and well cultivated. Bits are selected and planted as shown in Fig. 17.15 at about 30 cm spacing, requiring 3–3·5 tonnes of fresh bits per hectare. Compound fertiliser should be applied at 3 months. The leaves yellow and wilt at maturity 10–12 months after planting. Harvesting should be done not later than 12 months; the rhizomes are dug up (Fig. 17.16), cut off, washed, crushed and sieved in water to extract the starch. The starch is collected and dried. Yields of 2–4 tonnes of starch or about 10–20 tonnes of fresh rhizomes per hectare may be obtained.

Protection

Weeding should be carried out when needed while the crop is establishing. The leaf roller *Calpodes ethlius* causes defoliation. A fungal burning disease is controlled by proper drainage of moist soils.

Edible canna (*Canna edulis*)

Family Cannaceae

Other names Queensland arrowroot, purple arrowroot, achira, tous-les-mois (West Indies); adeira (Spanish, in Peru)

The purple-coloured rhizomes are eaten as a vegetable or used for starch extraction. The starch grains are very large and therefore easily digestible; it is useful as invalid or infant food. The leaves may be fed to livestock.

Botany and propagation

Edible canna is a herbaceous perennial 1–2 m tall with broad, pointed bronze-coloured leaves. The purple stems are in clumps each with 15–20 suckers and covered with short leaf sheaths. Flowers are orange-red, produced at the top of the plants. Rhizomes are fleshy, purple, somewhat cylindrical and up to 60 cm long, see Fig. 17.17. They are ringed by scale scars and the roots and suckers arise from them.

Propagation is from the terminal portions of the rhizomes.

fig. 17.17 Young shoots and rhizomes of edible canna

Climate and soil
It is a native of the West Indies and Central America and is commercially grown in Queensland, Australia. It needs a fertile well-drained soil rich in organic matter.

Cultivation and management
The rhizome divisions or 'seeds' are planted at a spacing of 1 m apart. The crop needs to be fertilised to yield well. It matures in 8–12 months, yielding about 38 tonnes of rhizomes and about 50 tonnes of tops per hectare.

Harvesting and processing
Starch is extracted by crushing and washing, as for arrowroot (p. 217). The whole rhizomes may be eaten baked or boiled.

Protection
As for arrowroot p. 217.

18 Spice, flavouring, perfume and colouring crops

Major crops

Pepper (*Piper nigrum*)

Family Piperaceae
Other names poivre (French); pimenta do reiro (Portuguese); pimienta (Spanish)

Commercial pepper is obtained from the fruits of the perennial climbing vine. It is one of the oldest and most important of the spices used extensively to flavour food, preserves and sauces. Two forms of the spice are available, black pepper and white pepper. The black type consists of the whole dried fruit, while white pepper is the dried fruit from which the skin has been removed.

Pepper originated in the south-western coastal region of India from where it spread to many parts of Asia, Africa and South America. Today the Southeast Asian countries of Malaysia, India and Indonesia together with Brazil produce about 90 per cent of the world's pepper.

Botany
The pepper plant has a fairly thin dark green stem with many slightly swollen nodes from which arise the leaves, adventitious roots which anchor themselves on to supports, and the inflorescence spikes (Fig. 18.1). In the wild state, the plant reaches a length of 10 m or more, but its growth under cultivation is controlled to ease harvesting. Its spike is characteristic of the members of the family Piperaceae to which pepper belongs and consists of 50–100 small flowers, yellowish-green in colour and without any visible stalk. The flowers are usually bisexual, although male flowers are sometimes produced at the tips of the spikes. In wild plants, unisexual flowers are produced. Pollination is usually brought about by wind although rain and ants may also act as pollinating agents. The fruit changes from dull green to bright red and at maturity consists of the peppercorn surrounded by a reddish and fleshy skin. Pepper contains the alkaloids piperine and piperidine. The level of the former, present in

fig. 18.1 Berries of pepper plant

much larger quantities (5–8 per cent) is an indication of the purity of pepper. The pungency of pepper is due to the resin chavicine which is most abundant in its fleshy skin; for this reason black pepper is more pungent than white pepper. The aroma is produced by volatile oils. Although wild pepper shows considerable variation, especially in the size of leaves and fruits, only a small number of cultivars have been selected on the basis of their adaptability to soil and climatic conditions, resistance to foot rot (caused by *Phytophthora palmivora*) and evenness of ripening. India has produced the greatest number of cultivars, of which 'Balancotta' and 'Kalluvali' are the most important. 'Balancotta' commonly produces high and regular yields and has large fruits. In Sumatra the most commonly cultivated cultivar is 'Belantung' which is fast growing, has an upright, cylindrical stem and is somewhat resistant to foot rot. It is grown mainly for black pepper. On the Indonesian island of Bangka, 'Lampong' and 'Muntok' ('Bangka') are widely grown mainly for the production of white pepper. 'Lampong' produces large light-coloured seeds of uniform size but it is quite susceptible to foot rot, while 'Muntok' bears dark-coloured seeds of irregular size and is only moderately susceptible to foot rot. In the Sarawak and Johore regions of Malaysia, the large-leaved high-yielding 'Kuching' is extensively grown, although it is generally susceptible to foot rot. The Indonesian variety 'Djambi' which has long fruiting spikes with small fruits which mature somewhat irregularly is only moderately susceptible to foot rot, and has been shown to yield almost as well as 'Kuching'. A less common variety in Sarawak is 'Sarikie' which normally gives 15–20 per cent less yield than 'Kuching' and produces small leaves. The main yields of some of the above cultivars of pepper are shown in Table 18.1.

Table 18.1 Yields of some pepper varieties in Sarawak, Malaysia

Source country	Cultivar	Yield of fresh fruits (27 per cent is white pepper); figures are the mean of 7 years (1962–69) in kg per hectare
Malaysia	'Kuching'	20 268
Indonesia	'Djambi'	19 965
Indonesia	'Bangka' ('Muntok')	16 804
Indonesia	'Belantung'	16 714
India	'Ithirancotta'	13 205
India	'Kaluvalli'	14 394
India	'Balancotta'	14 315
India	'Cheriakaniakadan'	8 856

Climate and Soil

Pepper needs a hot and uniformly wet climate with about 3 000 mm of rain annually; the plant is unable to withstand prolonged dryness. It thrives best near the equator, although it can be successfully grown as far as 20° latitude north or south of the equator. The vine can be grown at altitudes of up to 1 000 m, but low elevations give the best results. Level ground is the most suitable, provided there is no flooding. Hill slopes may also be used but here prevention of soil erosion by terracing or mulching should be carried out.

Pepper succeeds best on well-drained soils rich in organic matter and low in soil acidity. Waterlogged conditions result in poor growth and a high incidence of foot rot. Organic matter supplies plant nutrients and ensures good water-holding properties which become important during dry spells. Pepper has been successfully grown on a variety of soils. In Malaysia, it is planted on reddish brown or yellow latosols and in Indonesia on sandy clay soils.

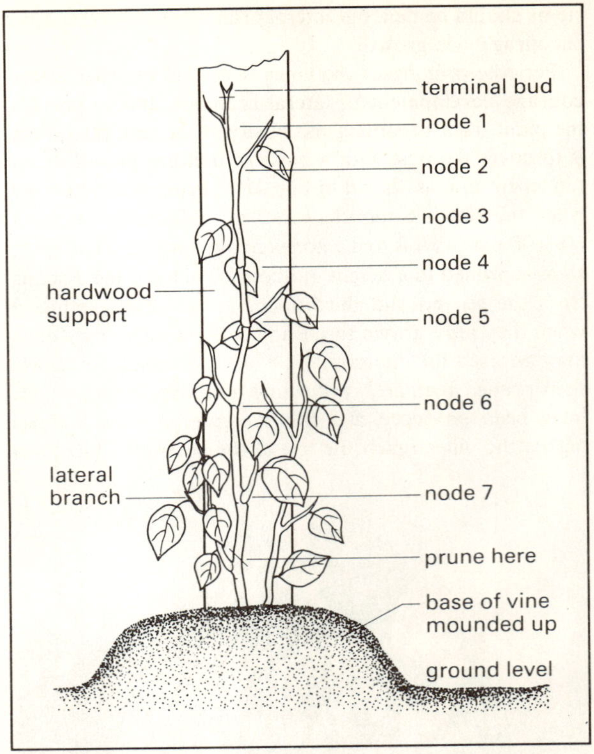

fig. 18.2 A pepper vine about 6 months after planting. The three main stems must be pruned back to encourage side branching and to obtain cuttings for further planting. The terminal bud is pinched out and the leaves or side branches on nodes 3 to 7 are removed; 7–10 days later the stem is pruned at the point indicated. (*Source*: Ministry of Agriculture pamphlet, Malaysia.)

Cultivation and management

Pepper is usually propagated by terminal cuttings taken from 1 to 2 year old plants. Propagation from seeds is seldom practised as the vines take a longer time to mature. Fruiting starts only 5 to 7 years after planting, compared with about 3 years for cuttings. Terminal cuttings are taken from parent vines. The parent vines are selected for high yields, vigorous growth, abundant rooting and absence of disease about 6–8 weeks before planting. The cuttings are raised in a nursery before planting out in the field. Cuttings should not be taken from lateral branches. Three to 4 cuttings may be taken at intervals of about 3 months from a vigorous vine before it is allowed to flower freely. Before taking a cutting, the terminal bud of the selected shoot is pinched off and the leaves and small branches between the 3rd and 7th nodes from the apex are stripped off. When a new terminal bud has regenerated, normally about 10 days later, the shoot is cut below the 7th node (Fig. 18.2).

The most satisfactory way of raising the cuttings is in nursery beds which should be sited on well-drained soils in a sheltered position. The cuttings are normally planted obliquely with 3 to 4 nodes below the soil surface, 15–20 cm apart in rows. They are usually shaded and frequent watering is necessary to maintain a high humidity. After 6 to 8 weeks in the nursery the rooted cuttings are ready for transplanting in the field. They are planted on mounds of soil at an angle of 20–40° with the rooted nodes underground. The aerial portions should be tied on to temporary supports. Intensive shading is necessary at this stage. The shade is gradually reduced and completely removed after 8 weeks. The plant spacing should be 2–3 m. The temporary supports must be replaced by permanent posts 4 m long and planted 60 cm deep in the ground about 15 cm from the pepper vine not later than 4 months after transplanting. Hardwood and sometimes concrete posts are used since the economic life span of pepper is about 15 years. The vines should be tied to the posts at frequent intervals; each internode on 3 or 4 main stems should be tied, but lateral branches should be left to encourage side growth.

Periodic pruning of the vines is needed in order to encourage development of lateral branches and to prevent the plant from becoming too bushy at the top; the object is to cover the post with a mass of uniform growth from top to ground, as shown in Fig. 18.3. Pruning is first done when the vine is 8 months to 1 year old and its height is 60–100 cm (with 8 to 12 nodes above ground). The main stem is pruned to a height of about 20 cm and the 2 stems from the second and third nodes are cut back similarly when they have grown to the same height. (These cuttings may be used for planting.) The top 3 climbing stems are then pruned regularly, each time a further 8 to 9 nodes have been produced and this is repeated 7 or 8 times before the vines reach the top of the supports. After this the terminal buds are removed periodically to prevent further vertical growth. Side branches and leaves near the ground are removed and during the first 2 years, any flowering spikes which form are also stripped off to prevent premature fruiting. Selective leaf plucking is sometimes practised to increase side branching.

fig. 18.3 A mature pepper plant

Weeding may be done manually or with herbicides, e.g. paraquat. Mounding which improves rooting may be carried out at the time of weeding and the weeds and other materials may be used as a mulch which improves pepper growth by reducing moisture loss and giving protection from high temperature.

Pepper has a high requirement for nutrients. Organic fertilisers have been traditionally used, e.g. bat guano, shrimp waste, farmyard manure, etc., but there is a growing use of well-balanced inorganic fertilisers. The rates of a compound fertiliser (N:P:K:Mg) shown in Table 18.2 may be recommended for application every 3 months.

Table 18.2 Recommended rates of compound fertiliser for pepper

	kg per hectare	g per plant
First year	250	100
Second year	500	200
Later years	750	300

Dolomitic limestone and minor elements are also recommended and the compound fertilisers such as nitrophoska blue special may be recommended at a rate of 2 kg per vine, particularly on acid soils.

The vines should not be allowed to flower freely until they are about 2 years old. Since it takes 8 to 9 months for the fruits to mature, the first harvest is possible only at about 3 years after planting. The whole spike is picked after some of the fruits have become red and the rest are yellow or green. The pepper must be picked at least once a week. Yields increase from the first picking to a maximum between the 5th and 10th year from planting after which yields decline gradually.

Harvesting and processing

Black pepper is produced from fruits picked when mature but still green. They are dried in the sun for 3–4 days and the stalks removed by rubbing against a wire mesh and winnowing. The pepper may be scalded in boiling water before drying to give a better product. About 33 kg of black pepper is produced from 100 kg of green pepper.

White pepper is prepared by soaking the ripe berries.

The fruit spikes are crushed slightly and then soaked in sacks under running water for 7 to 10 days. They are then trampled to loosen the stalks and skins which are removed by washing and rubbing with the hands in sieves. The pepper is then resoaked for 1 or 2 days and washed again to ensure a good quality product. The peppercorns are dried on mats in the sun, for 3–4 days. 100 kg of newly picked green pepper gives only 27 kg of white pepper.

Protection
The fungus *Phytophthora palmivora* causes foot rot, the most serious disease of pepper. The underground stem and roots are destroyed causing the vines to droop, the leaves turn yellow and drop off. The disease is spread by snails, water, plant materials and contaminated soils. As there is no cure, precautions must be taken to prevent the disease. Other fungal diseases which are fairly common are pink disease (*Corticum salmonicolor*) and white root disease (*Fomes lignosus*). Other less serious diseases are foot rot (*Pythium splendens*), anthracnose (*Colletotrichum piperis*) and black berry caused by a parasitic alga. Control of these diseases is through crop hygiene, removal of infected plants, foliage sprays or collar painting with fungicides such as Bordeaux mixture and PCNB.

The pepper weevil (*Lophobaris serratipes*) damages the fruits and the bugs *Dasynus piperis* and *Diconocoris hewetti,* cause premature fruit fall. Other important pests include the nematodes *Meloidogyne javanica* which causes root-knots, and *Radopholus similis* which burrows into the roots.

Minor crops

Allspice (*Pimenta dioica*)

Family Myrtaceae
Other names Jamaican pepper, pimento

Allspice is considered to combine the flavours of cinnamon, cloves and nutmeg.

Botany and Propagation
It is a tree reaching 7–10 m in height with smooth grey bark. Leaves grow in two opposite rows and are spear shaped. The greenish white flowers are carried on short branches. The flowers are functionally either male or female although bisexual in structure. A tree may bear entirely male or female flowers. The fruits are round and reddish brown, with two seeds.

Allspice is grown from seed in nursery beds. Seedlings may be transplanted in bags at 2–4 leaf stage and in the field when 25–40 cm high (i.e. about 1 year old).

Climate and soil
Allspice grows best in hot and fairly wet climates on islands with elevation of less than 1 500 m and rainfall of 2 000–3 000 mm per annum. Limestone soils appear most suitable. Soils must be well drained.

Cultivation and management
Three seedlings are planted per point at spacings of 8–10 m square and thinned down to 1 at first bearing. Three kg per tree of a complete fertiliser should be applied every year. Pimento starts fruiting in 5–6 years and maximum yields are obtained at 15 years. Fruits are harvested before fully ripe, and dried. Yields per tree of up to 45 kg per year may be obtained. In some places the leaves are harvested for oil extraction.

Protection
Pimento leaf rust due to the fungus *Puccinia psidii* causes defoliation; control by spraying with zinc or copper oxychloride. Termites and wood borers are sometimes serious pests and can be discouraged by the use of dieldrin around the tree base.

Annatto (*Bixa orellana*)

Family Bixaceae
Other names bixa; roucou (French); achiote (Spanish)

A red dye is extracted from the fruit; it is used for colouring cosmetics, confectionery and food.

Botany and propagation
Annatto is a tall shrub or tree 2–4 m in height, branching from the base of the trunk. The leaves are heart-shaped growing to 10–15 cm long. White or pink flowers grow in clusters. The heart-shaped, hairy, red fruit produce seeds containing the red pigment, annatto.

It may be grown from seed in nursery beds or bags, but is sometimes propagated by cuttings.

Climate and soil

Native to Central America and the West Indies it grows well in hot humid climates at 0–1 200 m elevation and 1 750–2 500 mm annual rainfall. Deep, well-drained soils are ideal.

Cultivation and management

Annatto is usually planted at a spacing of 3–4·5 m square but lower densities may be used on better soils. It is often grown as a hedge. Bearing starts at about 2 years. The capsules are harvested when brown. Dried seeds are macerated in water when the red dye bixin or annatto from the seed pulp settles to the bottom and is dried into a cake. The dye content of the seed is 5–6 per cent.

Protection

No major pests or diseases.

Bay (*Pimenta acris* or *P. racemosa*)

Family Myrtaceae

Bay oil, distilled from the bay leaves is a valuable essential oil containing 50–75 per cent eugenol. The oil is used mainly to make bay rum, useful in the manufacture of toiletries, hair tonic, etc.

fig. 18.4 Leaves and flowers of bay

Botany and propagation

The bay tree reaches about 8 m in height with scaly bark, leathery leaves each about 7 cm long and small white flowers carried on the branches. The fruits are black when ripe (Fig. 18.4).

It is usually grown from seed in containers and later transplanted in the field.

Climate and soils

The tree is native to the West Indies and Central America; it is only important in the West Indies. It requires a well-drained soil of good fertility.

Cultivation and management

The trees are usually planted close together and pruned to a height of about 1·5 m so that the leaves may be picked easily. Leaves are harvested 2 or 3 times a year and partially dried before baling for sale.

Protection

The crop is quite free from serious pests and diseases. Weeding is needed regularly.

Cardamom (*Elettaria cardamomum*)

Family Zingiberaceae

The dry fruits are used as a spice.

Botany and propagation

Cardamom is a herbaceous perennial with underground rhizomes. The aerial shoots form clumps up to 3 m tall (Fig. 18.5). The leaves with leaf sheaths and spear-shaped blades may grow up to 1 m long. The flowering stem tends to be horizontal. Flowers are similar to those of ginger, pale green in colour. The fruit or capsule is oblong and brown (Fig. 18.6).

It is propagated from seeds or rhizomes.

Climate and soil

Native to India and Sri Lanka, India now produces 80 per cent of the world's supply, cardamom grows best under light shade in free-draining soil rich in humus. It requires altitudes of 750–1 500 m and rainfall exceeding 2 500 mm yearly.

fig. 18.5 A clump of cardamom plants

fig. 18.6 Fruits of cardamom

Cultivation and management

There are two varieties; Indian type (malabar cardamom) and Sri Lanka cardamom. It is generally planted at a spacing of around 2 m. Planting holes should be given top soil and organic manure. Light shade is sometimes given. N:P:K fertilisers and organic manures are needed. Bearing starts in the 4th year; it must be replanted after the 15th year. Capsules are picked when nearly ripe 5 or 6 times a year. Yields of 100–350 kg dried capsules per hectare may be obtained. The capsules are dried and cured.

Protection

A virus disease ('foorkey' or 'katte') occurs in India and is spread by aphids. Affected clumps should be killed by injecting 40 per cent 2,4–D. Spraying with insecticides may be used in control. Thrips and caterpillars are controlled by insecticides.

Cinnamon (*Cinnamomum zeylanicum*)

Family Lauraceae

Other names true cinnamon or Sri Lanka cinnamon. Other species are *C. cassia* (Chinese cinnamon); *C. tamala* (Indian cassia) and *C. burmanii* (padang cassia).

The spice is obtained from the bark of this evergreen tree (Fig. 18.7).

Botany and propagation

The leaves are purplish when young. The small, yellow flowers grow in bunches. The fruits are black, about 1·5 cm long and pointed.

Cinnamon can be grown from cuttings but is usually grown from seeds sown in beds or polythene bags and transplanted at about 6 months.

Climate and soil

It is native to Sri Lanka and south India and commercially grown on white sandy soils in Sri Lanka and also in the Seychelles, Madagascar and Brazil. It is suited to hot and humid Tropics where rainfall is 2 000–2 500 mm yearly.

Cultivation and management

Planting is at a spacing of 2–2·5 m between trees. Stems are cut back after 2 or 3 years and later shoots are

fig. 18.7 A young cinnamon tree at 3 years

thinned down to 6 per tree. Harvesting starts at between 4 to 5 years; stems (1·5–5 cm thick and 2–3 m long) are cut close to the ground during the rainy season; it is harvested every 2 or 3 years. The bark is removed and the outer corky layer stripped and rolled into 'quills' (Fig. 18.22, p. 232). The best yields exceed 150 kg of dried quills per hectare, these occur at about the 10th year.

Protection

Cinnamon is free from major pests and diseases. Weeding is needed until shade is complete.

Citronella Grass (*Cymbopogon nardus*)

Family Graminae
Other names serai wangi (Malayan)

It is grown for the citronella oil contained in the leaves.

Botany and propagation

This vigorous grass reaches 1–2 m in height (Fig. 18.8), the leaves arise from thin stems growing up from short rhizomes.

It is planted using rhizome divisions.

fig. 18.8 A clump of citronella grass

Climate and soil

It is native to Sri Lanka. Citronella grass may be grown on a wide variety of soils in the humid Tropics. It grows at low elevations.

Cultivation and management

There are two groups of varieties; Sri Lanka type which is hardier, with lower oil content and narrower leaves; and the Java type which is less hardy and yields twice as much oil as the Sri Lanka type. Spacing is 0·5—1 m apart. Harvesting is carried out when the plants are 1–1·5 m tall, it is first cut at 6–9 months and later every 3–4 months. Yields of 25–45 tonnes of fresh grass per hectare per year may be obtained. The fresh grass contains 0·3–0·8 per cent oil. The oil is extracted by steam distillation, yielding up to 300 kg of oil per hectare.

Protection

Relatively free from pests and diseases.

Cloves (*Eugenia caryophyllus*)

Family Myrtaceae
Other names clou de girafle (French); cravoda India (Portuguese); clavero giroflé (Spanish)

The major product is the spice obtained from the dried flower buds. The stems and leaves are used for the production of clove oil.

Botany and propagation

This is a medium-sized tree growing 10–15 m high (Fig. 18.9), the leaves are oval and opposite, with oil glands. Pink flowers are produced in groups of 3 on short stalks (Fig. 18.10).

It is grown from seeds raised in a nursery for 1 year; seedlings emerge in about 5 weeks and are transplanted when 15–25 cm tall after 1 or 2 months of hardening.

Climate and soil

Native to the Moluccas, cloves are adapted to the humid Tropics; they do best with 1 200–2 500 mm annual rainfall near the sea and at around 300 m elevation. Soils should be well drained, rich in humus, deep and loamy.

fig. 18.9 A clove tree

Cultivation and management

Densities of 100–165 trees per hectare are usually adopted. Trees start flowering after 5 years but give economic yields only at 7–10 years, reaching a maximum at about 20 years; lifespan is 60–100 years. Organic manures and nitrogenous fertilisers are needed. Mulching is beneficial. Flowers are picked before blooming when just pink and the flower buds dried (Fig. 18.11). Clove oil is extracted from the stems and leaves. Each tree yields 3–12 kg of cloves per year.

Protection

Young plants are usually circle weeded. The fungus *Valsa eugeniae* causes 'sudden death' and *Cryptosporella*

fig. 18.10 Clove flowers growing at the tips of branches

fig. 18.11 Cloves drying in the sun on a woven mat

eugeniae causes die-back. Diseased trees should be removed and burned. In heavy soils, drains should be dug. Pests include a parasitic alga, termites and the trunk borer *Chrysotypus mabilianum.*

Ginger (*Zingiber officinale*)

Family Zingiberaceae
Other names gingembre (French); jingibre (Spanish)

Ginger oil, dried ginger, preserved ginger, etc. are made from the underground rhizomes. Ginger is also used to flavour foods and to produce drinks.

Botany and propagation

Ginger is a perennial herb growing up to 100 cm tall (Figs 18.12 and 18.13). Leaf sheaths cover the stem. The leaves are spear-shaped and the plant rarely flowers. The inflorescence consists of a cone of bracts with pale yellow flowers. Rhizomes are short, branching and scaly.

It is grown from small portions of rhizomes, 3–5 cm long.

Climate and soil

It probably originated in Asia and grows in the humid Tropics with at least 1 500 mm of rain yearly and a short dry season. It is planted from sea level to 1 500 m elevation. Soils should be loamy, rich in humus and well drained.

fig. 18.12 A field of young ginger plants

fig. 18.13 A clump of ginger

Cultivation and management

A number of clones exist in India, Malaysia, Jamaica and China. Soil should be well cultivated and ridged. Plant spacing should be 50 cm by 25 cm (900–1 500 kg of rhizomes per hectare). Lime, organic manure and inorganic N:P:K fertilisers should be applied. The crop should be harvested at 6–10 months after planting when leaves turn yellow and wilt. Yields of up to 30 tonnes of fresh ginger per hectare may be obtained.

Protection

Bacterial wilt *Pseudomonas solanacearum* is the most serious disease. Soft rot of rhizomes is caused by *Pythium* spp. fungi. The rhizomes should be treated with fungicides before planting and only healthy material should be planted. In the Philippines, red rot and black rot are common. Leaf spot fungi can be controlled by regular fungicide sprays.

Stem borers and root-knot nematodes are sometimes a problem.

Lemon grass (*Cymbopogon citratus*)

Family Graminae
Other name serai (Malayan)

fig. 18.14 A clump of lemon grass

Lemon grass is used as a flavouring for food in Southeast Asia and elsewhere. Its botany and cultivation resemble those of citronella grass, see p. 225. Lemon grass is smaller (Fig. 18.14) than citronella grass and is planted at a spacing of 1 m by 0·5 m. It should be cut 4–8 months after planting and then every 3–4 months; the yield of grass is 20–25 tonnes per hectare; the yield of oil is 40–110 kg per hectare.

Mint (*Mentha* spp.)

Family Labiatae
Other names peppermint, spearmint

The leaves are used as a pot-herb and as a source of essential oils. These oils are used to give the mint flavour in chewing gum, toothpaste, confectionery, medicine, etc.

Botany and propagation

The two major mint plants grown in the humid Tropics are *M. piperita* or peppermint and *M. spicata* or spearmint. They are small perennial herbs which produce underground stems (Fig. 18.15). Spearmint is lighter green in colour and has longer leaves. The crop is usually propagated from seeds or vegetatively from stem cuttings.

fig. 18.15 A bush of mint

Cultivation and management

The mint crops prefer rich, moist soils and may be grown in containers under partial shade. They are planted at a spacing of about 30 cm by 15 cm. The shoots are cut and the leaves collected. They may be used fresh, dried for storage or distilled to extract their essential oil.

Protection

Rust diseases of the leaves sometimes occur; spraying with sulphur is useful to keep the disease in check.

Nutmeg and mace (*Myristica fragrans*)

Family Myristicaceae

Two spices are obtained from this plant. The seed is known as nutmeg and the lacy scarlet aril around the seed is known as mace.

Botany and propagation

The cultivated tree grows up to 10–15 m tall; it has a short trunk and spreading branches. The leaves are 10 cm long, spear-shaped and alternate. Male and female flowers occur on different trees; the flowers are light yellow, bell-shaped, small and single or in bunches. The round fruit are 5–7 cm in diameter and yellow when ripe (Figs 18.16

fig. 18.16 Freshly plucked nutmeg fruits

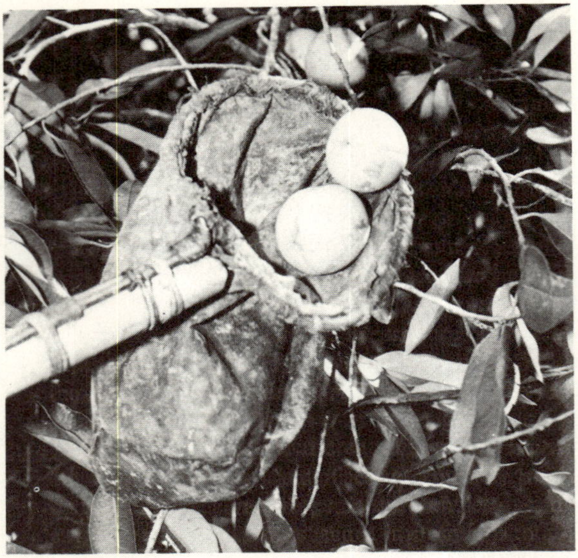

fig. 18.17 Plucking nutmeg fruits

fig. 18.18 Mace from nutmeg fruits

and 18.17). Inside the pericarp is the red mace covering the brown seed. (Fig. 18.18).

Climate and soil

A warm moist climate and deep loamy soil is required. It grows best in sheltered valleys near the sea with a rainfall of 1 500–2 500 mm per year and protection from wind. Soil must be well drained.

Cultivation and management

Nutmeg is planted with 3 seedlings per point in a triangular pattern 9 m apart. Two trees are removed from each point at first flowering, in 3–5 years this gives about 80 per cent female trees. Male trees in excess of 10 per cent are replaced. Manuring is carried out as for cloves, see p. 226. Bearing starts at 5–7 years, declines after 15–20 years and continues for about 50 years. Yields per year of 75 kg dried mace per hectare, 500 kg dried seeds per hectare and 3 000 kg of the fresh fruit flesh per hectare may be expected. The flesh can be used to make sweet pickle, and oil can be extracted from the fruits.

Protection

Nutmeg is relatively free from serious pests and diseases.

Patchouli (*Pogostemon cablin*)

Family Labiatae

Oil distilled from the leaves is used in making perfumes and for flavouring.

Botany and propagation

Patchouli is a small herb reaching 1·5 m. The stems are square and much branched; the leaves are opposite, with toothed margins. Small, purplish-white flowers grow in bunches at the ends of the branches.

Propagation is from semi-mature, 30–45 cm long stem cuttings planted directly in the field (Fig. 18.19).

Climate and soil

Native of the Philippines, patchouli is now grown in many tropical regions including Madagascar, West Africa and Brazil. It grows well in hot humid climates with even rainfall. Fertile soil rich in organic matter is most suitable.

fig. 18.19 A patchouli cutting taken for planting. (*Source*: Ministry of Agriculture leaflet No. 33, Malaysia.)

fig. 18.20 Rhizomes of turmeric

Cultivation and management

Cuttings are planted 1 m apart during the rainy season with 3 nodes below the soil surface; 300 kg of ammonium sulphate(VI) per hectare per month should be applied during establishment when planting on poor soils.

Harvesting starts at 6–7 months and is done every 4–6 months for 2–3 years. The plant is cut to 8 cm above ground. Branches and leaves are dried for 2–3 days. Yield from the first cutting is about 1 000 kg dried leaves per hectare; subsequent cuttings give 500–1 000 kg per hectare. Complete high-nitrogen fertilisers should be given after every cutting (about 300 kg per hectare) to maintain yields.

Protection

Regular weeding is needed. Patchouli is relatively free from pests and diseases. Die-back occurs on poor soils but this can be overcome with fertilisers.

Turmeric (*Curcuma domestica*)

Family Zingiberaceae

A bright orange spice is obtained from the underground rhizomes.

Botany and propagation

The plant grows 100–150 cm tall and is similar to ginger. The leaves are spear-shaped with broad clasping leaf stalks; the flower bunch is similar to ginger with greenish white to pink flowers.

It is propagated from rhizomes with one or two buds; 1 700 kg of rhizomes are needed to plant 1 hectare (Fig. 18.20).

Cultivation and management

A number of cultivars occur in India. Climate and soil, land preparation, spacing and fertilisation as for ginger. Turmeric should be harvested at 9–10 months when leaves become yellow. Yields of 13–35 tonnes per hectare of fresh turmeric are obtained.

Protection

Leaf spot can be controlled by fungicides. A shoot-boring caterpillar is the most serious pest; it is controlled by spraying with insecticides early in life.

Vanilla (*Vanilla fragrans* or *V. planifolia*)

Family Orchidaceae
Other names vanille (French); bannitta (Portuguese); vainilla (Spanish)

An essence called vanillin is extracted from the pods. This is widely used in confectionery.

Botany and propagation

Vanilla is a climbing vine-like perennial growing up to 5 m tall (Fig. 18.21). The stems are green and fleshy, they are supported by aerial roots. The leaves alternate and are also fairly fleshy. Greenish-yellow flowers in bunches arise from the leaf axils. The fruit is a 3-angled capsule 15–28 cm long, called a 'bean' and containing many small seeds. The beans are dark brown, oily and smooth in appearance (Fig. 18.22).

fig. 18.21 Vanilla vines growing on wooden stakes in Seychelle Islands. The plants have been mulched with coconut husks

fig. 18.22 Bundles of dark brown dried vanilla pods in the foreground with quills of rolled cinnamon bark in the background (see p. 225)

It is propagated from cuttings 80–100 cm long.

Climate and soil

Vanilla originated in Central America; it thrives in well-drained soils rich in organic matter. It grows from sea-level to 200 m elevation in humid Tropics and requires a rainfall of 2 000–2 500 mm yearly. A brief dry period is needed during fruit-ripening.

Cultivation and management

It is planted at a spacing of 0·9–1·3 m by 1·8–2·5 m; cuttings should be tied to stakes or prop-plants such as

Casuarina trees, they require support for 10–12 weeks. Mulching is beneficial. Organic manure and a foliar spray of 1 per cent urea should be given. Beans are produced from the third year; maximum yields can be achieved in 7–8 years. Artificial pollination should be carried out in the morning to increase yield. Bean formation can also be brought about by spraying with plant hormones, e.g. 2,4–D, IAA and IBA. Beans are harvested manually 4–9 months after pollination when they turn from green to yellow. Yields of about 120 kg of dried beans per hectare per annum may be obtained.

Beans are cured by successive periods of drying and sweating, until the aroma of vanillin develops.

Protection

There are no major pests or diseases but sometimes wilt and fungal root rot occurs. These can be discouraged by improving drainage.

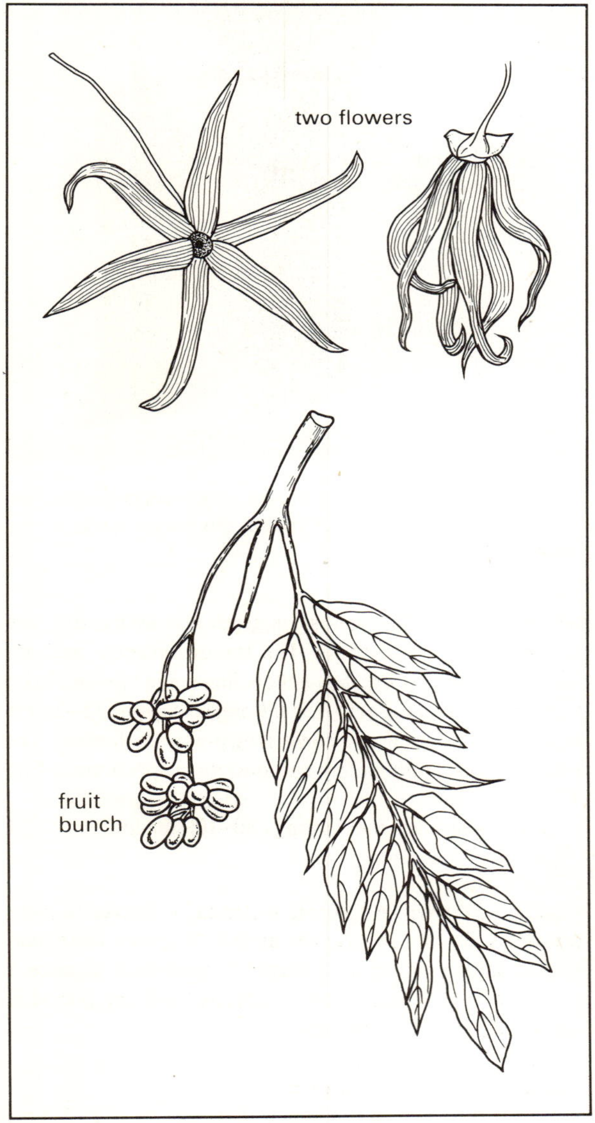

(a)

Ylang-ylang (*Cananga odorata*)

Family Annonaceae
Other names makosoi (Fijian); cadmia (Spanish); ilang-ilang (Tagalog, in the Philippines)

Two essential oils used in perfumery are distilled from the flowers: ylang-ylang oil is the more valuable, cananga oil is used to make cheaper perfumes or scents.

Botany and propagation

It is a large tree up to 30 m tall in the wild state. Leaves are 15–20 cm long, dark green above and somewhat hairy below. Flowers are yellowish green, appearing in

(b)

fig. 18.23 (a) *Cananga odorata*
(b) Flowers and leaves of ylang-ylang

hanging branches as shown in Fig. 18.23. Fruits are blackish when mature, oval and about 3 cm long. Each flower bunch produces 4–13 fruits each containing several black seeds. The tree flowers throughout the year.

Propagation is normally from seed. Selected types may be propagated from cuttings or graftings.

Climate and soil

Native to Southeast Asia, the bulk of the world's supply comes from Reunion, an island east of Madagascar. The crop grows well in the lowlands of the humid Tropics. It is not exacting in soil requirements but soils must be well drained. It is shallow rooting and therefore does not require deep soils.

Cultivation and management

Cultivated plants are usually topped to about 3 m to produce more branches and ease harvesting. Branches may also be pegged down to ground level, giving the tree a distorted shape. They start to flower when 1 or 2 years old. Flower yields increase with time, e.g. 9–10 kg per tree after 10 years. Equal proportions of ylang-ylang oil and cananga oil are usually obtained from the fresh flowers (about 1 per cent). Flowers are picked by hand before dawn because the scent is lost as the day gets warmer. The oils are extracted by distillation.

Protection

Similar to other members of Annonaceae—see p. 139.

19 Sugar crops

Major crops

Sugar cane (*Saccharum* spp.)

Family Graminae

Other names canne à sucre (French); cana de acucar (Portuguese); caña de azúcar (Spanish)

The generic name of sugar cane is *Saccharum* and the oldest cultivated species is *S. officinarum* which is a thick-stemmed juicy species that includes many varieties. These have been cultivated and selected at village level over thousands of years and the early commercial plantation industries of the nineteenth century were based on varieties of *S. officinarum* which are called 'noble canes'. The noble canes are of excellent quality and they are still widely grown throughout the tropical world for chewing, juice production and the manufacture of brown unrefined sugar. Fig. 19.1 shows a large yellow cane which is grown for juice in Southeast Asia. All of these noble canes are very handsome plants with broad spreading leaves and thick stems ranging in colour from yellow and green to red and glossy black. A black-stemmed noble cane was an item of trade with the Portuguese in Malacca, Malaysia, in the fourteenth century.

The main problems with the noble canes, which led to them being abandoned in plantation growing were susceptibility to certain diseases, relatively low yields when grown in large fields, and poor response to fertilisers. The noble canes were selected for growth in small clumps in back-yard agriculture, and with their broad spreading leaves they do not produce well at high densities (cf. p. 34).

The modern plantation sugar industry is based on hybrids between *S. officinarum* varieties and other varieties (chiefly *S. spontaneum*). Many hybrids have been produced which are resistant to various diseases such as 'sereh', 'Fiji-disease' and 'leaf-scale'. However, even hybrid varieties will succumb to diseases after prolonged cultivation and that is why new varieties are introduced every 10 or 20 years in sugar-producing regions. Hybrids have also been bred which are tolerant of drought, responsive to fertilisers (these have relatively erect leaves), and with shorter maturity. Short maturity types are very suitable for equatorial countries in which planting is based on a 10–12 month growing period.

Sugar cane is now grown widely throughout the subtropical and warm temperate regions of the world and it is only relatively recently that cane industries have spread back to the equatorial and humid regions of the Tropics. There are some problems to be overcome in the humid Tropics because sugar cane ideally requires a dry or a cold spell for maximum yields of sugar. However, the prospects for successful cane industries in the humid Tropics are good because the climate makes for a shorter growing period. The problem revolves around finding canes which will develop high sugar (brix) levels under relatively humid conditions.

fig. 19.1 *Saccharum officianarum*, one of the original noble canes still grown as a back-yard crop in Asia

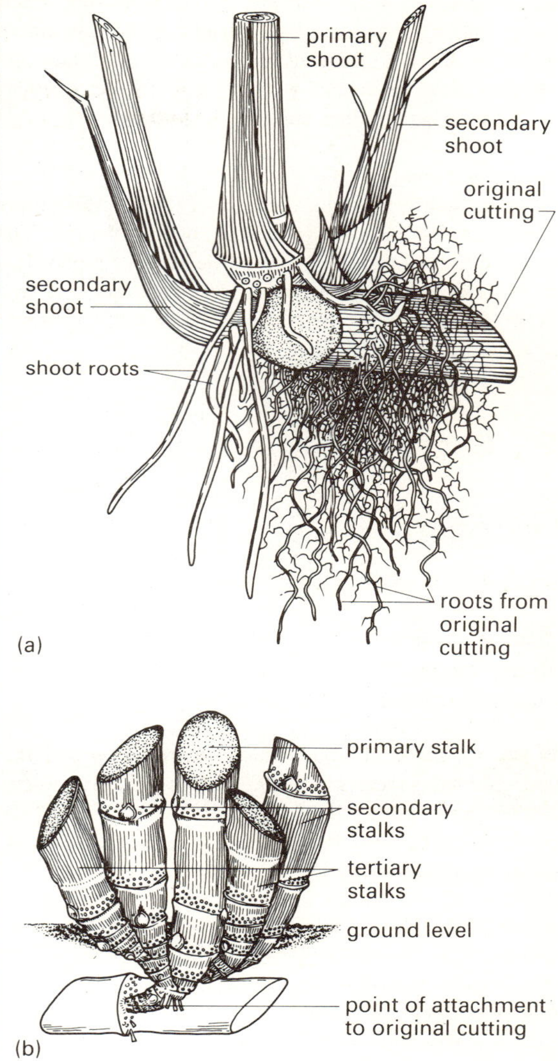

fig. 19.2 (a) When a cane sett is planted, shoots develop from the buds and roots grow from the root band. Later, secondary shoots and shoot roots develop from the primary shoots
(b) The underground portion of a developed cane stool showing the primary, secondary and tertiary stalks. (*Source*: King, Mungomery and Haws, *Manual of Cane Growing.*)

Botany

Sugar cane is one of the largest-sized members of the grass family. It is grown from stem cuttings which produce shoots and buds from the nodes as shown in Fig. 19.2. At first these shoots are leafy and the production of branch shoots occurs.

After a few months the stem begins to grow and the leafy canopy is raised (Fig. 19.3). Growth of stem continues to the time of harvesting which under equatorial conditions may be after 10–12 months. Very often, during the initial tillering phase, more tillers are formed than can reach maturity and death of up to 50 per cent of the early-formed tillers may occur before harvesting, at

fig. 19.3 A cane, 'Q 66', 16 weeks from planting showing the beginning of stalk formation

the normal planting distances. Thus, the growth of sugar cane involves first germination of the cuttings, tillering over the first few months, then growth in height of the cane with death of excess tillers. Certain varieties with vertical leaves will support a higher population of stems and will, in general, be higher yielding. If planting distance is too wide, tillering may continue late into the stem growth phase, instead of ceasing after the first few months. This will lead to low quality cane because late tillers will form on the cane stalks and reduce sugar content, and produce a cane unsuitable for handling and milling. Similarly, if cane lodges (falls over), due to wind or excessive fertility of the soil combined with the use of varieties not resistant to lodging, late tillers will form and reduce quality and sugar yield.

The early tillering phase is most sensitive to moisture stress. This is also when the greatest response to fertilisers occurs; therefore fertilisation is carried out early in growth of the crop; nutrient deficiencies cannot be made good by later applications of fertilisers. Ideally, the cane should be harvested before flowering takes place because flowering lowers sugar yields, but in practice flowering sometimes occurs, particularly when harvesting is delayed. Sugar cane is photo-periodic and flowering is promoted by short or declining daylengths.

The root system of sugar cane is sometimes quite extensive. Superficial absorbing roots occur in the top 50 cm of soil and spread laterally between the rows for at least 1·5 m. Deeper anchoring roots may occur down to about 2 m in friable soils.

Varieties and yields There are many varieties of sugar cane and sugar-producing countries are continually producing new varieties. However, aside from back-yard cultivation, successful sugar industries are at the present time almost all located in sub-humid regions. For successful sugar-production in humid regions, a rapidly maturing cane is required, and also one which is not susceptible to lodging. Recent attempts to revive a sugar industry in Malaysia have been based on the variety 'NCO 310', which develops a high sugar content early in development, is strong rooting, and is high tillering under warm humid conditions. The latter characteristic is important because canes tend to shoot too soon in very humid regions. The tendency for canes to grow too tall also exists in the humid Tropics leading to a high degree of lodging. The varieties 'POJ 2878', POJ 2883' and others were developed for tropical conditions in Indonesia in the Dutch colonial days and are still widely planted. They require warm conditions. More recent Indonesian varieties such as 'POJ 3016' are claimed to be superior to 'POJ 2878'. Variety 'F 134' from Taiwan has also been used in the humid Tropics. The variety 'Q 66' from Australia is a relatively short variety which does well on rich soils and under most conditions where other varieties may lodge. Certain newer varieties such as 'Q 78' and 'Q 83' are very fast maturing and therefore may be suitable for regions with a one-year crop.

However, the development of really successful sugar industries in the humid Tropics will depend on the breeding and selection of canes specifically for such areas, and in this regard attention must be given to the following properties:

1 good tillering;

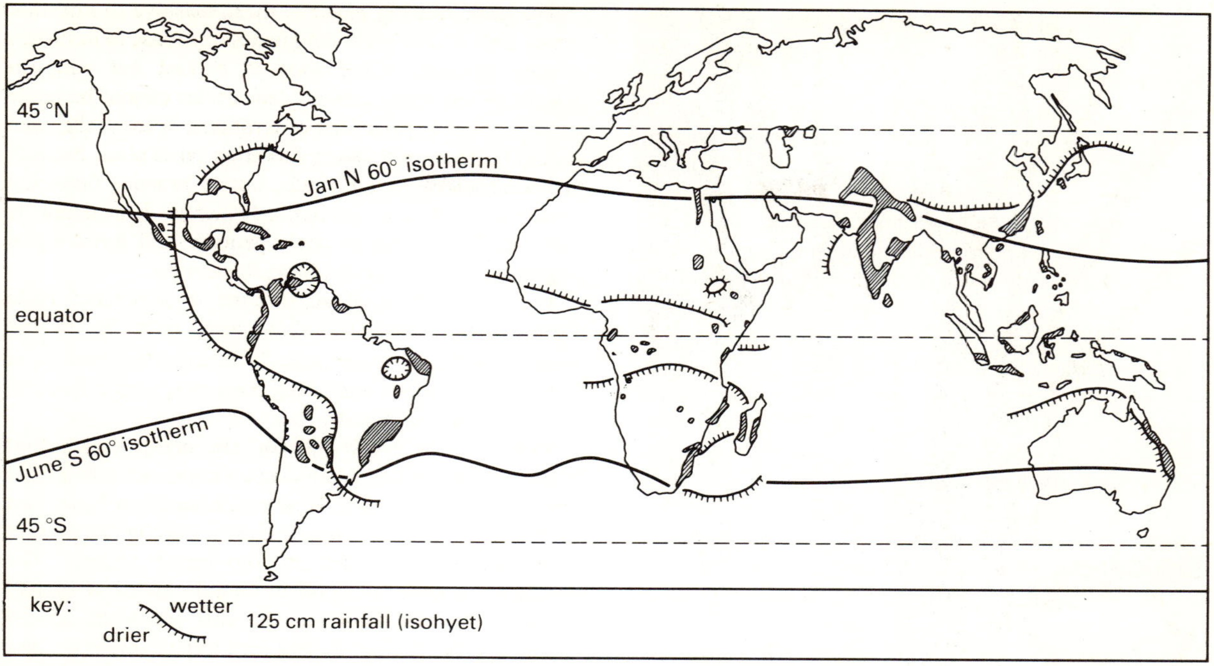

fig. 19.4 Production of sugar cane throughout the world. (*Source:* **Humbert, R. P.** (1968). *The growing of sugar cane.* Elsevier: London.)

2 freedom from lodging;
3 high sugar levels under warm humid conditions; and
4 no flowering.

Under the best conditions of husbandry, rather lower sugar yields may be expected in the humid Tropics than in the dry Tropics. This is because the build-up of sugar in the stems to a high level takes place under drier conditions, although the growth of the canes is satisfactory under humid conditions. For milling the sugar level should be as high as possible and the sugar content of properly matured canes should allow a recovery of at least 14 per cent of sugar.

Climate and soil

The main regions of sugar cultivation are shown in Fig. 19.4. They are mainly on the fringe of the humid tropical zone. Estimates of minimum rainfall needed are in the range of 1 000–1 200 mm per year, with a dry season to assist maturing of the cane. During the establishment phase, however, even slight moisture stress reduces growth.

Mean optimal temperatures in the main growing period are in the range 25–30°C for different varieties and about 20°C is considered the minimum for cane growth. Extension growth of the canes is very sensitive to temperature so that in the pre-maturity phase the growth of canes takes much longer in sub-tropical climates than in the Tropics. However, a lower temperature at maturity is beneficial in that, like drought, it helps to build up the sugar levels in the cane.

As stated before, sugar cane is photo-periodic and breeding and selection work is aimed at producing canes which do not flower in the region for which they are developed. This is another reason why the real success of sugar cane in the humid equatorial Tropics will depend on local breeding programmes. Some of the original noble canes which commonly grow in very humid regions seldom appear to flower.

Sugar cane also requires high sunlight to give good yields, so that cloudy hill sites are unsuitable. The best locations within the humid Tropics are probably those coastal regions in which rain often occurs as late afternoon or evening thunderstorms and the days are sunny.

Sugar cane is generally sensitive to frost although some

varieties can tolerate light frosts. Heavy frosts cause death of the main shoot and subsequent growth of stem tillers, reducing quality of the cane.

Soils Like most pan-tropical crops, sugar cane is grown on a wide range of soils. Drainage is required for development of the root system to at least 60 cm and heavy soils need to be broken up by deep cultivation.

Acid soils should be limed both to reduce the acidity and to provide calcium which stimulates cane growth. Lime also helps to improve the texture of heavy clay soils. The crop should benefit from lime when soil pH is below 5·0.

Saline soils are used for sugar cane in some regions. Certain cane varieties are more tolerant to saline conditions.

Cultivation and management

Planting material Before sugar cane can be grown, sufficient planting material must be obtained. If cane is being harvested at the time, the leafy tops may be planted. If not, special seed cane must be grown. If planting material is really in short supply one-node pieces should be used for planting seed fields and in these a planting distance of 60 cm by 60 cm should be used. Seed cane will be ready after about 6 months; older material is less vigorous. Fields of seed cane should be well fertilised with an N:P:K:Mg mixture fertiliser at a rate of 600 kg per hectare and ammonium phosphate(V) at 400 kg per hectare. Seed pieces which have been well fertilised germinate and grow better.

Field preparation For large-scale growing the field should be well cleared so that the seed bed can be prepared mechanically. Cane can conveniently be planted in rows 1·5–2 m apart and the planting furrow should be made about 25 cm deep. The whole field must be well drained.

Planting Seed pieces of 2–3 nodes are placed from 40 to 60 cm apart (closer if there is plenty of planting material). If planted by hand, the shoots develop better if the buds are placed upwards or sideways. Top pieces with the sheaths of the upper leaves present develop more slowly but will eventually grow and tiller strongly.

Prior to planting seed pieces should be treated with a mercurial fungicide to reduce diseases and pests, and lime and fertiliser should be added to the furrow.

After planting, under humid conditions, the cane pieces are lightly covered with soil. Planting should be done in a sufficiently moist season when soil moisture is fairly high, but avoiding very wet seasons and waterlogged conditions.

After about two weeks the shoots germinate and the furrow may be filled in a little at a time. This operation also controls weeds. The filling is done on a number of occasions to keep pace with the growth of the shoot until the furrow is completely earthed up.

Before complete earthing up, replacement planting should be done for seed pieces which have not germinated or are not growing well.

Fertilisers Rich alluvial soil may require little fertiliser for the initial crop but with continuous cultivation fertiliser use will become essential.

Certain varieties are also suited for highly fertile or less fertile conditions. The Australian variety 'Q 66', for example, grows well on rich soils while 'Q 75' is adapted to poorer soils.

In general some fertilisation is necessary, and the following recommendations are based on average conditions.

1 Before planting and during field preparation: If the soil is below pH 5·5 use one or more tonnes of magnesium limestone per hectare. This may also be ploughed in prior to furrowing. Superphosphate at 200 kg per hectare and potassium chloride at 200 kg per hectare should be placed in the furrow.

2 After germination: Nitrogen fertiliser in the form of ammonium sulphate(VI) should be supplied in two dressings of 200 kg per hectare at about 6 weeks and 12 weeks, during earthing up. One of the above dressings may be replaced by urea at 100 kg per hectare. After this no nitrogenous fertilisers should be used.

Generally fertilisation of this order will be satisfactory for the main crop of cane. Micronutrient deficiencies also occur in sugar cane, including deficiencies of copper, iron, manganese and zinc. Sugar cane has been found to respond to applications of silicon.

Irrigation Irrigation of cane is not practised in the humid Tropics.

Weed control During the earthing up procedure, mechanical weed control can be practised. After this, weed control in inter-row areas may be achieved with 2,4-D and diuron (at 1·1 and 2 kg per hectare respectively) for broad-leaf weeds, or grammoxone/diuron for grass weeds. Because excess tillers are usually formed with correctly-planted cane, destruction of a small proportion of the cane with herbicides is acceptable when spraying weeds between the rows.

(a)

Ratoon cultivation and rotation After the initial crop it is normal to grow 2 or 3 ratoon crops which are regenerated from the initial crop after harvesting. Since the cost of establishing the ratoon is less expensive this is generally a more profitable operation. The ratoons, however, will seldom be as weed-free and high-yielding as the initial crop.

After harvesting, trash from the first crop is raked up and the cane rows are straightened by running a disc up either side. This also opens up the soil and cuts off old roots and stimulates deep buds to germinate. The inter-rows should also be deep cultivated to loosen the soil and permit new root development in heavy soils. Fertilisers for a ratoon crop should be about 50 kg nitrogen per hectare higher than that used for the initial crop.

After several crops, the initial and 2 or 3 ratoons, it is sometimes advisable to plant alternate crops on the field to reduce build-up of cane diseases and pests. Suitable alternate crops are groundnuts, soya beans and cowpeas.

Harvesting and processing

Cane should, ideally, be harvested after a period of dry sunny weather when the sugar content is high. In the

fig. 19.5 (a) Bundles of sugar cane stems ready for processing. (b) A piece of sugar cane split to show the pith

(b)

equatorial Tropics, based on a one-year or shorter growing period, it is clear that sugar levels as high as those obtained under sub-tropical conditions, will not in general be possible. However, advantage should be taken of short dry spells when these occur.

Prior to harvesting, fields are sometimes burned to reduce the quantities of trash on the field and to flush out snakes and rats.

Harvesting of cane in developed countries is achieved mechanically. In other tropical countries the crop is harvested by hand (Fig. 19.5).

The preparation of refined sugar is carried out in large industrial plants, but brown sugar may be prepared by evaporating the juice in heated pans. By-products of cane include molasses and alcohol.

Protection

Important virus diseases are mosaic, chlorotic streak and ratoon-stunting disease. They are controlled by planting resistant varieties and by practising quarantine procedures to eliminate infected plants. Treatment of cuttings with hot water can be used to combat ratoon-stunting and chlorotic streak; for the former at 50°C for three hours and the latter at 52°C for 30 minutes. For mosaic, which causes a mottled pattern on the leaves of infected plants, the use of disease-free planting material and the rogueing of diseased plants is necessary. Mosaic is also spread (chiefly by aphids) from wild grasses so that good weed control is necessary. 'Sereh' is a virus disease serious in Java and elsewhere. The symptoms are leaf bunching and sprouting of roots on the cane stalks. Certain varieties, notably the POJ series, have resistance to this disease.

Pineapple disease is caused by a soil fungus which invades cuttings in the soil. Infected stems have a smell of pineapples. It is prevented by treating the cuttings at planting with mercurial fungicides at a strength of about 0·05 per cent. Red rot is another fungal disease causing red coloration in the stem tissue. Infection occurs by means of stem borers.

Two bacterial diseases which are sometimes difficult to control under humid conditions are leaf scald and red stripe. The former causes white or creamy streaks and patches and the latter red streaks between the veins. Some varieties of cane are more resistant, but where the disease occurs it is important to use uninfected planting material. Sterilisation of planting equipment with methanal (formaldehyde) solution reduces spread.

The most serious pests of cane are stem borers, and various soil grubs of beetles that attack the roots. Vigorous tillering varieties are more likely to overcome these pests. The insecticide BHC can be used to control borers and soil grubs but the long period that cane is in the field makes this expensive. Certain varieties are more resistant to borers.

Minor crops

Sugar palm (*Arenga pinnata* or *A. saccharifera*)

Family Palmaceae
Other names gemuti palm

The black fibre at the leaf bases is a source of ropes and twines particularly useful to fishermen. Sugar and molasses are prepared from the juice tapped from the flower bunches. The stem contains considerable amounts of starch which may be extracted in the same way as sago.

Botany and propagation

It is a palm growing up to 12 m tall. It has large ascending leaves with many leaflets which are whitish beneath and with stout leaf-stalks covered with a black fibre at their bases (often called gemuti fibre). Flowers first appear from the leaf axils and flowering continues downwards. Flower bunches are large, branched and unisexual in cycles. After two years of flowering the palm usually dies. Fruits are green when young and yellow at maturity; they contain many black seeds, each about 2·5 cm long (Figs 19.6 and 19.7).

Propagation is usually from seed.

Cultivation

The palm grows wild in Southeast Asia but is sometimes cultivated. Its requirements of climate, soil and cultivation are similar to those of the sago palm see p. 182.

Harvesting and processing

The palms flower at 7–10 years of age. The stalk of each male flower bunch is tapped for 2–3 months. Each palm yields about 3·5 litres of sap daily containing about 15

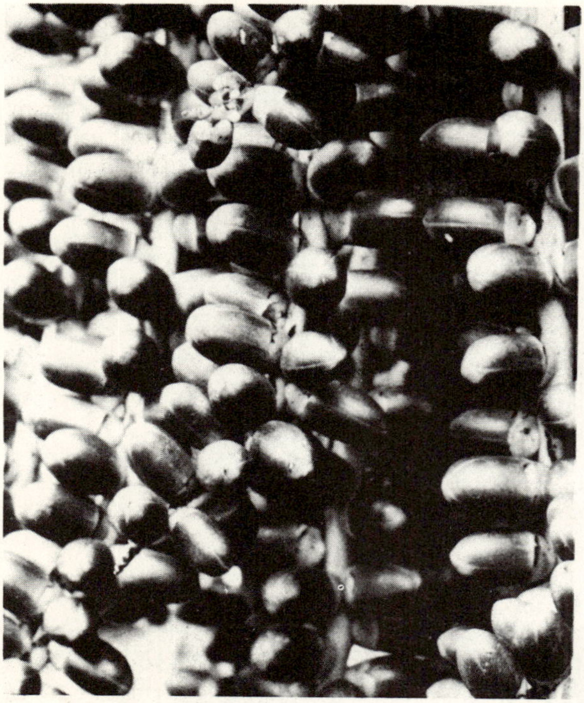

fig. 19.6 Flower bunches of sugar palm

per cent sugar. The sap is boiled to a brown sticky syrup which solidifies on cooling.

Protection

The palm is free from serious pests and diseases.

fig. 19.7 A sugar palm plant

20 Forage grasses

Major crops

Napier grass (*Pennisetum purpureum*)

Family Graminae
Other names elephant grass

Napier grass originated in Africa but has now spread throughout the Tropics and sub-tropics where it is used primarily as a forage grass. It has also been used to a limited extent as a soil cover for erosion control of steep land, for example in tea estates, but its vigorous nature makes it generally too difficult to control. Napier grass is the most high-yielding of all the perennial tropical grasses and it responds to very high levels of fertilisation both in terms of dry material and quality as expressed by protein content. However it is rather fibrous compared to other perennial forage grasses such as guinea grass and Guatemala grass when allowed to grow for more than about 50 days. At this age it has a digestible nutrient content

fig. 20.1 Established Napier; on the left regrowth after 30 days, on the right newly-cut Napier. It has been cut with a reciprocating scythe but can be cut with a forage harvester

of about 60 per cent and a fibre content of 30–35 per cent. Napier is used for producing dried grass pellets for compounding animal feeds and for fresh forage production. It can be silaged under tropical conditions. Napier recovers from grazing but yields less than one-third of the yield under cutting unless used in the fence-grazing system (p. 244).

Botany

Napier is a perennial grass producing large shoots and is only moderately tillering. Consequently it tends to form tussocks, or when planted in rows to maintain itself in the rows without colonising the spaces between the rows. Variation exists in the extent of tillering in any particular climate, some varieties will produce relatively few large shoots and others a larger number of smaller shoots. There is also a considerable variation in leafiness and vigour between Napier varieties in the humid Tropics and generally it is best to use locally-adapted varieties in planting this crop or varieties from a similar climatic area. Table 20.1 shows a summary of yields obtained in a variety trial carried out in North Borneo which is an equatorial location typical of the humid Tropics. The local variety and a variety, 'Mardi', from nearby Malaysia show the best performance.

Napier is best planted from stem cuttings that sprout rapidly from the nodes. Initially tillering occurs and after a time, with repeated cutting, a massive stool of shoot stalks will form from which new tillers develop following each cut. After a cut, or on establishing cuttings, predominantly leafy growth occurs for about the first 30

Table 20.1 Yields (in kg per hectare) of fresh forage from a number of varieties of Napier cut at 50-day intervals, from planting to establishment

	Harvest					
	1st	2nd	3rd	4th	5th	average
North Borneo	8 170	44 200	29 300	22 000	30 800	27 000
Malaysia	18 100	41 100	31 300	23 200	30 100	28 900
Cameroon	19 600	39 300	28 700	23 400	28 400	27 900
Uganda	15 300	34 200	16 500	11 900	25 100	20 600
New Guinea	—	—	—	—	24 500	24 500
Queensland	—	—	—	10 700	23 300	17 000

(a)

(b)

fig. 20.2 (a) Napier grass at 50 days after cutting
(b) Flowering shoot of Napier grass. Flowering is seasonal and varies with the variety

days as shown in Fig. 20.1, which is Napier about 30 days after a cut at 15 cm as shown on the right-hand side of the photograph. After this, stem growth takes place from internode expansion, and although leaf growth continues for up to about 70 days under good conditions, stem growth predominates and forms the bulk of the dry matter produced. With heavy fertilisation, leaf growth is stimulated and leaf area index (p. 34) may reach values of about 7 with leafy varieties. Although growth will continue for more than 100 days under good conditions, growth rate declines sharply after about 70 days and the best and highest yield of forage is obtained with cutting frequencies in the range 50–70 days (Fig. 20.2). For very high-grade forage, as for example for making grass pellets, cutting is done at shorter intervals of about 4–5 weeks.

Variation also exists between varieties in the extent of flowering; it seems that short days and decreasing daylengths promote flowering. Varieties should be chosen that do not flower or only flower to a limited extent.

Climate and soil

Different varieties of Napier are grown over a wide range of climates from the humid equatorial regions to frost-free temperate regions. Very high yields may be obtained in the humid Tropics from climatically-adapted strains with high sunshine levels. Sunshine levels below about 200 h per month produce marked decline in yields. Napier requires good drainage and yields poorly in heavy soils. Heavy soils should be deep-cultivated and built into cambered beds for satisfactory yields. Root growth in friable soils is vigorous and roots will follow the water-table over a dry period producing good yields in regions with a dry season of up to about 3 months.

Cultivation and management

Establishment Napier should be grown in rows 1 m apart which will permit cultivation between the rows. Since harvesting is best done by forage harvester, high ridging should be avoided as this will impede the entry of the machinery. On heavy soils cambered beds should be made as shown in Fig. 20.3. On free-draining soils planting should be on the flat. Cuttings of up to three nodes should be sown in furrows about 6 cm deep end to end; if plenty of planting material is available two rows may be sown. Longer cuttings should not be used because they will bend upwards at the nodes and become exposed. Under good conditions Napier will take about 4 months to establish with cuts at about 50 days, peak yields may not be obtained before about 8 months. However, once established, the sward is virtually perennial and may be cut for many years with correct management.

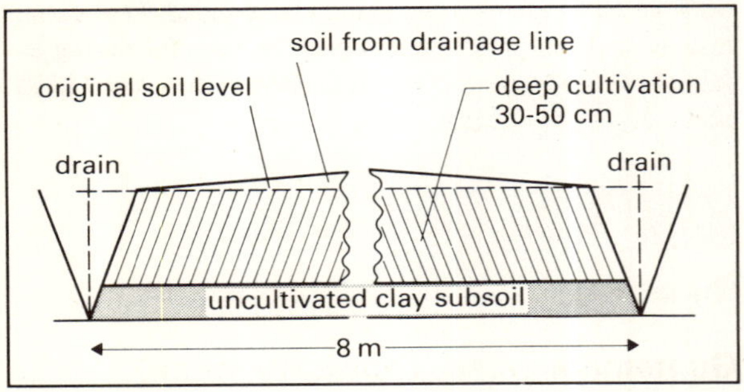

fig. 20.3 Construction of a cambered bed suitable for Napier grass in heavy soils

Management Management involves weeding and cultivation in inter-row areas; on heavy soils, twice-yearly; on friable soils, once every 2 years. Weeding may be done using the herbicides paraquat and simazine. Paraquat should be sprayed immediately after a cut at a rate of 1 litre per hectare. Napier is not affected by this herbicide providing spraying is not delayed beyond 24 hours. After this, new buds develop that will be killed by the herbicide. After paraquat, simazine at 2 kg per hectare should be applied. Generally this herbicide treatment will give at least 4 months free from weeds (two 60-day cuts).

Management for grazing Very high stocking rates can be obtained exceeding 12 steers per hectare when Napier is grazed by the fence-grazing method. The grass is established in double rows 1·2 m apart with fenced grazing corridors of width 1·5 m between the double rows as shown in Fig. 20.4.

fig. 20.4 Successive stages of fence-grazing in Napier grass

A rotation system is established with intensive grazing on each area for about 4 days, followed by a 55–60 day recovery period.

Fertilisers It has been found that Napier responds to very high levels of fertiliser. Fertilisers should be balanced since response to high levels of nitrogen does not take place without potassium and other nutrients. Over a long period the fertiliser rates shown in Table 20.2 have been found to give high yields under humid tropical conditions. They should be applied after each 60-day cut. Fertilisers should not be applied during dry spells or floods.

Table 20.2 Recommended rates for application of fertiliser to Napier grass (in kg per hectare)

Urea	300
Potassium chloride	100
Kieserite	60
Phosphate	60

On acidic soils below about pH 4·5 Napier responds to lime at rates up to 1 tonne per hectare.

fig. 20.5 Napier is best harvested using a forage harvester that breaks up the stalks. Material harvested in this way will be quite acceptable to livestock. If the stalks are not broken the animals will eat only the foliage leaves

Harvesting and processing

For feeding to livestock such as cattle, buffaloes and goats, Napier should be cut between 50 and 70 days with a forage harvester (Fig. 20.5). Yields in excess of 60 tonnes per hectare of dry material or 300 tonnes per hectare of fresh forage may be obtained under good physical conditions and with high rates of fertiliser use. When Napier is used for producing grass pellets for non-ruminants (poultry and pigs) cuts should be made at 4 week intervals and yields will be only half the above rates. However, with heavy fertilisation, protein levels exceeding 18 per cent may be obtained. For 4-week cuts a closer row spacing of 60 cm is recommended.

Protection

Napier is extraordinarily free of serious pests and diseases. Occasional leaf-spot fungi infect the crop to a minor extent and certain varieties are susceptible to a die-back virus. These varieties should be avoided. Leaf-eating insects such as grasshoppers may be harmful during establishment and should be controlled with a suitable insecticide when necessary.

Minor crops

Guatemala grass (*Tripsacum laxum*)

Family Graminae

This grass is used as a fodder crop; it is not suitable for grazing.

Botany and propagation

It is a tall, strong-growing erect perennial grass which reaches 3–4 m in height. Its leaves are broad and glossy with slightly serrated margins. It produces very little stem material.

Propagation is usually by plant divisions which are actually underground stems. Aerial stem cuttings may be used to grow the crop but root divisions are preferred as they take 2–3 weeks to strike compared to 2 months for stem cuttings.

Cultivation and management

Climate and soil as for Napier grass. It is more drought

fig. 20.6 Guatemala grass about 50 days after cutting

fig. 20.7 Regeneration growth of Guatemala grass after a cut at 40 cm

resistant than guinea grass. The crop is planted in 1 m rows at a spacing of about 30 cm. It should be cut at a height of 30–40 cm at intervals of 50–70 days (Fig. 20.6). It should not be cut below 30 cm as this affects regeneration (Fig. 20.7). Fertilisers at the following rates can be recommended, after each 50–70 day cut apply these quantities per hectare: urea – 150 kg, potassium chloride – 40 kg, kieserite – 25 kg, phosphate – 25 kg (on acidic soils below pH 4·8 ground rock phosphate may be used instead of super). Yields up to about 100 tonnes of fresh grass per hectare may be obtained annually.

The crop is protected as for guinea grass.

Guinea grass (*Panicum maximum*)

Family Graminae

A fodder grass grown for stall-feeding of ruminants and for manufacture of dried grass pellets for animal feed mixtures. It produces a fine forage with very little fibre. The fine-leaved varieties are also suitable for grazing purposes.

Botany and propagation

It is a perennial grass growing up to 1·5 m tall when mature. It spreads by tillering and grows into large clumps or stools (Fig. 20.8). The flower bunch is a large spreading panicle.

fig. 20.8 A plot of guinea grass

The grass is usually grown from divisions of mature plants.

Climate and soil

It will produce well in the humid tropical climates or in frost-free sub-tropical or temperate regions. It grows at elevations up to 1 500 m in the Tropics but is best suited to the plains. It does not produce well during dry periods and does not tolerate waterlogging. Rich and deep alluvial loams are preferred. Flowering seems to be enhanced by short or decreasing daylength.

Cultivation and management

Many cultivars have now been developed, e.g. 'Sigor' and 'Coloniao', varying in height, tillering, etc. Plant divisions (tillers) should be planted in double rows of 60 cm by 30 cm with spaces of 1 m between each double row. This gives better recovery from cutting than evenly-spaced rows of 60 cm. In the humid Tropics the grass may be first cut 3 months after planting and thereafter at intervals of 4–5 weeks, at a height of 15 cm.

The following fertiliser rates per hectare may be recommended after a 35-day cut: urea – 150 kg, potassium chloride – 50 kg, kieserite – 25 kg and superphosphate – 25 kg. Yields of up to 120 tonnes of fresh grass per hectare can be obtained annually. The grass tends to degenerate with time when yields diminish somewhat. This is primarily due to failure of tillers to recover after cutting if the stand becomes too dense. Delaying of cutting to more than 40 days enhances this reaction. Replanting should be done when necessary.

Protection

Guinea grass is relatively free from serious pests and diseases. Weeding is necessary initially and after every harvest. Cultivation of inter-row areas should be done periodically.

Pangola grass (*Digitaria decumbens*)

Family Graminae

Pangola is a pasture grass grown for grazing livestock. It is seldom used as a fodder grass. Well-adapted varieties can carry a stocking rate of 4 steers per hectare with correct management.

fig. 20.9 Pangola grass fertiliser trial, the plot in the foreground shows about 25 days regrowth

Botany and propagation

It is a low-growing leafy grass of high nutrient value to livestock (Fig. 20.9). There are many varieties. It is a perennial which regenerates by means of long rooting stolons.

Propagation is usually from stem cuttings or mowings.

Climate and soil

This grass is native to South Africa and is suited to the lowlands of the humid Tropics. It needs a high rainfall but is able to survive drought as well as flooding. The best growth is observed on well-drained soils.

Cultivation and management

Aerial shoots are broadcast and worked into the soil at the rate of 1·25–2·50 tonnes per hectare. The pasture should be topped or grazed lightly 6–8 weeks after planting to induce tiller growth. Regular applications of fertiliser are needed and the following rates per hectare can be recommended every 6 weeks: urea – 50 kg, potassium chloride – 15 kg, kieserite – 20 kg, phosphate – 10 kg. It can be grazed at the rate of about 4 steers to the hectare with continuous grazing. However, rotation grazing every 4 weeks is recommended.

Protection

The grass may be attacked by a virus; resistant varieties should be planted.

Para grass (*Brachiaria mutica*)

Family Graminae
Other names Mauritius grass

The grass may be used as a fodder crop or for grazing.

fig. 20.10 Para grass about 30 days after grazing

Botany and propagation

It is a spreading perennial grass (Fig. 20.10). The stems are hairy, particularly at their nodes. There are many varieties.

Propagation is usually from stem cuttings or mowings of three-node length.

Cultivation and management

Para grass requires climate and soil similar to those for guinea grass. It is usually planted at a spacing of about 30 cm square at the start of the rainy season. Two nodes are buried in the soil and one node is exposed. Fertilisers and cattle manure are usually needed; they are applied before planting and after each cut or grazing spell. The following rates of fertiliser per hectare may be recommended for grazing,

fig. 20.11 Signal grass about 30 days after grazing

applied after grazing: urea – 50 kg, potassium chloride – 15 kg, kieserite – 20 kg, phosphate – 10 kg. Twice these rates may be used for forage production. The grass may first be harvested or grazed 3–4 months after planting when the flowers are appearing and thereafter every 40–50 days. A yield of about 60 tonnes per hectare of fresh grass may be obtained yearly. Replanting is often necessary every 3 years.

The crop is protected as for guinea grass.

Signal grass (*Brachiaria decumbens*)

Family Graminae

This grass may be used for grazing or as a fodder crop.

It is a spreading perennial grass with considerable persistance and resistance to grazing in some varieties (Fig. 20.11). Propagation, cultivation and protection as for para grass. Climate and soil conditions are the same as for guinea grass.

Further reading

Allen, B. M. (1967). *Malayan fruits.* Donald Moore: Singapore.

Arnon, I. (1975). *Mineral nutrition of the maize.* Int. Potash. Inst.: Switzerland.

Avocado (1974). Department of Primary Industries: Queensland, Australia.

Berger, J. (1962). *Maize production and the manuring of maize.* Centre d'etude de l'azote: Zurich.

Brown, N. H. (1951–8). *Useful plants of the Philippines,* 3 vols. Tech. Bull. No. 10. Department of Agriculture and Natural Resources, Manila.

Burkill, I. H. A. (1966). *Economic products of the Malay Peninsula,* Vols 1 and 2. Ministry of Agriculture: Malaysia.

Child, R. (1964). *Coconuts.* Longman: London.

Cobley, L. S. revised by **Steele, W. M.** (1976). *An introduction to the botany of tropical crops.* Longman: London.

Coconut palm. Agricultural leaflet No. 41: Malaysia.

Conell, D. S. (1953). Vanilla. *Econ. Botany* **7,** pp. 291–358.

Coursey, D. G. (1968). The edible aroids. *World Crops* **20,** pp. 25–30.

(1977). *Yams.* Longman: London.

Djikman, M. J. (1951). *Hevea.* University of Miami Press: Florida.

Eden, T. (1965). *Tea.* Longman: London.

Gade, D. W. (1966). Achira the edible canna. *Econ. Botany* **20,** pp. 407–15.

Greenway, P. J. et al (1953). *The pawpaw.* Tanganyika Department of Agriculture. Pamphlet No. 52.

Grist, D. H. (1955). *Rice.* Longman: London.

Haarer, A. E. (1963). *Coffee growing.* OUP: London.

Harler, C. R. (1966). *Tea growing.* OUP: London.

(1963). *Tea manufacture.* OUP: London.

Hartley, C. W. S. (1977). *The oil palm.* 2nd Ed. Longman: London.

Humbert, R. P. (1968). *The growing of sugar cane.* Elsevier: London.

Johnson, R. M. and **Raymond, W. D.** (1957). *The sago palm.* Tropical Products Institute: London.

Kay, D. E. (1973). *Root crops.* Tropical Products Institute: London.

Kirby, R. H. (1963). *Vegetable fibres.* Leonard Hill: London.

Koh, A. H. (1975). *A guide to nutmeg cultivation.* Department of Agriculture: Malaysia.

Koh, A. H. and **Ding, S. M.** (1973). *A guide to clove cultivation.* Department of Agriculture: Malaysia.

Macadamia nut culture. (1976). Department of Primary Industries: Queensland, Australia.

Maize. Agricultural Leaflet No. 33: Malaysia.

Mango growing in Queensland. (1975). Department of Primary Industries: Queensland, Australia.

Massal, E. and **Barran, J.** (1956). *Food plants of the South Sea Islands.* South Pacific Comm. Tech. Paper No. 94.

Ng Siew Kee (1972). *The oil palm, its culture, manuring and utilisation.* Int. Potash Inst.: Switzerland.

Osman, M. N. (1972). *Cashew nut in Trengganu.* Agricultural Leaflet No. 51: Malaysia.

Patchouli. Agricultural Leaflet No. 33: Malaysia.

Purseglove, J. W. (1968). *Tropical crops: Dicotyledons.* Longman: London.

(1972). *Tropical crops: Monocotyledons.* Longman: London.

Rice Production Manual. (1967). University of the Philippines.

Simmonds, N. W. (1966). *Bananas.* Longman: London.

Smart, J. (1978). *Tropical pulses.* Longman: London.

Thompson, W. T. (1970). *Agricultural chemicals* Book II, *Herbicides.* Thompson Publications: USA.

Wee, Y. C. and **Tay, T. H.** (1968). *Pineapple cultivation in West Malaysia.* Pineapple Board Leaflet: Malaysia.

Wood, G. A. R. (1975). *Cocoa.* Longman: London.

Appendix

A note on pests and pest control by D. J. McCrae

It is impossible to deal comprehensively with this aspect of agriculture within a short space. This is because the crops grown and the pest species that attack them vary with region, locality and season. Pesticides now in common use may be replaced by others quite soon. It is possible to control many pest species with some precision, although there is still plenty of room for improvement and refinement of technique.

Good control of pests depends on close observation and timely treatment. It is impossible to cure damage once caused and a treatment should never be presumed effective unless observation proves it to be so. Until or unless a grower is familiar with pest species and the correct measures to be taken against them, it is wise to seek advice. This can usually be obtained from government or university advisory services, but if not, commercial firms dealing in pesticides are often able to give advice.

Pest species themselves present a very wide variety of organisms, ranging from large mammals such as wild pigs, through rats and birds, down to insects and mites. The circumstances requiring control of pests vary in their turn, from crops in the field to seed and all kinds of produce in storage. For crops of major importance, whole books have been written on the pests that affect them, e.g. rice, rubber, oil palm, coffee, bananas, etc. These are always worth consulting because they indicate what pest species are likely to do damage in a given region and the nature of that damage.

Unfortunately, no one pesticide exists which will deal with all pest species in practical terms. Mammals such as pigs and rats can be dealt with—not always satisfactorily—by the use of poison baits. Good control of rats in The field and in stores can be achieved by using anticoagulant baits, though instances have arisen of rats becoming resistant to this type of toxicant. Bird control is more difficult to achieve and to date no really satisfactory method has been discovered, despite the multiplicity of bird scarers and attempts at poisoning.

Insects are responsible for most losses to crops both in storage and in the field. Control of insects in the field usually involves the application of an insecticide, which if ingested or contacted by insects, proves lethal to them over a short or long period. It is as well to remember that insecticides are poisonous to man and livestock to some degree and it is quite wrong to think that they confine their action to insects. The impression given that only insects are affected by them is solely due to the fact that insects are relatively small, thus requiring a small amount of insecticide in relation to their small body weights to poison them. It follows that safety precautions must be followed when handling insecticides and these are usually detailed on the container label.

The mode of action of insecticides is not important to the ordinary farmer as long as the insecticides used are effective. It is as well to avoid using insecticides of long persistence, such as DDT or dieldrin because they eliminate natural enemies of pest species and so may well bring about more serious attacks. The systemic insecticides, i.e., those which are absorbed by the plant and make its juices toxic to insects (i.e. methomyl or phosphamidon) are usually not systemically effective against large insects. Their action is however most useful against newly hatched larvae, where only a minute quantity is enough to poison such small organisms. This is not to say that the contact action of such insecticides is not effective against fully adult insects exposed on foliage. It should also be remembered that very often two or more applications of an insecticide are required before full control of a pest can be obtained. Few insecticides will kill insect eggs, particularly if these are embedded in plant tissues, so further applications are necessary to kill the insects that hatch out from the eggs.

The English names of insects cause much confusion because they are imprecise. This confusion can only be dispelled by experience, which of course is impossible to teach. The following notes may prove helpful, but really have to be backed by experience.

Aphids These are insects which feed on the juices of plants and are incapable of consuming solid food. Many species give rise to new individuals without any egg stage. Some forms have wings and others, even of the same species, do not. They are all small soft-bodied insects, usually infesting the terminal shoots and growing points of the plant. They may be black (blackfly), or green (greenfly), others are yellow. Some species feed below ground, on roots or at the collar of the plant. Many species can breed up to large numbers very quickly, forming dense clusters on terminal shoots. They are important economically as many species transmit plant viruses, but this occurs more in temperate than in tropical regions. They are fairly easy to control, even with a spray of soapy water. Systemic and contact insecticides are also effective, especially if they contain a wetting agent.

Mealybugs These are sucking insects totally incapable of taking solid food. Mealybugs are small flattened, usually soft-bodied, white insects with a covering of floury wax which is secreted through pores in the skin. The males are minute winged insects, not usually seen, and do no harm. Thus the term 'mealybug' applies to the female. In severe infestations, all parts of the plant will harbour them, but they are usually found in groups on tender parts of the plant and below leaves. Some species feed on roots below the ground. They are rather immobile insects, they secrete honeydew and are often attended by ants. Excess honeydew falls on foliage and gives rise to black sooty mould on leaf surfaces. This can be seen from a distance and indicates the presence of mealybugs, aphids or scale insects. Serious infestations lead to leaf fall and die-back of affected plants. Some species also transmit plant virus diseases.

Control can be achieved with systemic insecticides: direct sprays are often not effective as they do not penetrate the waxy deposit on the insects. If attendant ants are denied access to mealybugs by spraying around the plants this allows predatory insects (e.g. ladybirds) and natural parasites better access to them.

Scale insects They are related to mealybugs and again confined to liquid food. Many species are small, flattened and scale-like, hence the name. Some are very injurious and many plant species are attacked by them. Usually, it is only the females which are seen on the plant. They are often completely immobile and feed in one position for most of their lives. Many species secrete honeydew like mealybugs and aphids and are often attended by ants for that reason. Sooty mould again is an indication of their presence. Not all species produce honeydew, however. They usually occur in groups along twigs, on the main stem or beneath leaves. They are more susceptible to direct sprays than mealybugs, but several applications are necessary to deal with individuals subsequently hatching from eggs.

Planthoppers and leafhoppers These vague terms collectively apply to a range of small active insects which jump and fly when disturbed. Like aphids, mealybugs and scale insects, they feed exclusively on plant sap. Their eggs are embedded in plant tissues and hatch into active young which eventually develop wings. Heavy infestations often result in the death of the host plant and very serious damage can be done by some species to important crops such as rice and sugar cane. Sooty mould often accompanies heavy attacks. There are many species, not all injurious. Some species are virus vectors, e.g. tungro virus of rice is spread by the green leafhopper in Asia.

Bean fly This is a real two-winged fly, but the damage is done by its maggots. The actual fly is very small; eggs are laid on leaf surfaces and the maggots bore down the leaf stalk and down the stem of young bean plants. The damage done to the plant by this mining often kills it. The maggots of other related species confine their activities to the leaves of the host plant and produce wavy mines.

Leafminers These are small caterpillars or maggots which tunnel in leaves. Leafmining caterpillars (the immature stages of moths) usually make blotch mines—irregular dead patches on leaves. The maggots of leafmining flies usually make serpentine mines. Heavy attack by leafminers results in defoliation and hence loss of yield.

Fruit-flies They are fairly large flies, often with smoky patterns on the wings and the body with bright yellow markings. The female lays her eggs in fruit and the resulting maggots feed within, making the fruit unsaleable and uneatable. There are many species and a very wide range of fruit suffers from attack by fruit-flies—from watermelon and cucumber to citrus, mango and guava to mention but a few. Certain systemic insecticides will control fruit-fly but have to be applied before the fruit is attacked so that the deposit is ready to kill the newly hatched maggots before they can do damage.

Ladybirds These are small, almost hemispherical beetles, often brightly coloured with black spots on the wing cases. Most ladybirds and their larvae are predatory and carnivorous, feeding on aphids, scale insects and mealybugs. Some species, however, are plant feeders. These are larger beetles, often yellowish, but again with black spots. The larvae are bristly and feed on leaf surfaces, producing characteristic leaf skeletons and scarification.

Borers A term applied to any insect which damages the plant by boring. Most borers are the larvae of moths or beetles. Some are very serious pests of important crops such as rice, maize and sugar cane. Most borers are difficult to control and no single method is effective for all of them.

Grasshoppers These are insects which jump and fly on being disturbed. The hindmost pair of legs is enlarged and enables the insect to jump. Locusts, among the most serious of all insect pests, are large grasshoppers. There are many species, some small, some large. They feed, for the most part, on grasses, but by no means exclusively and there are many species with widely differing habits and habitats.

Index